CREATING RESILIENT LANDSCAPES IN AN ERA OF CLIMATE CHANGE

This book delivers a realistic and feasible framework for creating resilient landscapes in an era of anthropogenic climate change.

From across six continents, this book presents fifteen case studies of differing socio-cultural, economic, and biophysical backgrounds that showcase opportunities and limitations for creating resilient landscapes throughout the world. The potential to create socio-ecological resilience is examined across a wide range of landscapes, including agricultural, island, forest, coastal, and urban landscapes, across sixteen countries: Argentina, Australia, Brazil, Denmark, Finland, Greece, Guatemala, Japan, Mexico, Norway, Samoa, South Africa, the United States, Turkey, Uruguay, and Vanuatu. Chapters discuss current and future issues around creating a sustainable food system, conserving biodiversity, and climate change adaptation and resilience, with green infrastructure, nature-based architecture, green-tech, and ecosystem services as just a few of the approaches discussed. The book emphasizes solution-oriented approaches for an "ecological hope" that can support landscape resiliency in this chaotic era, and the chapters consider the importance of envisioning an unpredictable future with numerous uncertainties. In this context, the key focus is on how we all can tackle the intertwined impacts of climate change, biodiversity loss, and large-scale land-cover conversion in urban and non-urban landscapes, with particular attention to the concept of landscape resiliency. The volume provides that much-needed link between theory and practice to deliver forward-thinking, practical solutions.

This book will be of great interest to students, researchers, practitioners and policymakers who are interested in the complex relationship between landscapes, climate change, biodiversity loss, and land-based conversion at local, national and global scales.

Amin Rastandeh is a landscape analyst, with experience in the United States, New Zealand, and Iran, working on human-environmental interactions and climate change in evolving multifunctional landscapes. He worked as a postdoctoral researcher in the Department of Sustainability and Environment at the University of South Dakota, United States. He specializes in multi-scale design and management of landscapes for safeguarding biodiversity and human communities in the face of climate change.

Meghann Jarchow is chair and associate professor in the Department of Sustainability and Environment at the University of South Dakota, United States. Her expertise includes taking a place-based approach to working toward greater sustainability including serving as chair of Greening Vermillion, president of Spirit Mound Historic Prairie, and board member of EcoSun Prairie Farms.

Routledge Studies in Conservation and the Environment

This series includes a wide range of inter-disciplinary approaches to conservation and the environment, integrating perspectives from both social and natural sciences. Topics include, but are not limited to, development, environmental policy and politics, ecosystem change, natural resources (including land, water, oceans and forests), security, wildlife, protected areas, tourism, human-wildlife conflict, agriculture, economics, law and climate change.

Threatened Freshwater Animals of Tropical East Asia
Ecology and Conservation in a Rapidly Changing Environment
David Dudgeon

Conservation Effectiveness and Concurrent Green Initiatives
Li An, Conghe Song, Qi Zhang, and Eve Bohnett

Religion and Nature Conservation
Global Case Studies
Edited by Radhika Borde, Alison A. Ormsby, Stephen M. Awoyemi, and Andrew G. Gosler

Jackals, Golden Wolves, and Honey Badgers
Cunning, Courage, and Conflict with Humans
Keith Somerville

Case Studies of Wildlife Ecology and Conservation in India
Edited by Orus Ilyas and Afifullah Khan

Creating Resilient Landscapes in an Era of Climate Change
Global Case Studies and Real-World Solutions
Edited by Amin Rastandeh and Meghann Jarchow

For more information about this series, please visit: www.routledge.com/Routledge-Studies-in-Conservation-and-the-Environment/book-series/RSICE

CREATING RESILIENT LANDSCAPES IN AN ERA OF CLIMATE CHANGE

Global Case Studies and Real-World Solutions

Edited by Amin Rastandeh and Meghann Jarchow

Routledge
Taylor & Francis Group
LONDON AND NEW YORK

earthscan
from Routledge

An electronic version of this book is freely available, thanks to the support of libraries working with Knowledge Unlatched (KU). KU is a collaborative initiative designed to make high quality books Open Access for the public good. The Open Access ISBN for this book is 9781003266440. More information about the initiative and links to the Open Access version can be found at www.knowledgeunlatched.org.

Designed cover image: Getty Images

First published 2023
by Routledge
4 Park Square, Milton Park, Abingdon, Oxon OX14 4RN

and by Routledge
605 Third Avenue, New York, NY 10158

Routledge is an imprint of the Taylor & Francis Group, an informa business

British Library Cataloguing-in-Publication Data
A catalogue record for this book is available from the British Library

Library of Congress Cataloging-in-Publication Data
Names: Rastandeh, Amin, editor. | Jarchow, Meghann, editor.
Title: Creating resilient landscapes in an era of climate change : global case studies and real-world solutions / Edited by Amin Rastandeh and Meghann Jarchow.
Description: New York, NY : Routledge, 2023. | Includes bibliographical references and index.
Identifiers: LCCN 2022033767 (print) | LCCN 2022033768 (ebook) | ISBN 9781032210377 (hardback) | ISBN 9781032210384 (paperback) | ISBN 9781003266440 (ebook)
Subjects: LCSH: Landscape ecology. | Vegetation and climate. | Biodiversity—Climatic factors.
Classification: LCC QH541.15.L35 C74 2023 (print) | LCC QH541.15.L35 (ebook) | DDC 577.5/5—dc23/eng/20220831
LC record available at https://lccn.loc.gov/2022033767
LC ebook record available at https://lccn.loc.gov/2022033768

ISBN: 978-1-032-21037-7 (hbk)
ISBN: 978-1-032-21038-4 (pbk)
ISBN: 978-1-003-26644-0 (ebk)

DOI: 10.4324/9781003266440

Typeset in Bembo
by codeMantra

CONTENTS

FIGURES

NOTES ON CONTRIBUTORS

Pippin Anderson, University of Cape Town, South Africa
Pippin Anderson is an associate professor in the Department of Environmental and Geographical Science at the University of Cape Town. While primarily a plant ecologist, her research focuses on urban ecology and the way in which the city both filters and constructs its inhabitant flora. She is interested in understanding the rules in cities and in particular how we might use this knowledge to restore and create more green spaces in cities that speak to the dual purpose of human wellbeing and plant conservation.

Laura Astigarraga, Universidad de la República, Uruguay
Laura Astigarraga is a professor in the Department of Animal Production and Pastures at the Faculty of Agronomy (Universidad de la República, Uruguay). She works on livestock grazing systems, with an emphasis on sustainability. Her approach is wide, studying pasture management, animal nutrition, production systems, and greenhouse gases emission. She is a leader author in the Working Group II of the Sixth Assessment Report of the IPCC.

Birsen Kesgin Atak, İzmir Demokrasi University, Turkey
Birsen Kesgin Atak is an associate professor in landscape architecture at İzmir Demokrasi University. After completing her bachelor's degree (2002) at Ege University, Department of Landscape Architecture, she received her M.Sc. (2007) and Ph.D. (2013) degrees from Ege University, Department of Landscape Architecture. At the same time, she completed the postgraduate program of environmental management in the Mediterranean Agronomic Institute of Chania, Crete, in Greece in 2005. Her research interests lie in the field of detection of land-use/cover change, analyzing spatial patterns on the landscape, and modeling future trends of landscape structure based on GIS and RS.

Paul Blaschke, Victoria University of Wellington, New Zealand
Paul Blaschke is an independent environmental researcher and consultant based in Wellington, Aotearoa New Zealand. With a background in landscape ecology and environmental policy work, Paul has wide-ranging experience in environmental management and applied ecology throughout Aotearoa New Zealand and several Pacific Island countries. More recently he has been particularly active in applied urban ecology and ecological restoration, and nature-based climate change adaptation. He is an honorary research fellow at Te Herenga Waka Victoria University of Wellington and the University of Otago.

Victoria Chanse, Victoria University of Wellington, New Zealand
Victoria Chanse is a community designer and planner at Victoria University of Wellington, where she is a senior lecturer and landscape architecture program director. She focuses her research, teaching, and practice on solving problems associated with sea level rise, flooding, and stormwater. Her work focuses on participatory, community-based approaches to develop local responsive designs that consider community needs and landscape changes under different scenarios of sea level rise and stormwater management. Her recent work includes a chapter on Island Bay, Greater Wellington Region, New Zealand, in the edited book, *Landscape Architecture for Mitigating Sea Level Rise: Innovative Global Solutions*, and "Engaging Stakeholders in the Sea Level Rise Design Process" in *The International Journal of Climate Change: Impacts and Responses.*

Rachael Ann Cox, EarthEmpower
Rachael Ann Cox is a specialist in socially inclusive and sustainable development. She founded EarthEmpower, a women-led internationally focused consulting firm, in 2017 and leads projects focused on rural development, women's empowerment, supply chain management, and agricultural research. Rachael holds a B.S. in agronomy and an M.S. in crop production and natural resources and ecology management from Iowa State University. Rachael was a National Science Foundation research fellow from 2010 to 2012 and was a Fulbright World Learning professional specialist in 2021 in Bucaramanga, Colombia. She has worked in Kenya, Uganda, Rwanda, Mexico, Guatemala, Ecuador, Costa Rica, Colombia, and the United States.

Timothy E. Crews, Land Institute, USA
Timothy E. Crews is a soil ecologist and director of the international program at The Land Institute in Salina, Kansas. His research career has focused on ecological processes and attributes that underlie the sustainable productivity of natural ecosystems and how grain producing agroecosystems can be improved with the integration of attributes such as plant diversity and perenniality.

Wenche E. Dramstad, Norwegian Institute of Bioeconomy Research, Norway
Wenche E. Dramstad is a landscape ecologist with a strong interest in the process of landscape change, including what drives these changes and the effects they have on biodiversity, ecology, and landscape perception. Her main focus has

been on agricultural landscapes, and one key activity has been researching, monitoring, and analyzing aspects of agricultural landscape change. More recently her emphasis has been on sustainable development and ecosystem services, still focused on agricultural landscapes.

Katsue Fukamachi, Kyoto University, Japan
Katsue Fukamachi is an associate professor at the Graduate School of Global Environmental Studies, Kyoto University. Her specialty is landscape ecology and planning, and research topics include determining key factors and changes in the relationship between people and nature in satoyama landscapes. She is interested in the close links between the mountainous and coastal areas, which are not only significant in everyday life but were also important under extraordinary circumstances that necessitated a response. She has been involved in the Central Environmental Council of Japan, and responsible for several projects for the conservation and productive use of cultural landscapes.

Scott Hawken, University of Adelaide, Australia
Scott Hawken is an urban designer, landscape architect, and landscape archaeologist with 20 years' experience in professional design practice and academia. His research and teaching bring together these three disciplines in creative ways. He is a strong supporter of the kind of transdisciplinary thinking necessary to tackle the large problems of our time. He directs the Urban Design and Planning program at the University of Adelaide, and his research focuses on long-term approaches to achieve urban sustainable transitions.

Meghann Jarchow, University of South Dakota, United States
Meghann Jarchow is chair and associate professor in the Department of Sustainability and Environment at the University of South Dakota. Jarchow's expertise includes taking a place-based approach to working toward greater sustainability including serving as chair of Greening Vermillion, president of Spirit Mound Historic Prairie, and board member of EcoSun Prairie Farms.

Pia Heike Johansen, University of Southern Denmark, Denmark
Pia Heike Johansen is an associate professor at The Danish Centre for Rural Research at the University of Southern Denmark. Pia's research interest lies on the rural-urban dichotomy in relation to landscape and cultural life, with a specific focus on the everyday life, entrepreneurship, and the political conditions for these.

Gabriel Luke Kiddle, Victoria University of Wellington, New Zealand
Gabriel Luke Kiddle's research and teaching at Victoria University of Wellington is focused on Pacific urban issues, informal settlements, the urbanization and climate change interface, and wellbeing. Luke has also worked for many years in international development, with focus on the Pacific region, including long-term experience in Solomon Islands while working for the New Zealand Aid Programme.

Simon Kilbane, Deakin University, Australia
Simon Kilbane is landscape architect and urban designer with diverse experience across public, private, and academic sectors. He leads the environmental consultancy, Rhizome, and currently teaches planning, urban design, and landscape architecture at Deakin University, Geelong, Australia. Simon's work focuses on the intersection of people, place, and ecology at large and small scales through Green Infrastructure as a specific planning and design approach that articulates ecological science and policy intent and intertwines community aspirations toward the creation of more healthy, resilient, and inclusive landscapes and cities.

Marta Moura Kohmann, University of Florida, United States
Marta Moura Kohmann is a postdoctoral associate at the University of Florida's Range Cattle Research and Education Center. The work she is currently developing under the U.S. Department of Agriculture's Long-Term Agroecosystem Research initiative focuses on evaluating sustainability of rangelands managed with prescribed fire at various frequencies and in association with mechanical control. Marta received a Ph.D. in forage management from the University of Florida's Agronomy Department in 2017, where she evaluated ecosystem services associated with the inclusion of legumes in grass-based beef cattle systems, including nutrient cycling, greenhouse gas emissions, forage production and nutritive value, and animal diet selection.

Silviya Korpilo, University of Helsinki, Finland
Silviya Korpilo, Ph.D., is a postdoctoral researcher and studies human-nature connections in cities from a socioecological-technological systems perspective and particularly how participatory GIS methods can support planning and management of urban green spaces. In her current research, she is examining the links between landscape and soundscape quality and psychological restoration.

Brian Kronvang, Aarhus University, Denmark
Brian Kronvang is professor in catchment science and environmental management in the Department of Ecoscience at Aarhus University. He has more than 30 years of experience with research projects related to pressure-impact analysis on freshwater and use of nature-based and technological solutions to improve the environment.

Anita Latai-Niusulu, National University of Samoa, Samoa
Anita Latai-Niusulu is a geography lecturer and the head of the Department of Social Sciences at the National University of Samoa. Her research activities have focused on explorations of islanders' resilience and survival strategies, climate change, and other environmental challenges affecting islanders, sustainability, environmental governance/management, and urban and children's geographies. Her hobbies include spending time in her garden, playing with her son, and having a laugh with her colleagues at the Faculty of Arts' Round Table.

Elisa Lähde, Aalto University, Finland
Elisa Lähde is an assistant professor of landscape architecture in Aalto ARTS. She actively develops new tools to integrate green infrastructure approach into urban planning through research. Her expertise is related to co-creation methods that enable sustainability change, systems thinking and deeper understanding of ecological processes in urban context.

Emily Leyden, University of Adelaide, Australia
Emily Leyden is Research Fellow at the University of Adelaide, with research interests in soil geochemistry, coastal acid sulfate soils, geochemical and hydrological modelling, environmental resource management, and sea level rise. Her research primarily examines geochemical changes and complexities in coastal soils and landscapes due to sea water intrusion.

Andrew MacKenzie, University of the South Pacific, Vanuatu
Andrew MacKenzie is associate professor and director of the University of the South Pacific, Emalus Campus, Vanuatu. His research includes land-use regulation, landscape governance, and the adaption of urban landscape systems in response to disasters. Andrew is a registered landscape architect and a member of the International Federation of Landscape Architects Education Capacity Building Working Group.

Magdalena Reynoso Martinez, EarthEmpower
Magdalena Reynoso Martinez is an agricultural value chain researcher who specializes in rural development and social inclusivity in Latin America. She co-founded EarthEmpower Consulting, which focuses on providing socially inclusive consulting services in agricultural research, monitoring and evaluation, and value chain development. Magda holds a B.S. in agricultural engineering and animal sciences from Monterrey Technological University. She has led research and consulting projects with the Nature Conservancy, the Food and Agricultural Organization, the Economic Commission for Latin America and the Caribbean, and the World Wide Fund for Nature in Mexico, Cuba, Guatemala, El Salvador, the Dominican Republic, Honduras, and Belize.

Tania Eulalia Martinez-Cruz, Université Libre de Bruxelles, Belgium
Tania Eulalia Martinez-Cruz is an Ëyuujk Indigenous woman from Tamazulápam del Espíritu Santo, Mixe, Oaxaca, Mexico. She is a multidisciplinary scientist with a B.S. in irrigation engineering from Chapingo Autonomous University, an M.S. in agricultural and biosystems engineering from the University of Arizona, and a Ph.D. in social sciences from Wageningen University. Currently, she is a research associate at Laboratoire d'Anthropologie des Mondes Contemporain, Université Libre de Bruxelles. Tania was a Fulbright scholar from 2010 to 2012 and was awarded the National Youth Prize for Academic Achievement by the Mexican President in 2016.

Fezile Mathenjwa, South African National Biodiversity Institute, South Africa and University of Cape Town, South Africa

Fezile Mathenjwa is a biodiversity data technician at the South African National Biodiversity Institute. Her work focuses on managing biodiversity data. She is a trained environmental scientist who holds a master's degree in environment, society, and sustainability from the University of Cape Town. Her research interest is in sustainable management strategies for biodiversity conservation in urban landscapes.

Thomas P. McKenna, University of Kansas, United States
Thomas P. McKenna is an assistant research professor at the Kansas Biological Survey, a University of Kansas Designated Research Center. McKenna's research focuses on plant-soil interactions in natural and agricultural systems, with an emphasis on ecological intensification in perennial agriculture.

Pablo Modernel
Pablo Modernel is a Uruguayan agricultural engineer working in sustainable livestock farming systems. He completed his Ph.D. at Wageningen University in 2018 studying the impact of intensification strategies on the sustainability of the livestock production in Uruguay, with an emphasis on the native grasslands species diversity. His aim is that research can be used for policymaking, by applying strategies of intensification that consider both the farmer's livelihoods and the environment they are working on in a context of global change.

Luke Mosley, University of Adelaide, Australia
Luke Mosley is an environmental scientist with over 15 years of experience in academia and government. The overarching aim of his research is to understand and solve complex environmental problems across multi-disciplinary boundaries (water, soil, ecology, and climate). Previously he was a principal scientist (water quality) at the Environment Protection Authority (EPA), South Australia, where he won the 2011 Premier's Award for "Assessing and managing risks to water quality during extreme drought in the Lower River Murray and Lakes."

Elina Nyberg, Finnish Environment Institute, Finland
Elina Nyberg, M.Sc., works currently as a researcher at the Finnish Environment Institute. She holds a master's of science in technology in the field of spatial planning and transport engineering. Elina's research interests lie in the development of sustainable and green urban environments and participatory planning approaches.

Patrick O'Farrell, University of Cape Town, South Africa and United Nations University, Institute for Integrated Management of Material Fluxes and of Resources, Germany.
Patrick O'Farrell has worked in applied research exploring the role of nature in development for the past 22 years. His work at the interface of people and nature positions him well in making different forms of contributions toward sustainable, equitable, and just development. Notably, he provides scientific support for local, national, and international decision-makers and managers in working

toward sustainable development. He has led research work through multiple large, internationally funded multi-disciplinary projects focused on resilience, ecosystem services, the Sustainable Development Goals, and understanding the plural values of nature and human well-being.

Jakob Vesterlund Olsen, University of Copenhagen, Denmark
Jakob Vesterlund Olsen is senior adviser and holds a Ph.D. in agricultural economics from the University of Copenhagen. His research is in several aspects of agricultural economics primarily based on farm-level analysis related to productivity analysis, investment behavior, and policy assessments. His research is based on profound knowledge about the institutional organization in the agricultural sector, and he is especially interested in policy evaluation and in farmers' incentives and responses to policy changes.

Soledad Orcasberro, University of Wisconsin–Madison, United States, and Universidad de la República, Uruguay
Soledad Orcasberro is a doctoral student in Agronomy at the University of Wisconsin–Madison. She joined Dr. Picasso's Lab to research the sustainability of perennial forage systems. She holds a degree in agronomy and a master's degree in agriculture (animal science option) from Universidad de la República (Uruguay), where she studied methane emissions and ecological intensification of cow-calf systems in native grasslands. Her research interests include sustainable grazing beef and dairy systems, enteric methane mitigation strategies, resource use efficiency, and perennial forages.

Maibritt Pedersen Zari, Auckland University of Technology, New Zealand
Maibritt Pedersen Zari is an associate professor at Huri Te Ao, The School of Future Environments, Auckland University of Technology, Aotearoa New Zealand. Her research area of regenerative design redefines sustainable architecture and urban design through emulating ecosystems, working with ecologies and nature, and integrating complex social factors into architectural and urban design. Pedersen Zari's expertise includes biomimicry, biophilia, urban ecosystem services, nature-based solutions, urban climate change adaptation, urban biodiversity, and climate sensitive vernacular architecture. Her most active current research stream relates to urban climate change adaptation in Oceania. She leads a complex and diverse team aiming to co-design nature-based urban design solutions, rooted in Indigenous knowledges that support climate change adaptation and individual and community wellbeing in different contexts across Oceania (including Aotearoa). Pedersen Zari is author of *Ecologies Design: Transforming Architecture, Landscape, and Urbanism* (2020), and *Regenerative Urban Design and Ecosystem Biomimicry* (2018).

Valentín D. Picasso, University of Wisconsin–Madison, United States, and Universidad de la República, Uruguay
Valentín D. Picasso is an associate professor at the University of Wisconsin–Madison, and an adjunct professor at Universidad de la República, Uruguay. He holds a degree in agronomy from Universidad de la República, Uruguay,

and a Ph.D. in sustainable agriculture from Iowa State University. He researches forages and perennial grain crops, resilience to climate change, and ecological intensification of dairy and beef systems. He has identified ecological intensification trajectories to inform policymakers and farmers for beef, dairy, and silvopastoral systems. He leads international transdisciplinary research projects to link science and policy for sustainability in perennial diverse forage systems, Kernza, carbon footprint, and grazing systems.

Kerstin Potthoff, Norwegian University of Life Sciences, Norway
Kerstin Potthoff is a landscape geographer interested in landscape change and drivers of change. She has been working on transformation of different types of agricultural landscapes and the different factors that impact on their transformation. A main interest has been not only in mountain landscapes, changes in land use, changes in climate and elevational treeline, and forest line changes but also in how 'driving forces' can be used as an analytical tool.

Alexandra D. Solomou, Institute of Mediterranean Forest Ecosystems, Hellenic Agricultural Organization DEMETER (ELGO DIMITRA), Greece
Alexandra D. Solomou is a researcher at Plant Biodiversity in Mediterranean Ecosystems in the Institute of Mediterranean Forest Ecosystems, Hellenic Agricultural Organization DEMETER (ELGO DIMITRA). Her interests are focused on the monitoring, assessment and conservation of plant diversity in Mediterranean ecosystems. She has published more than 40 articles in international peer-reviewed journals and more than 50 articles in national and international conferences. She was also a member of the National Committee of Protected Areas "NATURE 2000" and a National Focal Point of FAO organization for Biodiversity for Food and Agriculture.

Søren Præstholm, University of Copenhagen, Denmark
Søren Præstholm is cultural geographer who holds a Ph.D. degree about changes in the agricultural landscape due to new types of land owners and changing societal demands to farming and landscapes. His research has focused on landscape functions and balancing between different demands – recent years mainly within recreational behavior, user groups, recreational opportunities, constraints, recreational planning and people's nature connection/perception. Søren is a special consultant in the Department of Geosciences and Natural Resource Management at the University of Copenhagen, and he is head of the cross-institutional partnership Children & Nature – Denmark, hosted by the university.

Amin Rastandeh, University of South Dakota, United States
Amin Rastandeh is a landscape analyst, with experience in the United States, New Zealand, and Iran, working on human-environmental interactions and climate change in evolving multifunctional landscapes. He worked as a postdoctoral researcher in the Department of Sustainability and Environment at the University of South Dakota, United States. He specializes in multi-scale design

and management of landscapes for safeguarding biodiversity and human communities in the face of climate change.

Christopher Raymond, University of Helsinki, Finland and Swedish University of Agricultural Sciences, Sweden
Christopher Raymond is a professor and an interdisciplinary scientist who creates new policy and management options for engaging diverse actors in the design and implementation of nature-based solutions. His expertise is related to the development and application of conceptual approaches for assessing human-nature relationships, knowledge co-creation processes, and analytical methods in the social valuation of land use and ecosystem services under global change.

Jesper Sølver Schou, Secretariat of the Danish Climate Council, Denmark
Jesper Sølver Schou holds a Ph.D. in Environmental Economics and has more than 25 years of experience in research, research management, and research-based public-sector service work. Schou's research covers fields of agricultural and environmental economics· and the economics of nature management and provision of public goods, with a specific focus on application of cost-benefit analysis to support policy development. Jesper has received a large number of research grants from both private and public funds and has extensive experience with cross-disciplinary research projects.

Aubrey Streit Krug, Land Institute, United States
Aubrey Streit Krug is a writer, teacher, and researcher who studies stories of relationships between humans and plants. She is the director of ecosphere studies at The Land Institute in Salina, Kansas, where she leads transdisciplinary civic science communities that bring together researchers and a range of people around the United States to grow, study, and learn with new perennial grain crops. Streit Krug grew up in a small town in Kansas and loves limestone soils and rocky prairie hillsides.

Susana Taua'a, National University of Samoa, Samoa
Susana Taua'a is an associate professor of geography at the National University of Samoa in Apia. She teaches undergraduate and postgraduate courses in urban and rural geography, coastal processes, and land management issues in the Pacific. Her research interests include the informal economy and employment creation and prospects for smallholder farming in the Pacific.

Ebru Ersoy Tonyaloğlu, Aydın Adnan Menderes University, Turkey
Ebru Ersoy Tonyaloğlu is an associate professor in landscape architecture at Aydın Adnan Menderes University. After completing her bachelor's and M.Sc. degrees at Ege University, Department of Landscape Architecture, she continued her Ph.D. studies at the University of Sheffield, Department of Landscape. She is particularly interested in multifunctional urban planning approaches. Her current work includes research on landscape ecology and landscape planning as well as national and international projects on integrated landscape planning and

historic landscapes. Her international collaborations include projects in Turkey and the UK with landscape architects, landscape planners, urban ecologists, and archaeologists.

Kati Vierikko, Finnish Environment Institute, Finland
Kati Vierikko is a docent in urban ecology and works as a senior researcher at the Finnish Environment Institute (SYKE). Her current research relates to urban biodiversity, nature-based solutions and blue-green infrastructure. She is developing methods and tools for enhancing biodiversity in cities at different spatial scales. Her previous research topics have focused on sustainable forest use in Finland, biocultural diversity and cultural ecosystem services of urban blue-green infrastructure.

Kaihang Zhou, University of Adelaide, Australia
Kaihang Zhou has a degree in landscape architecture from Shenyang Jianzhu University in China and a master's degree in landscape architecture from the University of Adelaide, Australia. He is interested in the relationship between landscape design and ecology. His current research focuses on risks and opportunities to coastal wastewater treatment systems in response to the sea level rise in Australia and evaluating different landscape approaches to consider the consequence for different courses of action.

ACKNOWLEDGEMENTS

We deeply thank Pippin Anderson, Laura Astigarraga, Birsen Kesgin Atak, Paul Blaschke, Victoria Chanse, Rachael Ann Cox, Timothy E. Crews, Wenche E. Dramstad, Katsue Fukamachi, Scott Hawken, Pia Heike Johansen, Gabriel Luke Kiddle, Simon Kilbane, Marta Moura Kohmann, Silviya Korpilo, Brian Kronvang, Anita Latai-Niusulu, Elisa Lähde, Emily Leyden, Andrew MacKenzie, Magdalena Reynoso Martinez, Tania Eulalia Martinez-Cruz, Fezile Mathenjwa, Thomas P. McKenna, Pablo Modernel, Luke Mosley, Elina Nyberg, Patrick O'Farrell, Jakob Vesterlund Olsen, Soledad Orcasberro, Maibritt Pedersen Zari, Valentín D. Picasso, Kerstin Potthoff, Alexandra D. Solomou, Søren Præstholm, Christopher Raymond, Jesper Sølver Schou, Aubrey Streit Krug, Susana Taua'a, Ebru Ersoy Tonyaloğlu, Kati Vierikko, and Kaihang Zhou for their significant, impactful, and precious contributions to this book. Support received from the National Science Foundation through the EPSCoR Track II cooperative agreement OIA-1632810 and NSF DBI-1560048 is highly appreciated.

1

LANDSCAPE RESILIENCE IN THE FACE OF CLIMATE CHANGE

A call to transition from despair to hope

Amin Rastandeh and Meghann Jarchow

A dramatic change is here

The world is getting warmer, landscapes fragment, wildlife species perish, ecosystems collapse at alarming rates, wildfires, drought, and water scarcity engulf us more than ever before, more people are displaced as a result of extreme, unusual, and unseasonal climate events, desertification advances quickly, and food insecurity is on the rise in many parts of the world. In our landscapes, we are grappling with a wide range of socio-ecological challenges triggered by the interwoven impacts of land-use change and climate change. Fear, alarmism, guilt, and shame are all possible responses to these challenges, but they are unlikely to be effective in motivating the large-scale and long-term change needed to address these challenges. What we need now are effective ways to change including creating widespread hope.

Our climate has been relatively stable, and humans did not have a global influence until more recently. This has created a mismatch; where many of our societies developed during a time of relative stability, and we are now entering a period of rapid change – potentially catastrophic change. Hope, science, cooperation, and innovation are just some of the tools that we need to move toward greater resilience. In this book, we place our focus on the concept of landscape resilience, its applications, and implications at multiple scales, from small urban gardens in Vanuatu to vast grasslands in the Great Plains of the United States, to discuss how this change may occur on the ground.

In order to change, we need to be brave enough to see our world as it is and to create a transformed world that is able to support human and non-human species. We use the verb "to create" in the title of this book to draw attention to the importance of proactive actions toward the future that we want. Creating resilient landscapes is a critical component of that desired future. Crawford Stanley

DOI: 10.4324/9781003266440-1

(C.S.) Holling, the father of resilience research, defined the concept of resilience in ecological systems as "a measure of the ability of [complex ecological systems] to absorb changes of state variables, driving variables, and parameters, and still persist" (Holling 1973). Resilience is an approach to facilitate the process of re-thinking, re-organizing, and reshaping socio-ecological systems in dynamic and turbulent conditions (Reyers et al. 2018).

The change from dysfunctionality to multifunctionality; from defenselessness to resilience, requires a comprehensive, multi-dimensional transformation. Therefore, we need to increase our capacity for transformability (Folke et al. 2004) in ways that we can absorb unforeseen and unexperienced changes while ensuring critical ecosystem functions and services in our landscapes. Transformability is "the capacity to create a fundamentally new system when ecological, economic, or social (including political) conditions make the existing system untenable" (Walker et al. 2004; cf. Folke et al. 2004). One important requirement for this transformation is to restrain the current competitive consumerism and shift the global community to an equity-centered vision that ensures the resilience of socio-ecological systems (Chapin et al. 2022). Unprecedented transformative solutions that recognize the reorganization of complex socio-ecological systems can be an effective vehicle to increase socio-ecological resilience in this context (e.g., an era of anthropogenic climate change, widespread land-use change, and rapid urbanization; Elmqvist et al. 2019).

Currently, there is a plethora of information about the impacts of climate change, biodiversity loss, intensive agriculture, and rapid urbanization; and scientists have depicted and/or documented the magnitude and extensiveness of these impacts on the Earth's abiotic, biotic, and cultural resources. Despite this plethora of information, action and change are too slow. There is a need for more focus on innovative, feasible, and flexible solutions.

Because less focus has been placed on the question of "how to change" toward a more resilient future, we believe that time is ripe to utilize a forward-thinking "how-to-change" approach in order to adapt to the impacts of climate change, while slowing down the current trends in biodiversity loss and land-cover change. Here, we address the question of "how to change toward a better future through creating resilient landscapes" by providing a collection of place-based and science-backed solutions capable of responding to the current and future socio-environmental challenges in our landscapes.

In this book, we show that resilience is an evolving, plastic, and multidimensional concept with diverse and wide-ranging implications for how we might address creating more resilient landscapes. In this context, we define landscape resilience as the ability of landscapes, including their human communities, to adapt to and to mitigate the impacts of climate change; and to persist under changing and uncertain conditions such that safeguard biodiversity and nature's contributions to people (Díaz et al. 2018). Although creating resilient landscapes is one fundamental means for adaptation to the impacts of climate change, we show that tools, methods, expectations, and outcomes are, to some extent, region-specific, sensitive to socio-ecological conditions, and therefore subject to change over time. Despite this, we show that the root causes of the challenges and many general themes of the solutions are common. As such, the book suggests that characteristics of socio-ecological resilience in Australia, Japan, and South Africa differ; socio-ecological requirements to make agricultural landscapes of Turkey resilient are not the same as those of landscapes of Uruguay and Argentina; what Indigenous Peoples of Vanuatu, Samoa, Guatemala, and Mexico expect from socio-ecological resilience in their landscapes are, to some extent, divergent; and the creation of resilient landscapes in the Western countries like the United States, Denmark, Norway, and Finland requires its own local considerations. Accepting these differences, the book reflects foundational commonalities, as well. According to the case studies presented in this book, perhaps our biggest and most important commonality is that the global community wants to create resilient landscapes for its future; however, the book shows that tools, methods, and approaches differ from region to region. To empower the global community in the face of the impacts of climate change, we believe that accepting the existence of these differences is essential.

These case studies indicate that we have aggressively exploited biodiversity, vastly altered the structure of landscapes, and irreversibly destroyed ecosystems in order to respond to our increasing needs and demands for food, water, shelter, and recreation. Today, mega-scale, anthropogenic interventions in ecological patterns and processes, combined with climate change, unprecedentedly influence our communities and landscapes. At the same time, these case studies show that there are still innovative, locally acceptable solutions to leave the current "despair" and move toward a science-backed "hope".

Recognizing challenges and creating hope

We believe in the importance of hope, and through this, we are not advocating ignorance or avoidance. "Having hope within uncertainty is not unrealistic positivity or optimism – rather hope fuels meliorism; a belief in our agency to make the world a better place in spite of the turbulence we experience" (Strazds 2019). We need to understand the crises posed by climate change so that we may address them. Climate change is expected to cause a reduction of 11%–14% of the global economic output by 2050 (Swiss Re Institute 2021). The global productivity loss as a result of rising temperatures will be 2.2% of total worldwide working hours by 2030 (i.e., 80 million full-time jobs; ILO 2019). Of 31 million people displaced by disasters in 2020, about 98% were as a result of weather-related hazards (Internal Displacement Monitoring Centre 2021). By 2060, the world's human population in coastlines will reach to approximately 1.4 billion people (Dawson et al. 2018), and this, in turn, increases the vulnerability of a larger portion of the global community to sea level rise, flooding, and coastal storms while much of our infrastructure is still inadequate. By 2100, at least 200 million people will face the direct impacts of sea level rise, especially in Asia,

Brazil, and the United States (Kulp and Strauss 2019). Currently, 2.3 billion people live in water-stressed or high/critically water-stressed countries (UN-Water 2021). In drought-sensitive countries of Sub-Saharan Africa, in particular, the number of undernourished people has increased by 45.6% since 2012 (FAO, IFAD, UNICEF, WFP and WHO 2019). The Amazon rainforest has faced an unforeseen drought (Jiménez-Muñoz et al. 2016). More than 75% of this critical ecosystem has been losing resilience since the early 2000s (Boulton et al. 2022). In 2003, extreme heat waves killed ca. 72,000 people in Europe (UNISDR 2018). The Dixie Fire in the summer of 2021 destroyed almost the entire town of Greenville, California, depicted an apocalyptic, horrifying picture of the impacts of climate change on human communities and ecosystems in mountainous landscapes. Australia's 2019–2020 bushfires led to the displacement of about 65,000 people (du Parc and Yasukawa 2020), mostly Indigenous Australians who traditionally have been living in the bushfire-affected areas (Williamson et al. 2020). Bushfires also impacted a large number of wildlife species in about 11.46 million hectares of landscapes of the southeast, southwest, and northern Australia (WWF-Australia 2020; cf. Ward et al. 2020). Only ca. 3% of the world's terrestrial ecosystems are still intact (Plumptre et al. 2021). Vector-borne diseases are very likely to become more widespread as the climate continues to change (Tanser et al. 2003; Watts et al. 2021). The outbreak of malaria epidemics is linked to unusual and unseasonal rainfalls in West Africa (Diouf et al. 2022), mainly because malaria-carrying mosquitoes benefit from such conditions. The increasing rate of mental disorders can be partly attributed to climate change (Berry et al. 2010; Qi et al. 2015; Dumont et al. 2020).

Yet there is reason for hope. According to the Peoples' Climate Vote (UNEP and University of Oxford 2021), the largest survey of public opinion on climate change ever conducted with 1.2 million participants from 50 countries, 64% of people believed that climate change is an "emergency". The climate emergency was most recognized among peoples of the small island developing states (74%), where the impacts of climate change on landscapes and communities are widespread and devastating. According to this survey, 59% of people believed that "everything necessary" must be conducted to mitigate the impacts of climate change. Between 2013 and 2018, concerns about climate change have increased globally from 56% to 67%; and in 2018, at least 59% of Americans considered climate change as a major threat (PEW Research Center 2019). The Climate Strike, as a growing phenomenon, is an example of a global network of youth, with common goals, who strive to make positive change for their futures. Day by day, more people are changing their lifestyle, especially in terms of their choices for clothing and food consumption, to fight climate change. People are being informed of the fact that their diets directly influence their landscapes, and this, in turn, will have implications for the rate of greenhouse gas emissions, worldwide (Poore and Nemecek 2018). Today, we know that about 34% of global GHG emissions are attributed to human food systems (Crippa et al. 2021), mainly for the production of meat and dairy (Watts et al. 2021). The growing number of people who ethically/environmentally decide to change their diet and lifestyle in response to climate change can be one outcome of public awareness about the impacts of human food preferences on landscapes (Cooper 2018; Kortetmäki and Oksanen 2021). This social readiness is admirable and necessary, but is that enough?

To respond to climate change, the number of landscape restoration/conservation projects is on the rise around the world (Tucker and Simmons 2009; Franklin 2018; Goffner et al. 2019; Carver et al. 2021). Efforts to integrate land-sharing/sparing with land-use policies are rapidly increasing (Karner et al. 2019; Ibáñez-Álamo et al. 2020; Fischer et al. 2020). Our understanding of the importance of urban agriculture for food security, biodiversity, and landscape resilience has greatly expanded (Orsini et al. 2013; Clucas et al. 2018; Ferreira et al. 2018; Langemeyer et al. 2021). Today, more heed is given to the sustainable use of water for food production in water-stressed landscapes (Faramarzi et al. 2010; Maroufpoor et al. 2021).

As the climate continues to change, the role of nature-based solutions in creating resilience is being increasingly recognized, worldwide. According to Griscom et al. (2017) about 37% of the emission reductions needed by 2030 to keep global temperature increases fewer than 2°C can be provided by nature-based solutions. It is perhaps why nature-based solutions have recently received more attention for enhancing socio-ecological sustainability and landscape resilience at multiple scales (Lafortezza et al. 2018; Bush and Doyon 2019; Thorn et al. 2021). Humans have started to use landscapes for clean and renewable energy production. For example, the global capacity of wind power has substantially increased, currently helping the world avoid over 1.2 billion tons of CO_2 emissions per year (GWEC 2022). In 2020, global energy generation from renewable sources, mainly wind and solar energy, reached to a new record (358 TWh increase), while carbon emissions from energy use decreased by 6.3%, to their lowest level since 2011 (BP 2021).

A call to transition to a more resilient future

Ecological and social systems are similar in their adaptive and renewal cycles (Holling and Gunderson 2002). As a result, the concept of resilience has been evolving over the past five decades into a multi- and cross-disciplinary framework for understanding socio-ecological systems. Today, socio-ecological resilience is a recognized concept in a wide range of environmental and social sciences (Chambers et al. 2019) and can be measured, even in complex systems (Arani et al. 2021). However, we still observe a chasm between science and practice when addressing the possibility of creating resilient landscapes in the real world, perhaps because of a widespread conflict of interests among a diverse range of stakeholders, ranging from local people within a small village to nations of different colors, norms, traditions, and values, who want to use their landscapes for multifarious, and sometimes contrasting reasons/purposes. The existence of this divergence across multiple spatiotemporal scales requires us to take proper actions using different tools, methods, and approaches. Moreover, our priorities constantly change over space and time based on our socio-economic, political, and environmental conditions. Therefore, while we embrace the significance of resilience as an overarching goal in landscapes of different socio-ecological characteristics, we simultaneously need to find new processes through which the global community can effectively act upon all of the commonalities in spite of all of the existing differences and contradictions. In addition, we need to examine how low-cost and cost-effective solutions can put this intention into action in the real world. In response to this complicated challenge, the chapter authors employ a hopeful, future-facing attitude to bridge the current gap between science and practice by exploring feasible paths into the future. This book is a collective call to a science-informed transition from inaction to proaction; from dysfunctionality to multifunctionality; and from brittleness to resilience in our landscapes. The book consists of 17 chapters, including 15 case studies addressing possibilities for creating socio-ecological resilience in landscapes of Argentina, Australia, Brazil, Denmark, Finland, Greece, Guatemala, Japan, Mexico, Norway, Samoa, South Africa, the United States, Turkey, Uruguay, and Vanuatu.

Biggs et al. (2012) suggested seven principles for enhancing the resilience of ecosystem services in landscapes: (1) maintain diversity and redundancy, (2) manage connectivity, (3) manage slow variables and feedbacks, (4) foster an understanding of socio-ecological systems as complex adaptive systems, (5) encourage learning and experimentation, (6) broaden participation, and (7) promote polycentric governance systems. Various aspects of these seven principles have been addressed from different angles by international subject-matter experts whose forward-thinking case studies contributed to this book. The case studies explore landscapes at multiple scales with an in-depth understanding of the importance of interactions between society and nature. They suggest new ways for creating landscapes that are socially and ecologically tolerant and flexible, capable of supporting the quality of abiotic, biotic, and cultural resources in the face of the major drivers of environmental change, including climate change, land-cover transformation, biodiversity loss, and urbanization. In light of this, they suggest processes through which human communities learn how to use their landscapes to better interact and coexist with the natural capital (i.e., abiotic and biotic resources), while understanding the evolving nature of ecological patterns and processes under a changing climate.

Acknowledgment

This work was supported by the National Science Foundation through the EPSCoR Track II cooperative agreement OIA-1632810 and NSF DBI-1560048.

References

Arani, B. M., Carpenter, S. R., Lahti, L., Van Nes, E. H., & Scheffer, M. (2021). Exit time as a measure of ecological resilience. *Science*, 372(6547), eaay4895.

Berry, H. L., Bowen, K., & Kjellstrom, T. (2010). Climate change and mental health: A causal pathways framework. *International Journal of Public Health*, 55, 123–132.

Biggs, R., Schlüter, M., Biggs, D., Bohensky, E. L., BurnSilver, S., Cundill, G.,... & West, P. C. (2012). Toward principles for enhancing the resilience of ecosystem services. *Annual Review of Environment & Resources*, 37, 421–448.

Boulton, C. A., Lenton, T. M., & Boers, N. (2022). Pronounced loss of Amazon rainforest resilience since the early 2000s. *Nature Climate Change*, 12(3), 271–278.

BP. (2021). Statistical Review of World Energy, July 2021, 70th Edition. Available at: https://www.bp.com/en/global/corporate/energy-economics/statistical-review-of-world-energy.html

Bush, J., & Doyon, A. (2019). Building urban resilience with nature-based solutions: How can urban planning contribute?. *Cities*, 95, 102483.

Carver, S., Convery, I., Hawkins, S., Beyers, R., Eagle, A., Kun, Z.,... & Soulé, M. (2021). Guiding principles for rewilding. *Conservation Biology*, 35(6), 1882–1893.

Chambers, J. C., Allen, C. R., & Cushman, S. A. (2019). Operationalizing ecological resilience concepts for managing species and ecosystems at risk. *Frontiers in Ecology and Evolution*, 7, 241.

Chapin, F. S., Weber, E. U., Bennett, E. M., Biggs, R., van den Bergh, J., Adger, W. N.,... & de Zeeuw, A. (2022). Earth stewardship: Shaping a sustainable future through interacting policy and norm shifts. *Ambio*, 51, 1907–1920. https://link.springer.com/content/pdf/10.1007/s13280-022-01721-3.pdf

Clucas, B., Parker, I. D., & Feldpausch-Parker, A. M. (2018). A systematic review of the relationship between urban agriculture and biodiversity. *Urban Ecosystems*, 21(4), 635–643.

Cooper, L. L. (2018). A new veganism: How climate change has created more vegans. *Granite: Aberdeen University Postgraduate Interdisciplinary Journal*, 2(1), 16–24.

Crippa, M., Solazzo, E., Guizzardi, D., Monforti-Ferrario, F., Tubiello, F. N., & Leip, A. J. N. F. (2021). Food systems are responsible for a third of global anthropogenic GHG emissions. *Nature Food*, 2(3), 198–209.

Dawson, R. J., Khan, M. S. A., Gornitz, V., Lemos, M. F., Atkinson, L., Pullen, J., and Osorio, J. C. (2018). Urban areas in coastal zones. In Rosenzweig, C., W. Solecki, P. Romero-Lankao, S. Mehrotra, S. Dhakal, and S. Ali Ibrahim (eds.), *Climate Change and Cities: Second Assessment Report of the Urban Climate Change Research Network*. New York: Cambridge University Press, 319–362.

Díaz, S., Pascual, U., Stenseke, M., Martín-López, B., Watson, R. T., Molnár, Z.,... & Shirayama, Y. (2018). Assessing nature's contributions to people. *Science*, 359(6373), 270–272.

Diouf, I., Adeola, A. M., Abiodun, G. J., Lennard, C., Shirinde, J. M., Yaka, P.,... & Gbobaniyi, E. O. (2022). Impact of future climate change on malaria in West Africa. *Theoretical and Applied Climatology*, 147(3), 853–865.

du Parc, E., & Yasukawa, L. (2020). *The 2019–2020 Australian Bushfires: From Temporary Evacuation to Longer-Term Displacement*. Geneva: Internal Displacement Monitoring Centre.

Dumont, C., Haase, E., Dolber, T., Lewis, J., & Coverdale, J. (2020). Climate change and risk of completed suicide. *The Journal of Nervous and Mental Disease*, 208(7), 559–565.

Elmqvist, T., Andersson, E., Frantzeskaki, N., McPhearson, T., Olsson, P., Gaffney, O.,... & Folke, C. (2019). Sustainability and resilience for transformation in the urban century. *Nature Sustainability*, 2(4), 267–273.

FAO, IFAD, UNICEF, WFP and WHO (2019). *The State of Food Security and Nutrition in the World 2019. Safeguarding Against Economic Slowdowns and Downturns*. Rome: FAO.

Faramarzi, M., Yang, H., Mousavi, J., Schulin, R., Binder, C. R., & Abbaspour, K. C. (2010). Analysis of intra-country virtual water trade strategy to alleviate water scarcity in Iran. *Hydrology and Earth System Sciences*, 14, 1417–1433.

Ferreira, A. J. D., Guilherme, R. I. M. M., & Ferreira, C. S. S. (2018). Urban agriculture, a tool towards more resilient urban communities?. *Current Opinion in Environmental Science & Health*, 5, 93–97.

Fischer, L. K., Neuenkamp, L., Lampinen, J., Tuomi, M., Alday, J. G., Bucharova, A.,... & Klaus, V. H. (2020). Public attitudes toward biodiversity-friendly greenspace management in Europe. *Conservation Letters*, 13(4), e12718.

Folke, C., Carpenter, S., Walker, B., Scheffer, M., Elmqvist, T., Gunderson, L., Holling, C. S. (2004). Regime shifts, resilience, and biodiversity in ecosystem management. *Annual Reviews of Ecology, Evolution, and Systematics*, 35, 557–581.

Franklin, J. (2018). Can the world's most ambitious rewilding project restore Patagonia's beauty? *The Guardian*. Available at: https://www.theguardian.com/environment/2018/may/30/can-the-worlds-largest-rewilding-project-restore-patagonias-beauty

Goffner, D., Sinare, H., & Gordon, L. J. (2019). The Great Green Wall for the Sahara and the Sahel Initiative as an opportunity to enhance resilience in Sahelian landscapes and livelihoods. *Regional Environmental Change*, 19(5), 1417–1428.

Griscom, B. W., Adams, J., Ellis, P. W., Houghton, R. A., Lomax, G., Miteva, D. A., ... & Fargione, J. (2017). Natural climate solutions. *Proceedings of the National Academy of Sciences*, 114(44), 11645–11650.

GWEC. (2022). *Global Wind Report 2022*. Brussels: Global Wind Energy Council.

Holling, C. S. (1973). Resilience and stability of ecological systems. *Annual Review of Ecology and Systematics*, 4(1), 1–23.

Holling, C. S., & Gunderson, L. H. (2002). Resilience and adaptive cycles. In L. H. Gunderson and C. S. Holling (eds.), *Panarchy: Understanding Transformations in Human and Natural Systems*. Washington, DC: Island Press, 25–62.

Ibáñez-Álamo, J. D., Morelli, F., Benedetti, Y., Rubio, E., Jokimäki, J., Pérez-Contreras, T.,... & Díaz, M. (2020). Biodiversity within the city: Effects of land sharing and land sparing urban development on avian diversity. *Science of the Total Environment*, 707, 135477.

ILO. (2019). Working on a warmer planet: The impact of heat stress on labour productivity and decent work. International Labour Office, ILO: Geneva.

Internal Displacement Monitoring Centre. (2021). *Global Report on Internal Displacement 2021: Internal Displacement in a Changing Climate*. Geneva, Switzerland: Internal Displacement Monitoring Centre.

Jiménez-Muñoz, J. C., Mattar, C., Barichivich, J., Santamaría-Artigas, A., Takahashi, K., Malhi, Y.,... & Schrier, G. V. D. (2016). Record-breaking warming and extreme drought in the Amazon rainforest during the course of El Niño 2015–2016. *Scientific Reports*, 6(1), 1–7.

Karner, K., Cord, A. F., Hagemann, N., Hernandez-Mora, N., Holzkämper, A., Jeangros, B.,... & Schönhart, M. (2019). Developing stakeholder-driven scenarios on land sharing and land sparing–Insights from five European case studies. *Journal of Environmental Management*, 241, 488–500.

Kortetmäki, T., & Oksanen, M. (2021). Is there a convincing case for climate veganism?. *Agriculture and Human Values*, 38(3), 729–740.

Kulp, S. A., & Strauss, B. H. (2019). New elevation data triple estimates of global vulnerability to sea-level rise and coastal flooding. *Nature Communications*, 10(1), 1–12.

Lafortezza, R., Chen, J., Van Den Bosch, C. K., &Randrup, T. B. (2018). Nature-based solutions for resilient landscapes and cities. *Environmental Research*, 165, 431–441.

Langemeyer, J., Madrid-Lopez, C., Beltran, A. M., & Mendez, G. V. (2021). Urban agriculture—A necessary pathway towards urban resilience and global sustainability?. *Landscape and Urban Planning*, 210, 104055.

Maroufpoor, S., Bozorg-Haddad, O., Maroufpoor, E., Gerbens-Leenes, P. W., Loáiciga, H. A., Savic, D., & Singh, V. P. (2021). Optimal virtual water flows for improved food security in water-scarce countries. *Scientific Reports*, 11(1), 1–18.

Orsini, F., Kahane, R., Nono-Womdim, R., &Gianquinto, G. (2013). Urban agriculture in the developing world: A review. *Agronomy for Sustainable Development*, 33(4), 695–720.

Pew Research Center. (2019). Climate Change Still Seen as the Top Global Threat, but Cyberattacks a Rising Concern. Spring 2018 Survey February 10, 2019 Release.

Plumptre, A. J., Baisero, D., Belote, R. T., Vázquez-Domínguez, E., Faurby, S., Jędrzejewski, W.,... & Boyd, C. (2021). Where might we find ecologically intact communities?. *Frontiers in Forests and Global Change*, 4, 26.

Poore, J., &Nemecek, T. (2018). Reducing food's environmental impacts through producers and consumers. *Science*, 360(6392), 987–992.

Qi, X., Hu, W., Page, A., & Tong, S. (2015). Associations between climate variability, unemployment and suicide in Australia: A multicity study. *BMC Psychiatry*, 15(1), 1–12.

Reyers, B., Folke, C., Moore, M. L., Biggs, R., & Galaz, V. (2018). Social-ecological systems insights for navigating the dynamics of the Anthropocene. *Annual Review of Environment and Resources*, 43, 267–289.

Strazds, L. M. (2019). Radical hope: Transforming sustainability. *Journal of Sustainability Education*, 21. Retrieved from http://www.susted.com/wordpress/wp-content/uploads/2019/12/Strazds-JSE-December-2019-General-Issue-PDF.pdf

Swiss Re Institute. (2021). The economics of climate change: No action not an option. Available at: https://www.swissre.com/risk-knowledge/mitigating-climate-risk/no-action-climate-change-not-option.html

Tanser, F. C., Sharp, B., & Le Sueur, D. (2003). Potential effect of climate change on malaria transmission in Africa. *The Lancet*, 362(9398), 1792–1798.

Thorn, J. P. R., Aleu, R. B., Wijesinghe, A., Mdongwe, M., Marchant, R. A., & Shackleton, S. (2021). Mainstreaming nature-based solutions for climate resilient infrastructure in peri-urban sub-Saharan Africa. *Landscape and Urban Planning*, 216, 104235.

Tucker, N. I., & Simmons, T. (2009). Restoring a rainforest habitat linkage in north Queensland: Donaghy's Corridor. *Ecological Management & Restoration*, 10(2), 98–112.

UNEP and University of Oxford. (2021). People's Climate Vote Results. *United Nations Development Programme (UNDP).* Available at: https://www.undp.org/library/peoples-climate-vote

UNISDR. (2018). *Economic Losses, Poverty and Disasters 1998–2017.* Geneva, Switzerland: The UN Office for Disaster Risk Reduction.

UN-Water. (2021). *Summary Progress Update 2021 – SDG 6 – water and sanitation for all. Version: July 2021.* Geneva, Switzerland: UN-Water.

Walker, B., Holling, C. S., Carpenter, S. R., & Kinzig, A. (2004). Resilience, adaptability and transformability in social–ecological systems. *Ecology and Society*, 9(2). https://www.jstor.org/stable/26267673

Ward, M., Tulloch, A. I., Radford, J. Q., Williams, B. A., Reside, A. E., Macdonald, S. L.,... & Watson, J. E. (2020). Impact of 2019–2020 mega-fires on Australian fauna habitat. *Nature Ecology & Evolution*, 4(10), 1321–1326.

Watts, N., Amann, M., Arnell, N., Ayeb-Karlsson, S., Beagley, J., Belesova, K.,... & Costello, A. (2021). The 2020 report of the Lancet Countdown on health and climate change: Responding to converging crises. *The Lancet*, 397(10269), 129–170.

Williamson, B., Markham, F. & Weir, J. K. (2020). *Aboriginal Peoples and the Response to the 2019–2020 Bushfires*, Working Paper No. 134/2020, Centre for Aboriginal Economic Policy Research, Australian National University, Canberra.

WWF-Australia. (2020). *Australia's 2019–2020 Bushfires: The Wildlife Toll.* WWF-Australia. Available at: https://www.wwf.org.au/what-we-do/bushfire-recovery/in-depth/resources/australia-s-2019-2020-bushfires-the-wildlife-toll

2

MULTIFUNCTIONAL LAND CONSOLIDATIONS IN DENMARK

Rethinking the pattern of landownership to create resilient future landscapes

Søren Præstholm, Brian Kronvang, Jakob Vesterlund Olsen, Jesper Sølver Schou and Pia Heike Johansen

A vision for a multifunctional land consolidation

The sixth assessment report from the UN Intergovernmental Panel on Climate Change (IPCC) clearly states that human impacts have and will influence the climate (IPCC, 2021:1–41). There is a need for resilient and flexible landscapes to adapt to these changes to mitigate the negative societal consequences. On the other hand, urgent calls for reducing the greenhouse gases (GHG) emission factors that accelerate these climate changes can be addressed by land use changes in the very same landscapes. This is not least the case in Denmark, where more than half of the country's land surface is covered with agricultural crops in combination with intensive livestock production on many farms (Statistics Denmark, 2021). Today, the agricultural sector accounts for 22.4% (2017) of the Danish emission of CO_2 equivalents (Nielsen, 2019:14–16).

The present governmental and a parliamentary majority has recently agreed to find new measures to reduce approximately half of the emissions from the agricultural sector compared to former predictions for 2030 emission levels to reach a national reduction of 70% compared to 1990. One of the expected measures is abandonment of agricultural production and termination of drainage on 100,000 ha peat soils before 2030 (Ministry of Finance, 2021).

Reduction of GHGs is not the only societal demand that local agricultural landscapes and rural communities are facing. Other societal demands include enriching biodiversity; reducing the impact of nutrients on water quality; creating better outdoor recreation opportunities and supporting development of rural communities while still having food and fibers produced. Balancing demands can include both sharing and sparing land (Ben et al., 2011; Loconto et al., 2020). It was recently decided to spare land for better protection and improvement of biodiversity by designation of 15 new so-called Nature National Parks and to

DOI: 10.4324/9781003266440-2

abolish timber production in many state-owned forests (Ministry of Environment, 2020). However, promoting a land sharing strategy in terms of a more multifunctional agricultural landscapes has been encouraged for decades both in Danish legislation and in the EU's Common Agricultural Policy (CAP) (van G. Huylenbroeck & Durand, 2003; Primdahl, 2014).

In 2017, it was estimated that there is a total need for 30–40% additional land surface in Denmark if recent policy-stated goals and demands are to be fulfilled (Arler et al., 2017). Based on both expert knowledge, involvement of politicians and organizations and public hearings, the final report recommends both land sharing in terms of integration of multiple functions and prioritization of the national goal. It further proposes to strengthen the national planning and designations along with more freedom to finding local solutions and differentiation according to the local context, e.g. by better local landscape/village development strategies and plans based on participatory processes (Arler et al., 2017).

The need for finding better solutions for the future landscapes was also addressed in another Danish initiative under the headline of 'Collective Impact' in 2014. National stakeholders, representing both NGOs and authorities, committed each other to finding sustainable solutions to societal challenges. One of the themes is 'Future Sustainable Landscapes' and stakeholders within agriculture, forestry, nature conservation, recreation and regional authorities collaborate about multifunctional land consolidation as one of the means. The broad group of stakeholders emphasized that a heterogenic and complicated spatial pattern of land ownership is a severe obstacle to a transformation of landscapes to better pursuing recent and future societal demands. The concept of 'resilient landscape' was not explicitly given attention by the stakeholders. Nevertheless, it was central to the project approach to align ownership pattern better with ecosystem characteristics to coping with changing climate, environment or other conditions. Hence, the multifunctional land consolidations may be a relevant tool for creating future resilient landscapes.

Denmark has a long tradition for land consolidation processes. In the 1760s to 1780s, the so-called enclosure movement resulted in a land reform that over the next few decades completely changed the allotment pattern in Denmark: Land consolidations amalgamated the individual properties to one or few cadastral plots (Hartvigsen, 2014, 2015). This reform was successful in terms of fulfilling a societal demand for a more efficient agricultural production at that time. Inspired by the historic enclosure movement, the present group of stakeholders in the Future Sustainable Landscapes initiative formulated a vision for a contemporary land reform with a restructuring of the pattern of ownership to create landscapes suitable for fulfilling recent and future societal demands. To prepare the ground for a new land reform, pilot projects with multifunctional land consolidations were initiated in three local areas in 2016. A fourth project followed later.

The chapter presents the vision, the pilot projects and learnings from the process and the still ongoing research based evaluation (Johansen et al., 2020; Johansen, Ejrnæs, et al., 2018) of the multifunctional land consolidation initiative in Denmark. The last section broadens the perspectives from the lesson learned in Denmark by asking the question: Are multifunctional land consolidations – in the way they were done in the pilot projects – an applicable way to create resilient future landscapes in Denmark and abroad?

Land consolidation as a means for multifunctional landscapes

A scale mismatch in social-ecological systems is a well-known challenge (Cumming et al., 2006). For instance, water management on a large spatial scale might well optimize the total system while disrupting functions at a local scale where farmers have run out of water for irrigation of crops or have flooded parts of their land. A scale mismatch between societal demands for ecosystem services and the local social-ecological systems have also been linked to the Danish cadaster by Vejre et al. (2015). Their study shows how designations of e.g. potential wetlands, drinking-water supply and natural areas based on ecological parameters are poorly aligned with the spatial pattern of land ownership. The authors conclude that the cadastral divisions between individual parcels impede a sound management of the ecosystem services that the designations aim to support. The scattered allocation of agricultural plots is partly a reminiscence of the 18th century land reforms where strips of peat land were allocated to all farms and partly due to a rapid structural development with farm acquisitions of distant farmland to utilize economics of size.

The Danish land reform more than 200 years ago addressed a societal goal for a more efficient agricultural production. The land consolidation was compulsory in the sense that if at least one of the farmers in the village insisted on a re-allotment, the land consolidation was performed for all farms. (Barton, 1988). Land consolidation projects with the main purpose of improving conditions for agriculture continued – under changing legislation over time – until 2006 (Hartvigsen, 2014). But already by 1990, an amendment of the law introduced a possibility for landscape consolidations with multiple purposes. This allowed for e.g. nature restoration, afforestation and infrastructural constructions. The broader scope mirrored the development in other Western Europe countries such as the Netherlands and Germany (Hartvigsen, 2015; Van Huylenbroeck et al., 1996). Whereas it has been compulsory for the landowners to participate in many land consolidation projects in these countries, Hartvigsen (2015) emphasizes that it is voluntary for Danish landowners. Hence, a common understanding and commitment among the landowners is important if a land consolidation is to be successful in providing opportunities for large-scale changes. A single landowner might, for example, refuse a re-allotment and thereby obstruct a re-flooding of a reclaimed seabed or floodplain for nature purposes. In some cases, a legal basis in other legislation than the land consolidation act makes expropriation possible. However, authorities are generally hesitant in overruling private property rights if voluntary solutions are possible, and in some cases – e.g. afforestation projects – expropriation is illegal. Therefore, the local landowner's commitment is essential to ensure a successful implementation of land consolidation projects.

The vision by the Collective Impact stakeholders for a new land reform including multifunctional land consolidations to create a better match between patterns of ownership and present and future societal demands has gained much political interest. When a new Social Democrat government was inaugurated in 2019, the policy program included a vision for a land reform (Frederiksen, 2019). The former government had already provided additional funding to initiate more multifunctional land consolidation projects. As of 2021, four new projects have been approved (Ministry of Food Agriculture and Fisheries of Denmark, 2021).

The political visions for a land reform and multifunctional land consolidations seem to be developing on the national scene. Meanwhile, we will now focus on the local scale and the first four pilot projects in the Future Sustainable Landscapes initiative, as initiated by the Collective Impact. Firstly, we present the collaborative approach that was adopted in four pilot projects. Secondly, we describe the local context and some examples from the pilot projects. The final section then provides insights to some of the lessons learnt so far.

Finding a common agenda for multifunctional land consolidations

The four pilot projects for multifunctional land consolidation aimed at generating a common local agenda as a basis for the land consolidation process. As mentioned above, Future Sustainable Landscapes was part of the Danish Collective Impact initiative. The ambition is that stakeholders address societal challenges and together find suitable solutions based on the principles of the collective impact approach, as proposed by Kania and Kramer (2011). Finding a 'common agenda' is one of five important conditions for generating a collective impact. The common agenda is not a steady state; it constantly evolves and is redefined due to interaction between 'continuous communication' and 'mutually reinforcing activities'. Professionals facilitate and support the process – the so-called backbone support element – and 'shared measurements' during the process aim at creating a transparent and sound knowledge base for the communication, the actions and the common agenda. These five elements of the collective impact were adapted in the four pilot projects as illustrated in Figure 2.1 (Johansen, Ejrnæs, et al., 2018).

The collective impact process in the pilot projects was supported by professionals from the local municipality and, in some cases, consultants from private companies, NGOs or governmental bodies. The professionals had two roles: While on the one hand supporting the local dialogue and the ideas of the local stakeholders, they were also committed to inspire or propose how overarching societal demands might be addressed in the local context. Land consolidation consultants were also a part of the backbone support by negotiating with the individual farmers about their interests in the land consolidation. An interdisciplinary research group developed an indicator framework to guide the process by shared measurements of the results. These indicators constitute the framework

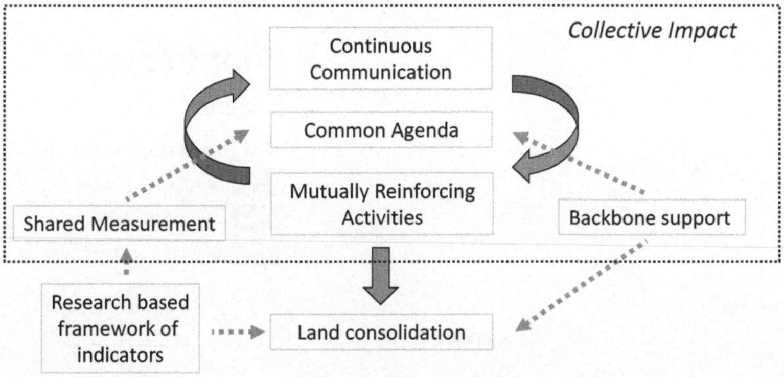

FIGURE 2.1 The five elements in the collective impact framework by Kania and Kramer (2011) were adapted into a figure that illustrates the approach in the Danish multifunctional land consolidation pilot projects (Johansen et al., 2018). The figure here is a slightly modified version.

for the evaluation of the first three pilot projects (Ejrnæs et al., 2019) and a forthcoming final evaluation of all four pilot projects.

Implementing a national vision in a local context

The national group of stakeholders in the Future Sustainable Landscape initiative negotiated and defined five overarching societal demands to be pursued in the pilot projects: (1) increasing biodiversity; (2) protecting the water environment including drinking water resources; (3) improving the farm economy on the active farm enterprises; (4) creating better possibilities for outdoor recreation; and (5) strengthening rural community development (climate adaption and mitigation of GHG emissions was later added as a sixth demand). The interdisciplinary research group operationalized the demands into a framework with a range of indicators (Ejrnæs et al., 2016; Johansen, Ejrnæs, et al., 2018). The indicator framework was used to evaluate potentials and results of the local pilot projects.

All Danish municipalities with substantial rural landscapes were encouraged to propose pilot projects in 2015. Five municipalities with promising projects proposals were invited to further develop an initial plan for the multifunctional land consolidation, including a description of the potentials within the five societal goals. The research group evaluated the plan by the indicator framework. The evaluation guided the group of national stakeholders' decision to initiate three pilot projects in early 2016. Eventually, the fourth pilot project was initiated in 2018 to further develop the process based on the preliminary experiences from the first projects. Figure 2.2 shows the four areas.

After approval of the initial plans, a further development of ideas was initiated in the three selected municipalities. Here, the local project managers guided the dialogue. Public meetings and workshops were held to find a common agenda. Parallel

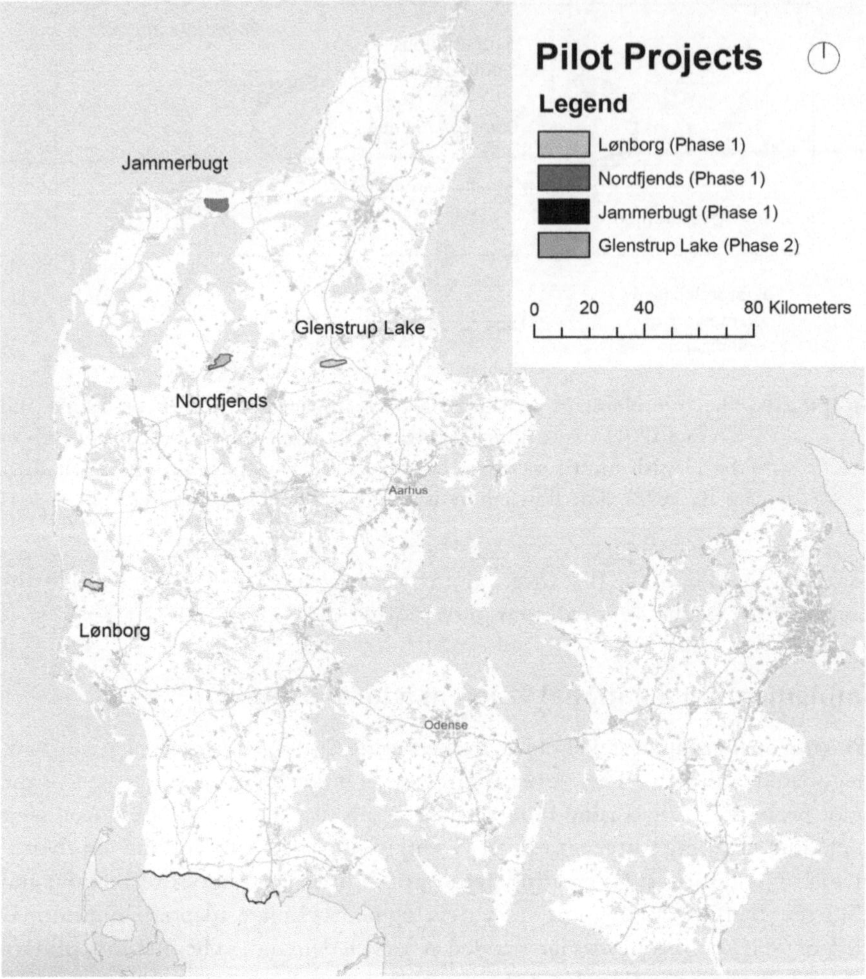

FIGURE 2.2 Location of the four pilot projects. The size of the areas varies from 1,700 ha to 9,900 ha.

to this process, a land consolidation consultant started having individual meetings with landowners to learn their thoughts about future development of their property, including wishes for purchase of additional land, swap of land or sale of land.

The process differed between the three first pilot projects. The overall results of a first round of land consolidation in the three first pilot projects are summarized in Table 2.1.

In Nordfjends, the process was well-prepared by the former landscape strategy making process, which was firmly anchored in the municipal administration and among the politicians in the council, contrary to the Lønborg Hede area (Figure 2.2). Here, the process was partly detached from the municipal

TABLE 2.1 Main characterization of the three first pilot projects and the results after the land consolidation, based on Ejrnæs et al. (2019)

	Jammerbugt	*Nordfjends*	*Lønborg*
Municipality	Jammerbugt	Skive	Ringkøbing
Backbone support	The project was headed by a municipal landscape planner, assisted by other officers in the municipal administration and a land consolidation consultant.	Headed by a municipal planner, the project enjoyed a strong commitment from politicians and several municipal officers who had been involved in the participatory planning process prior to the pilot project. A land consolidation consultant also participated.	The project was headed by a private consultant. Cooperation with the municipal administration was weaker than the other projects. A land consolidation consultant also participated.
The process	Many ideas came up during public workshops with many stakeholders. Less interest in participating among land owners, partly due to strong interest in hunting in a central part of the project area.	Synergy with another EU-financed initiative was not possible, which reduced the potentials for the pilot project. The creditors for some of the large dairy farms vetoed against land consolidation due to marked uncertainties and a high level of debts on the farms.	The politicians turned out to be against a substantial reduction of the area of farmland. Hence less willingness to support and encourage the process.
Pilot project area*	9,900 ha	2,200 ha	1,700 ha
Area of re-allotment	141 ha	79 ha	29 ha
Area with a cadastral clause (e.g. public access or no use of fertilizers)	90 ha	20 ha	5 ha
Other agreements or results due to the pilot project	New walking trail (800 m)	New recreational access/ trails (3,750 m)	Initiated a local dialogue for a possible 230 ha wetland project near the boundary of the project area.

* The project area represents the target area for initiatives, but re-allotments also included lots outside the borders of the project area, see Figure 2.3.

administration and undertaken by a private consultant. Furthermore, the politicians in the municipal council restricted the room to maneuver for the consultant when deciding that the land consolidation should not result in a substantial reduction of farmland in the selected pilot area.

Providing an example from the Nordfjends project area, Figure 2.3 shows a landscape strategy that was developed prior to the land consolidation pilot project as part of a participatory planning process in Skive Municipality (Kristensen & Primdahl, 2020; Skive Kommune, 2016). Besides the landscape strategy, the process revealed that 64 out of 94 farmers were motivated to participate in a land consolidation to either enlarge/reduce the size of their property or to change the location of the fields belonging to their farms. The landscape strategy was further elaborated into a common agenda with potential projects, which the land consolidation might pave the way for. Figure 2.4 provides a map showing the results of what actually came out of the process in Nordfjends – the same project area as in Figure 2.3.

Based on the experiences from the process in the three areas (Haldrup & Iversen, 2018; Mouritsen, 2017), the fourth pilot project was organized slightly differently.

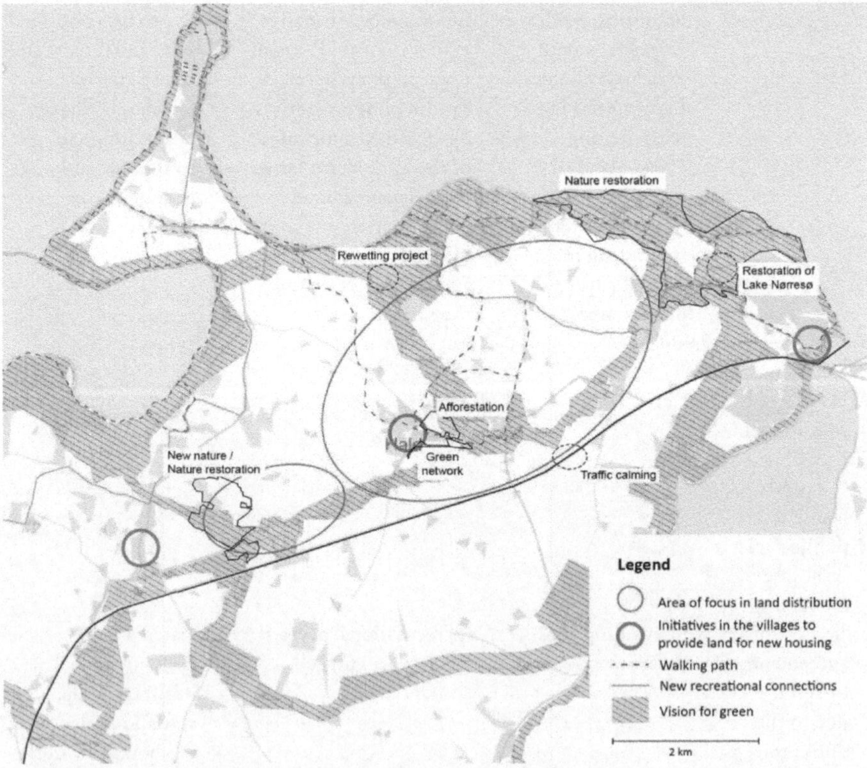

FIGURE 2.3 Map of the initial landscape strategy that was adapted in the in the multifunctional land consolidation pilot project in Nordfjends. Adapted from Pears et al. (2019) and Skive Kommune (2016).

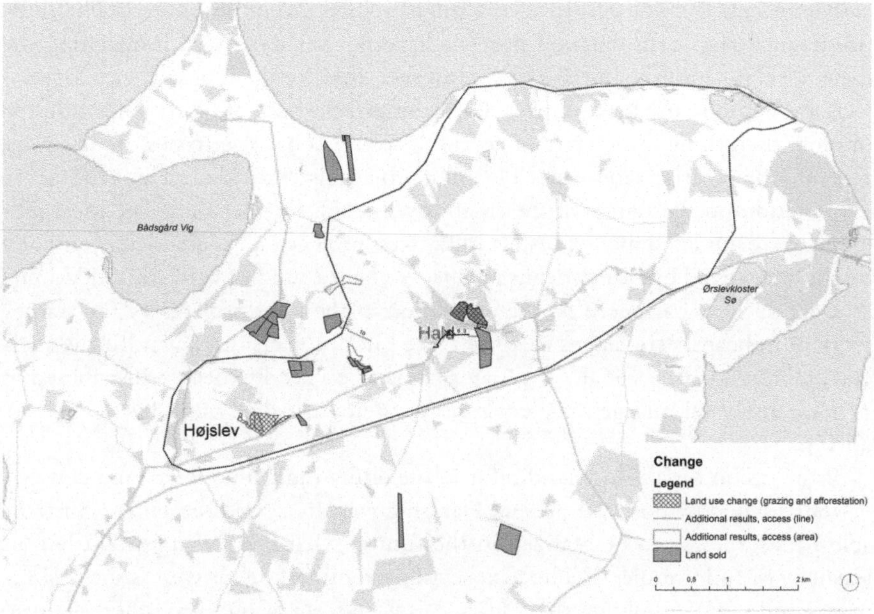

FIGURE 2.4 Changes due to the multifunctional land consolidation in Nordfjends. Adapted from Ejrnæs et al. (2019).

In the Glenstrup Lake area (Figure 2.2), a facilitator group consisting of representatives from both the farmers' advisory association and the land consolidation agency accompanied the principal project manager from the municipality administration. Secondly, a board of local stakeholders helped guide and encourage the local dialogue. As a third initiative, the interdisciplinary research group was asked to inspire the local process by spatially illustrating potentials in the local pilot area. They also gave their interpretation of where potentials for positive or negative interactions between societal goals were present in the local landscape (Johansen et al., 2020). These inputs represented an external perspective as a supplement to the local experts from the municipality or the facilitator group, who were mainly insiders.

So far, the outcome of the local process in the Glenstrup Lake area is a local development plan called 'Fælles forandring' [United Change] (Mariagerfjord Kommune, 2019). The plan presents a catalogue of possible initiatives and landscape changes that the land consolidation (and other measures) may be able to realize in the future. The negotiation with the landowners in the fourth pilot is not yet finished (primo 2022).

Lesson learned by the multifunctional land consolidations

A quick glance at Table 2.1 and a comparison between the initial visions in Nordfjends in Figure 2.3 and the outcome of the pilot project so far in Figure 2.4

clearly reveals that the results of the pilot projects are quite limited. The multifunctional land consolidation projects have not yet succeeded in reaching the many local goals of the initial plans. However, this does not mean that pilot projects are without any local impact. For example, the case of Nordfjends resulted in more than three kilometres of corridors for trails or other types of public access into the agrarian landscape. The pilot project also provided an area for future afforestation on the fringe of the small town Hald. Furthermore, the local dialogue has formulated many possible initiatives, projects, ideas etc. that eventually might be realized by other means or due to a later round of land consolidation.

Despite these perspectives, the evaluation of the pilot projects by the framework of indicators concludes that the three pilot projects did not fulfil the societal potentials (Ejrnæs et al., 2019), nor has land ownership been redistributed to better comply with large scale ecological systems as called upon by Vejre et al. (2015).

If the vision for a future land reform, including multifunctional land consolidation as stated by both the present Danish government and the Future Sustainable Landscapes, is to be realized in the future, what are the important lessons learnt by the pilot projects? The next section examines this, as well as the preliminary results, and combines these learnings with inspiration from other (Danish) projects and findings. We focus on four learning themes: (1) the multifunctional societal demands at the local scale; (2) the process of finding a local agenda; (3) combining measures and (4) the time perspective.

Multifunctional societal demands at the local scale

The five overall societal demands to be pursued in the pilot projects were, as previously mentioned, defined by the national stakeholders, who asked the interdisciplinary research group to measure the local potentials and outcomes of these societal demands in the pilot projects. The indicator-based framework was developed to score the (potential) outcome on each of the five societal demands and provide a total score for the pilot projects (Johansen, Ejrnæs, et al., 2018). However, the indicator-based approach did not provide any guidelines or measures for spatial interactions between the demands. For example, improving recreational access may have a negative impact on some aspects of biodiversity, but limited or no impact on other aspects. Furthermore, the interactions may vary across different local contexts.

The research group developed a guide for potential interactions inspired by the work of the International Council for Science on interactions between the UN Global Development Goals (International Council for Science, 2017). The guide provides guidance for potential synergy or conflicts between indicators (and thereby also between the different societal demands). The interactions are evaluated on a seven-point scale ranging from 'indivisible' (+3), i.e. the indicator fully depends on the other, to 'cancelling' (–3) meaning that potential of the indicator is fully levelled or cancelled by the other conflicting indicator. The guide was created by an iterative process. Each of the researchers initially interpreted the expected interaction between indicators of their own research fields (one of the societal demands) and the indicators of the other research fields based on literature. The initial interpretations were adjusted in an iterative discussion between the researchers within all five disciplines to overcome what Luhmann (2000) describes as *autopoietic social systems*, i.e. when disciplines tend to center around internal academic language and understandings while reducing the outer complexity (Noe & Alrøe, 2003).

The final guide (Johansen, Kronvang, et al., 2018) revealed both positive (synergy) and negative (conflict) interactions, many of which were characterized by a spectrum of values (e.g. interaction) ranging from −2 through 0 to acknowledge that the interaction is influenced by the local context. Hence, the guide aimed to qualify the discussions about the local agenda rather than providing the exact result. To further investigate the possibilities of connecting the national scale (the overall societal goals) with the local landscape, each of the five disciplines described and mapped potentials in the fourth pilot project area of Glenstrup Lake and used the guide in combination with background data and field visits to estimate interactions at the local scale. The result of the scientifically based proposal and analysis reveals potentials for developing a more multifunctional landscape with prevailing positive interactions (Figure 2.5; Johansen et al., 2020).

The process of finding a local agenda

Behind the result map in Figure 2.5 is a catalogue of suggested initiatives, and it is indeed not given that all of these are aligned with the attitudes and wishes of the local community, nor the landowners as discussed in Johansen et al. (2020:12). However, the catalogue served as an input from neutral 'outsiders' to inspire the local dialogue, as previously mentioned. A printed version of the catalogue was posted to the local inhabitants in the pilot area and researchers presented the catalogue at a local meeting. Afterwards, if further input from the researchers was requested by the locals to develop their project ideas, some of the researchers held thematic meetings in the field with small groups of local people. Eventually, the process spawned a great deal of input, which resulted in the 'United Future' plan. Whether the land consolidation will provide space for some of the proposals depends on the results of the ongoing negotiations with the individual landowners in the Glenstrup Lake area.

External researchers were also involved in Nordfjends. This was part of the participatory process of planning a landscape strategy (2014–2016) prior to the land consolidation project (Kristensen & Primdahl, 2020). The landscape strategies have been developed as collaborative processes involving both politicians, employees from the municipality, local stakeholders and researchers in a number of Danish rural case studies (Kristensen et al., 2015, 2019) based on inspiration from Patsy Healey's collaborative strategy making in urban settings

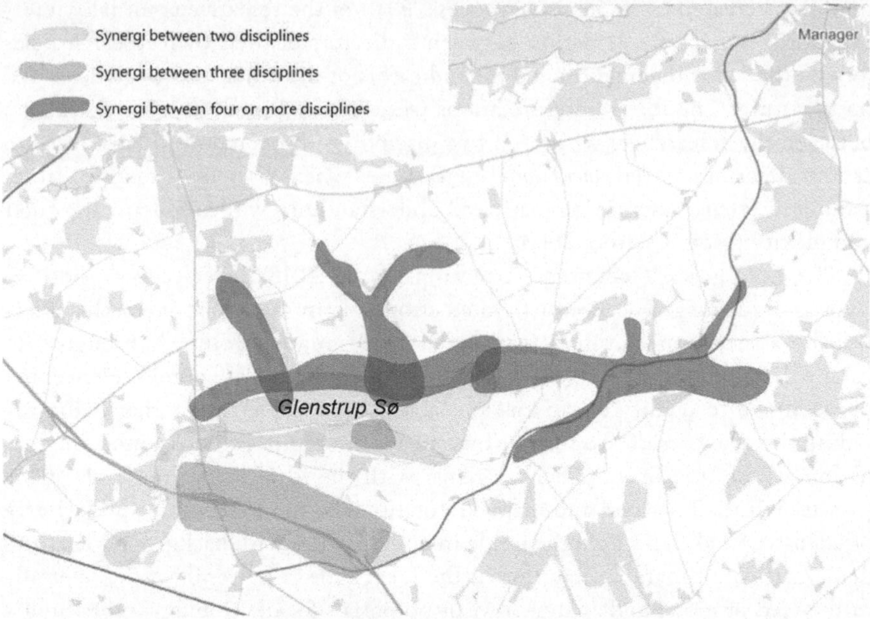

FIGURE 2.5 Positive interactions between societal demands based on modification of Johansen et al. (2020). There are a few areas with negative interactions (not shown on the map).

(e.g. Healey, 2007). The strategy-making process includes interaction between four elements, all of which support the collective impact framework (Figure 2.1): (1) mobilizing local resources; (2) generating interest and building confidence; (3) scoping formulation of vision and goals and (4) formulation of frames of prioritization (Kristensen & Primdahl, 2020). The outcome from these Danish processes has proved successful in some cases (Kristensen et al., 2022) while taking unforeseen turns in other cases (Primdahl et al., 2020).

The collaborative process is, however, not always straightforward (Brand & Gaffikin, 2007). Interviews with project managers/employees from the municipalities in the first three pilot projects show that it was challenging to pursue the multifunctional purpose. Some of the professionals felt it was hard to deliver on the high expectations from the Collective Impact stakeholders (Mouritsen, 2017). The land consolidation consultants emphasized that land consolidation projects usually have a clear purpose, such as creating space for new infrastructure projects or a specific nature conservation project. The multiple purposes complicated the dialogue with the individual landowners, especially when negotiations took place while the local multifunctional aims were still under discussion (Haldrup & Iversen, 2018:20). Even in Nordfjends, the result of the land consolidation was limited despite the landscape strategy process and a very high initial interest in re-allotment among the landowners (Table 2.1).

The Nordfjends case underlines that there are various other constraints to a successful multifunctional land consolidation than just the process (see also Haldrup & Iversen, 2018:13ff). For example, it proved to be impossible to combine the multifunctional approach with another EU-funded project in a part of the project area, because the EU only accepts that funding is spent on the very preapproved purpose of the measure.

Further, a challenging financial market situation combined with high level of debts among many of the large dairy farms further obstructed the process. The creditors perceived many of the farms as vulnerable investments at that time and consequently refused to participate in the land consolidation (Haldrup & Iversen, 2018). Finally, landowners might not be inclined to engage in a project if they are only compensated by the foregone income. Permanent projects could also appropriate the option value, which quite often is attached to land ownership e.g. the landowner has an option to develop the land for construction purposes.

The experiences from the pilot projects proved that finding a common agenda for the multifunctional land consolidations demands both time and a sound setup of the 'backbone support'(Kania & Kramer, 2011). Even with a successful process – as in Nordfjends or Glenstrup Lake – other factors might constrain the realization of the multifunctional land consolidation.

Combining of measures and funding

There are many possible means and strategies to transform landscapes to ensure better coping with changing and future societal demands (Pinto-Correia et al., 2018:198ff). The multifunctional land consolidation deals only with re-allotment of land. Many of the wishes for local landscape changes formulated in the landscape strategies or multifunctional land consolidation plans depend on more than just re-allotment. In Nordfjends, the land consolidation only provided space for afforestation and new recreational infrastructure. Planting forest or establishing pathways depends on additional means from other sources.

Some of the other local wishes had very little to do with land ownership, but would require other types of initiatives than land consolidation, e.g. improve communications about cultural heritage or development of new tourism products in the Jammerbugt area. Finally, the experiences from Nordfjends revealed that synergy is not always possible due to constraints imbedded in the measures, e.g. the rigid design of certain EU measures that explicitly prevents multifunctionality, as mentioned above.

Even if it was possible to combine funding from different sources, such as the EU, national schemes etc., most funding is allocated to direct measures to improve, for instance, water quality or to support the transaction costs when running a multifunctional land consolidation process. Very recently, with the new national scheme's targeted abandonment of agricultural production on peat land, a large sum of money was allocated to buy up agricultural land. Nevertheless, this funding is allocated to one specific purpose rather than to fulfil more concurrent goals for sustainable land use.

Time

The final learning from the pilot projects centers round the temporal scale of the land consolidations, not least the prior process of finding a common agenda. A collaborative process takes time. In the case of Nordfjends, the making of the landscape strategy lasted from 2014 to 2016. As also revealed in Figure 2.1, the local common agenda evolves based on interactions between dialogue and actions. Hence, the nature of the local dialogue is a long process. However, local processes are not enough. The projects also illustrated that there is a need for a general support from local government, and for having the projects anchored in the municipalities' overall policy for the development of rural areas.

Development of the plans for the multifunctional land consolidations in the three first pilot projects took place under a tight external deadline defined by the Collective Impact stakeholders. First, the municipality had to apply to be prequalified for the program. Next, the initial pre-plans in five potential municipalities were prepared in only a few months. The three municipalities then further developed the plan. However, the total time for creating the basis for the multifunctional land consolidation was limited and therefore negotiations with the individual landowners about re-allotment began while the plan including the aims to be fulfilled was still under discussion (Haldrup & Iversen, 2018). Referring to Figure 2.1, the negotiations about land consolidation were initiated without having a sound common agenda.

Land consolidation projects usually take two to four years and in large areas, successive stages of negotiations and implementation are often needed to successfully change the ownership structures. This was the case when 2,200 hectares of wetland and meadows were re-established in Skjern River Valley during the 1990s (Hartvigsen, 2014). Given the multifunctional purpose, it seems fair to conclude that the pilot projects could have gained from more available time than the approximately two years from being approved to the implementation of the land consolidation in the project areas.

Future perspectives of multifunctional land consolidation

We initially questioned whether land consolidations might be an applicable means to create resilient future landscapes in Denmark and abroad. The pilot projects in the Future Sustainable Landscapes initiative clearly illustrate that multifunctional land consolidation represents no shortcut to fulfilling societal demands or limiting scale mismatches in the social-ecological systems. However, the pilot projects yielded important learnings that are relevant to the recent discussions about a land reform and implementation of societal goals in local landscapes (Ejrnæs et al., 2019). Danish experiences might also contribute to the international debate on future role of land consolidation (Hartvigsen, 2015; Jiang et al., 2022; Pašakarnis et al., 2021).

It seems obvious that land consolidation potentially caters for social-ecological adaptions due to e.g. climate change and a need to reduce the emission of GHGs. Though the results of the pilot projects so far are limited, land consolidation proved valuable in the 2,200-hectare nature restoration project in the Skjern River Valley (Hartvigsen, 2014) and other Danish wetland projects (Haldrup, 2015). The recent political ambition to abandon farming on 100,000 ha peat soils and re-establish wetlands – equaling approximately 4% of the cropland in Denmark – seems to urge for land consolidation to a much higher extent than hitherto.

Despite the obvious potentials of multifunctional land consolidation, the question is whether the explicit focus on multifunctional land consolidation in the pilot projects has jammed an initial and broader discussion about local development. It might have proved fruitful to perceive the multifunctional land consolidation as only one of many tools to create landscape changes rather than the aim itself from the beginning of the project. The land consolidation consultants would have preferred that the plan for multiple functions was settled before negotiating with the landowners (Haldrup & Iversen, 2018). The local process actually generated many ideas that were not possible to obtain by land consolidation, or ideas that demanded additional actions, time and/or funding to obtain, and these were not available in the pilot projects. Also, some of the existing funding schemes were rigid and not possible to combine for multiple purposes. The explicit focus on land consolidation might also have blurred other ways of reaching the goals. For example, another Danish collaborative process proved that land consolidation is not necessarily needed, and concerted action and cooperation among existing landowners was successful in creating both richer natural values and recreational opportunities in the Odderbæk stream area (Kristensen et al., 2022).

The four pilot projects have, as in other Danish examples, produced local development plans or landscape strategies through collaborative processes. It is a relevant general concern whether these local development plans succeed in addressing overall societal demands in line with the ambition of the Future Sustainable Landscapes initiative. The work on interactions between societal demands by the interdisciplinary research group illustrated potentials for multifunctional initiatives in the Glenstrup Lake area (Figure 2.5; Johansen et al., 2020). Hence, input from outsiders and shared measurements (including the guide with interaction scores) might strengthen the local awareness about overall societal demands and bring in new perspectives in the local process. Attractive incentives, e.g. rural development programs anchored in the common agricultural policy, might also promote local initiatives aligned with societal demands (Pinto-Correia et al., 2018). However, it should be questioned whether the local collaborative initiatives and voluntary land consolidations (even with competent backbone support, such as the professional facilitator group in the Glenstrup Lake area) will be sufficient to create more flexible and resilient landscapes as well as address recent and future demands – or whether the collaborative local planning and actions should be accompanied by national compulsory goals or designations that must be complied with in the local solutions, as stressed by Arler et al. (2017). The future evaluation of the fourth pilot project and additional multifunctional land consolidations initiated at a later stage will hopefully help determine how to balance between local demands and societal needs for future multifunctional landscapes.

Multifunctional land consolidation is certainly not a quick fix to obtain resilient landscapes, but it may be a component in finding new pathways – not only in a Danish context. Current worldwide demands on solving at the same time the climate, biodiversity and environmental crisis requires that we develop novel paradigms for how to assist in creating future resilient landscapes.

Acknowledgment

The research was funded by Realdania as part of the Collective Impact initiative 'Future Sustainable Landscapes'. All figures were kindly generated/modified by the two research officers Marie Alstrup Jensen and Sara Folvig.

References

Arler, F., Jørgensen, M. S., & Sørensen, E. M. (2017). *Prioritering af Danmarks areal i fremtiden: Afsluttende rapport fra projektet*. Fonden Teknologirådet.

Barton, H. A. (1988). The Danish agrarian reforms, 1784–1814, and the historians. *Scandinavian Economic History Review*, 36(1), 46–61. https://doi.org/10.1080/03585522.1988.10408105

Ben, P., Malvika, O., Andrew, B., & E., G. R. (2011). Reconciling food production and biodiversity conservation: Land sharing and land sparing compared. *Science*, 333(6047), 1289–1291. https://doi.org/10.1126/science.1208742

Brand, R., & Gaffikin, F. (2007). Collaborative planning in an uncollaborative world. *Planning Theory*, 6(3), 282–313. https://doi.org/10.1177/1473095207082036

Cumming, G. S., Cumming, D. H. M., & Redman, C. L. (2006). Scale mismatches in social-ecological systems. *Ecology and Society*, 11(1). http://www.jstor.org/stable/26267802

Ejrnæs, R., Johansen, P. H., Kronvang, B., Olsen, J. V., Præstholm, S., & Schou, J. S. (2019). *Resultater af multifunktinel jordfordeling i tre pilotprojekter - dokumentation af effekter for landdistrikter, driftsøkonomi, natur, miljø og rekreation*. Center for Landdistriktsforskning, SDU.

Ejrnæs, R., Johansen, P. H., Kronvang, B., Præstholm, S., & Schou, J. S. (2016). *Muligheder for effektmåling af multifunktionel jordfordeling i fem projektområder - en tværfaglig vurdering*. Arbejdsrapport.

Frederiksen, M. (2019). *Retfærdig retning for Danmark. Politisk forståelse mellem Socialdemokratiet, Radikale Venstre, SF og Enhedslisten, 25. juni 2019*. Regeringen Mette Frederiksen.

Haldrup, N. O. (2015). Agreement based land consolidation – In perspective of new modes of governance. *Land Use Policy*, 46, 163–177. https://doi.org/10.1016/j.landusepol.2015.01.027

Haldrup, N. O., & Iversen, S. D. (2018). *Erfaringer fra tre demonstrationsjordfordelinger.* Orbicon.

Hartvigsen, M. (2014). Land consolidation and land banking in Denmark – Tradition, multi-purpose and perspectives. *Danish Journal of Geoinformatics and Land Management,* 47, 1–7.

Hartvigsen, M. (2015). *Experiences with land consolidation and land banking in Central and Eastern Europe after 1989.* Food and Agriculture Orgnization of the United Nations.

Healey, P. (2007). *Urban complexity and spatial strategies.* Routledge.

Huylenbroeck, van G., & Durand, G. (2003). *Multifunctional agriculture: A new paradigm for European agriculture and rural development.* Ashgate.

International Council for Science. (2017). *A Guide to SDG internactions: From Science to implementation.* ICSU.

IPCC. (2021). *Climate Change 2021. The Physical Science Basis. WG1 contribution to the Sixth Assessment Report.* WMO/UNEP.

Jiang, Y., Tang, Y.-T., Long, H., & Deng, W. (2022). Land consolidation: A comparative research between Europe and China. *Land Use Policy,* 112, 105790. https://doi.org/10.1016/j.landusepol.2021.105790

Johansen, P. H., Ejrnæs, R., Kronvang, B., Olsen, J. V., Præstholm, S., & Schou, J. S. (2018). Pursuing collective impact: A novel indicator-based approach to assessment of shared measurements when planning for multifunctional land consolidation. *Land Use Policy,* 73, 102–114. https://doi.org/10.1016/J.LANDUSEPOL.2018.01.040

Johansen, P. H., Ejrnæs, R., Kronvang, B., Olsen, J. V., Præstholm, S., Schou, J. S., & Johansen, S. K. (2020). Exploring the interdisciplinary potential of the Agenda2030—Interactions between five Danish societal demands for sustainable land use. *Land Use Policy,* 94, 104501. https://doi.org/10.1016/j.landusepol.2020.104501

Johansen, P. H., Kronvang, B., Præstholm, S., Ejrnæs, R., Schou, J. S., & Olsen, J. V. (2018). *Synergi og begrænsning imellem 25 effektmålepunkter for multifunktionel jordfordeling.* Center for Landdistriktsforskning, SDU.

Kania, J., & Kramer, M. (2011). *Collective impact.* Stanford Social Innovation Review. Leland Stanford Jr. University.

Kristensen, L. S., & Primdahl, J. (2020). Landscape strategy making as a pathway to policy integration and involvement of stakeholders: Examples from a Danish action research programme. *Journal of Environmental Planning and Management,* 63(6), 1114–1131. https://doi.org/10.1080/09640568.2019.1636531

Kristensen, L. S., Pears, D. Q., & Primdahl, J. (2022). Guiding multifunctional landscape changes through collaboration - experinces from a Danish case study. In D. Rizzo, E. Marraccini & S. Lardon (eds.), *Landscape Agronomy: Advances and Challenges of a Territorial Approach to Agricultural Issues.* (pp. 247–270). Springer Cham.

Kristensen, L. S., Primdahl, J., & Møller, K. H. (2019). *Fremtidens Landskaber: Visioner og Planer for det åbne land.* Bogværket.

Kristensen, L. S., Primdahl, J., & Vejre, H. (2015). *Dialogbaseret planlægning af det åbne land.* Bogværket.

Loconto, A., Desquilbet, M., Moreau, T., Couvet, D., & Dorin, B. (2020). The land sparing – Land sharing controversy: Tracing the politics of knowledge. *Land Use Policy,* 96, 103610. https://doi.org/10.1016/j.landusepol.2018.09.014

Luhmann, N. (2000). *Sociale systemer. Grundrids til en almen teori.* Hans Reitzels Forlag A/S.

Mariagerfjord Kommune. (2019). *Fælles Forandring. Lokal Udviklingsplan for Glenstrup Sø og Østerkær Bæk.* Mariagerfjord Kommune.

Ministry of Environment. (2020). *Natur- og biodiversitetspakken.* Ministry of Environment.

Ministry of Finance. (2021). *Aftale om grøn omstilling af dansk landbrug.* The Ministry of Finance.

Ministry of Food Agriculture and Fisheries of Denmark. (2021). Webpage. https://lbst.dk/landbrug/arealer-og-ejendomme/jordfordeling/multifunktionel-jordfordeling-mufjo/. Date 2021/11/30.

Mouritsen, A. K. M. (2017). En undersøgelse af dansk forhandlingsmetode og blokeringer for multifunktionelle jordfordelinger [Danish negotiation strategy and resistance in multifunctional land consolidation projects]. *Kart Og Plan*, 77(4), 321–332.

Nielsen, O. K. et al. (2019). *Denmark's national inventory report 2019. Emission inventories 1990–2017- submitted under the United Nations Framework Convention on Climate Change and the Kyoto Protocol.* Aarhus University, DCE – Danish Centre for Environment and Energy.

Noe, E. & Alrøe, H. (2003). Farm enterprises as self-organizing systems: A new transdisciplinary framework for studying farm enterprises?. *The International Journal of Sociology of Agriculture and Food*, 11, 3–14. https://doi.org/10.48416/ijsaf.v11i.325

Pašakarnis, G., Maliene, V., Dixon-Gough, R., & Malys, N. (2021). Decision support framework to rank and prioritise the potential land areas for comprehensive land consolidation. *Land Use Policy*, 100, 104908. https://doi.org/10.1016/j.landusepol.2020.104908

Pears, D. Q., Primdahl, J., Sørensen, E. M. & Herslund, L. H. (2019). Case D: Et Limfjordslandskab i udvikling – Nordfjends. In L. S. Kristensen, J. Primdahl & K. H. Møller (eds.), *Fremtidens Landskaber – visioner og planer for det åbne land.* Bogværket.

Pinto-Correia, T., Primdahl, J., & Pedroli, B. (2018). Landscape policy and planning – Managing conflicts and making places. In T. Pinto-Correia, J. Primdahl & B. Pedroli (eds.), *European landscapes in transition: Implications for policy and practice* (pp. 198–239). Cambridge University Press.

Primdahl, J. (2014). Agricultural landscape sustainability under pressure: Policy developments and landscape change. *Landscape Research*, 39(2), 123–140. https://doi.org/10.1080/01426397.2014.891726

Primdahl, J., Folvig, S., & Kristensen, L. S. (2020). Landscape strategy-making and collaboration. The Hills of Northern Mors, Denmark; A case of changing focus and scale. *Land*, 9(6). https://doi.org/10.3390/land9060189

Skive Kommune. (2016). *Procesplan for gennemførelse af multifunktionel jordfordeling. En del af projekt Liv og Limfjordslandskab i Nordfjends, Skive kommune.* Udarbejdet i samarbejde med KU. Skive Kommune og Københavns Universitet.

Statistics Denmark. (2021). Husdyrbestanden efter tid, areal, art og enhed. [Livestock population by time, area, species and unit] (www.statisbanken.dk/HDYR1).

Van Huylenbroeck, G., Coelho, J. C., & Pinto, P. A. (1996). Evaluation of land consolidation projects (LCPs): A multidisciplinary approach. *Journal of Rural Studies*, 12(3), 297–310. https://doi.org/10.1016/0743-0167(96)00024-1

Vejre, H., Vesterager, J. P., Andersen, P. S., Olafsson, A. S., Brandt, J., & Dalgaard, T. (2015). Does cadastral division of area-based ecosystem services obstruct comprehensive management? *Ecological Modelling*, 295, 176–187. https://doi.org/10.1016/j.ecolmodel.2014.09.027

3

RESILIENT FOOD PRODUCTION – RESILIENT LANDSCAPES

The role of heterogeneity and scale

Kerstin Potthoff and Wenche E. Dramstad

Norwegian agricultural history in a nutshell

Climate restricts yields and types of crops that can be grown in Norway due to its northern location (OECD, 2021). Only 3.5% of Norway is fully cultivated land, and arable land is a scarce resource (Statistics Norway, 2021a). To compensate for the lack of arable land and to tackle variations in, for example, climate, farms relied on a very diverse resource use, different productions and sources of income. Outfield areas such as mountains were used for grazing and hay production, branches and leaves as winter fodder and seaweed as fertilizer (Acksel et al., 2019; Lunden, 2004; Olsson, Austrheim & Grenne, 2000). Each farm typically produced grain, meat, and milk as basic products (Almås, 2004). Growing of potatoes did not become common until the early 19th century, and fruit and vegetable production has also been rather limited (Almås, 2004). In general, productions focused on farm self-sufficiency. Farming was commonly combined with other types of activities. Combinations differed among regions, however, to a large extent dependent on available resources and geographical location. In coastal regions, for example, fishery added to farm resources, in forest regions, forestry played the same role (Almås, 2004). Resource use was on the scale of the single farm and each farm had access to a proportion of different local resources.

As in other European countries, specialization and rationalization are ongoing trends within Norwegian farming since about the Second World War (Almås, 2004). Artificial fertilizer, commonly in use by the end of the Second World War (Gjerdåker, 2004), reduced the dependency on outfield resources by increasing infield yields. The rising production in infield areas made outfield resources increasingly marginal. At the same time, agricultural production followed an ideal of efficiency, i.e., increased output per invested unit of time, along the lines of the ideals of industrialization, as also described so well by Smiley (1997). To

DOI: 10.4324/9781003266440-3

become more efficient, production has been specialized, in general leading to a smaller number of productions, including crops grown, per farm. Specialization has been combined with the ideal of large units to benefit from the 'economies of scale'. In the landscape, these developments are visible in terms of larger individual fields and increased average field size (Potthoff, 2020; Stokstad & Krøgli, 2012). Similar to the development in many other countries (Eurostat, 2021), number of active farmers has been in decline since the Second World War in Norway, however. While Norway had 213,441 active farms (with more than 0.5 ha of land) in 1949, the number of active farms declined to 38,633 in 2020 (Norges Offisielle Statistikk, 1950; Statistics Norway, 2021b). The amount of farmland has changed to a lesser degree, though, due to an increase in land renting (Stokka, Dramstad & Potthoff, 2018). However, there is a limit to how large a farm can become, before farm size and land located at a distance will have an effect on productions. Farmers have a restricted amount of time and a restricted number of days with weather good enough for harvesting (Vik & Flø, 2017). During this time, farmers can manage a certain amount of farmland while the remaining farmland is dealt with during less optimal weather conditions. Carrying out farmland operations during non-optimal conditions means reduced productivity. While climate change may result in increased temperatures, an extended growing season and a larger time window for farm operations, other changes, such as occurrences of new weeds or diseases and cold hardening of plants challenged by increased autumn temperatures may negatively impact productions (Neset, Wiréhn, Klein & Käyhkö, 2019; Uleberg, Hanssen-Bauer, van Oort & Dalmannsdottir, 2014; Wiréhn, 2018).

'Economies of scale' have been accompanied and driven by a comprehensive technological development within the agricultural sector (Almås, 2004). Many labour-intensive agricultural practices have been almost entirely abandoned, such as haymaking and pollarding. To make hay, the grass was manually cut with a scythe and dried on wire systems mounted in the fields – called *hesjer* in Norwegian, a work typically involving the entire family, including children and elderly family members, during a couple of long working days when the weather and season was right (Almås, 2004). This practice changed as machine harvesting and storing grass as silage took over. Similar paths of development can be seen in a number of farming operations, e.g., feeding and milking. Automated milking systems, common since the year 2000, being one example of a more recent technological development (Rønningen, Fugestad & Burton, 2021). While technological development was a key driver for more efficiency in farming, it also provided opportunities for farmers to have 'normal' working hours and holidays. A case study from Southern and Central Norway even showed that intensifying production and investment in automated milking systems was driven by a desire for a better work-life balance (Burton & Farstad, 2020). Technological development can move productions towards reduced flexibility as the introduction of automated milking systems shows; it ties the livestock to the area close to the stable (Rønningen et al., 2021).

Since the Second World War, Norwegian agriculture has been linked to a selection of explicitly articulated aims, such as to upkeep settlement and employment in rural areas and to increase food production (Bjørkhaug & Rønningen, 2014). To ensure the best use of the scarce farmland and thereby maximize food production, what has been described as a canalization policy was introduced in the 1950s. The policy encouraged grain growing in areas most suitable for this kind of production (i.e., southern and eastern Norway) whereby livestock husbandry was concentrated in those areas less suitable for grain production (i.e., western and northern Norway) resulting in a strong regional differentiation of agricultural production (Almås, 2002; Jones & Rønningen, 2007).

Norway's current degree of self-sufficiency is 36% (data from 2018 and 2019), fish and imported concentrated feedstuffs excluded (Rustad, 2020). To sustain and preferably increase self-sufficiency – also taking into consideration expected population growth – the Norwegian government aims at increasing food production by 20% from 2011 to 2030 (Meld. St. 9 (2011–2012)). In 2016, the government confirmed the aim of increased production although no specific percentage was given (Forbord & Vik, 2017).

Agricultural production is carried out within a comprehensive legislative and regulatory framework and cost efficiency in productions has been encouraged by adjusting subsidies to benefit larger farms (Bjørkhaug & Rønningen, 2014; Forbord & Vik, 2017). At present, agricultural production is but one small element in a larger food system, a network of processes on national and global scales (Bjartnes, 2018; Nyström et al., 2019). In 2020, Norway imported c. 39% (784,350 t) of the raw material to be used for concentrated feed production (Landbruksdirektoratet, 2022). Although roughage is used in livestock productions, pig and poultry productions rely entirely on concentrated feed (Nysted, Uldal & Vakse, 2020). In addition to input transported to Norway from abroad, dependency on long-distance transport has increased also inside Norway to process and distribute products. The number of slaughterhouses is reduced, from 65 in 1996 to 39 in 2005, the number of mills accepting grain for further treatment has declined from 139 in 1998 to 70 in 2017, and at present only 11 locations accept animal wool and skins (Animalia, 2020; Hillestad & Bunger, 2019; Svin, 2018). The declining number of locations processing products result in comprehensive transport costs. Nortura drives ca. 17 million km to transport living animals to the slaughterhouse and eggs for packaging, while the main milk processing company in Norway (TINE) drives 55.5 million km to distribute its products (Hillestad, 2014). The coverage of other farm related services such as repair of machinery may also become patchier with a declining number of farms, further increasing the need for transport.

To analyse and discuss resilience in food production and landscapes, we selected three municipalities in different parts of Norway (Figure 3.1). The municipality of Rakkestad is located in Eastern Norway in an area with good conditions for grain production (Figure 3.2). Animal husbandry as well as vegetables are important productions in Time municipality, located in Western Norway (Figure 3.3).

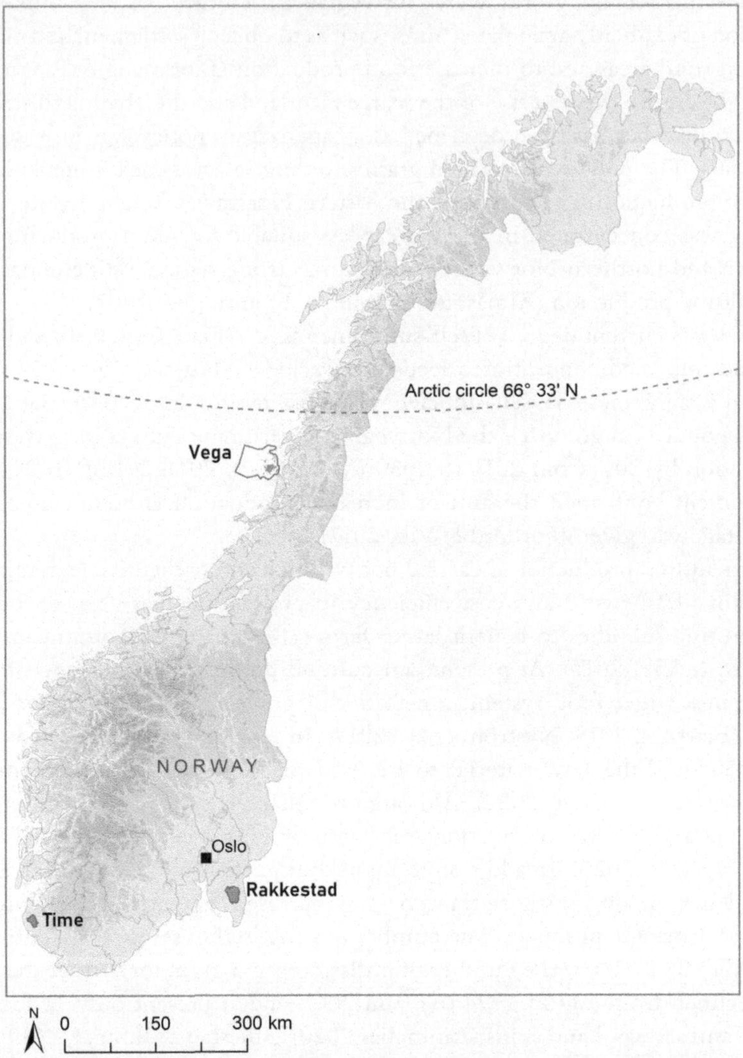

FIGURE 3.1 The municipalities of Rakkestad, Time and Vega, marked in grey. The grey line in Vega illustrates that the sea makes up a large part of the municipality (© NIBIO).

Due to its northern location, farming is constricted by climate in the municipality of Vega. Most important productions are animal husbandry.

Despite differences in the importance of productions and amount of agricultural land, agriculture production in all municipalities reflects the changes common for the whole country presented above. Productions have gone through an up-scaling with fewer farms managing an increasing amount of

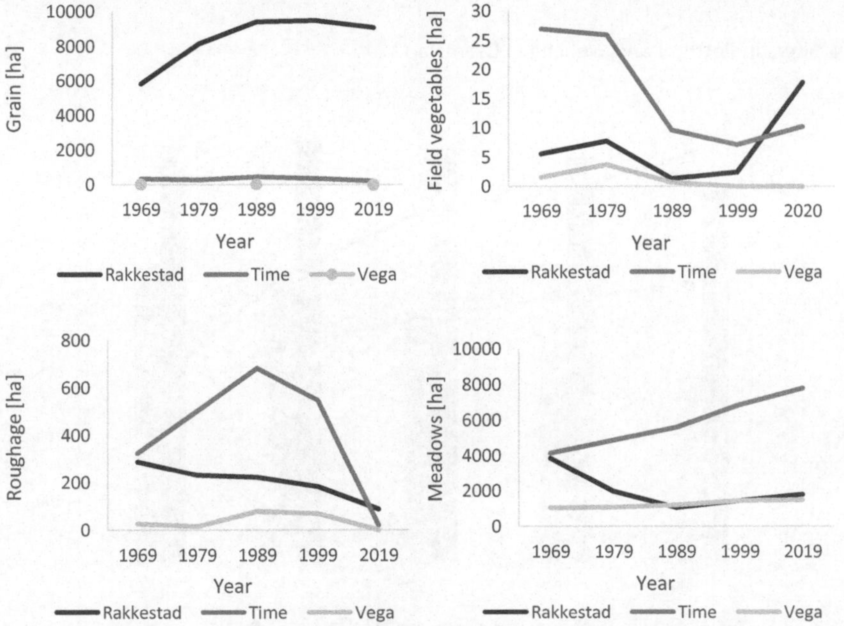

FIGURE 3.2 Changes in production of grain, roughage, field vegetables and meadow area from 1969 to 2019. Data for vegetables are lacking for 2019, data for 2020 have been used instead; data for grain production in Vega are 22 ha (1969), 26 ha (1989) and 0 ha (2019) (Statistics Norway, 2021d). (Note difference in scale on vertical axis).

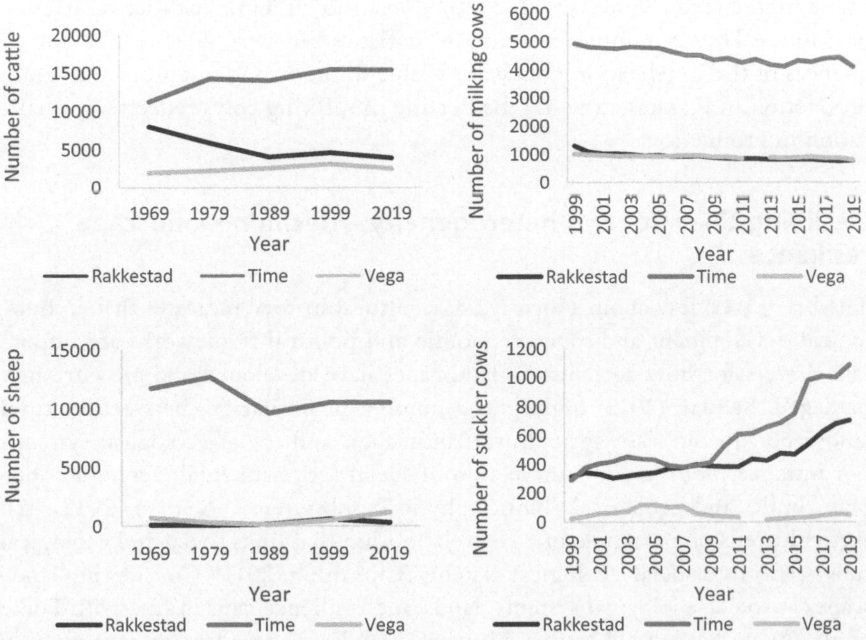

FIGURE 3.3 Changes in livestock productions from 1969 to 2019. Data separating milking and suckler cows is only available since 1999, thus annual data since 1999 have been used (Statistics Norway, 2021c). (Note difference in scale on vertical axis).

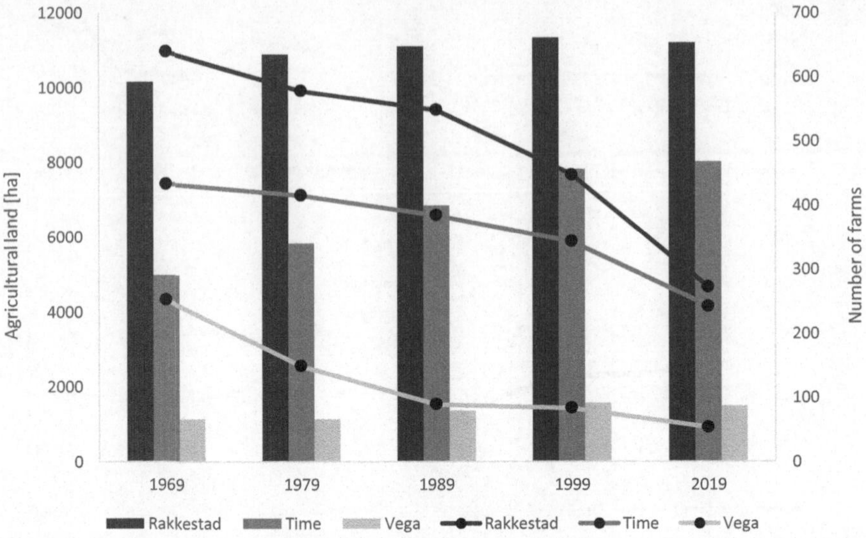

FIGURE 3.4 Change in agricultural land (columns) and number of farms (lines) from 1969 to 2019 (Statistics Norway, 2021e).

land per farm (Figure 3.4). Pig and poultry productions – relying entirely on concentrated feed (Nysted et al., 2020) – occur in all municipalities with Time having the largest number of animals (Statistics Norway, 2021c). The consequences of the canalization policy are visible in the strong importance of grain production in Rakkestad while the decline in milking cows reflects a specialization in production.

Declining diversity and heterogeneity – declining landscape resilience

The brief overview about Norway's agricultural history indicates that technological development and socio-economic and political frameworks are important drivers for how agricultural landscapes have developed and are currently managed. Selman (2012) highlights a number of similarities between cultural landscapes, in our case agricultural landscapes, and social-ecological systems. For instance, both 'are a combination of social (governmental, economic, human, built) and ecological (biotic, physical) subsystems' (Selman, 2012, 42). Agriculture is one example of a key relationship that links social and ecological subsystems into social-ecological systems (Cumming, 2011). Considering landscapes as social-ecological systems, landscape resilience can, in line with Folke et al. (2010), be defined as the ability of a landscape to adapt to continuously ongoing change and to tackle disruptions while at the same time retaining its essential function. From an anthropocentric point of view, a landscape's essential function can be defined as the provision of ecosystem services, such as food production. In the same way, as ecological resilience is improved by greater biodiversity and social resilience by diversification of livelihoods, an agriculture based on a diverse resource use enhances, according to our understanding, landscape resilience (Cumming, 2011). Thus, we argue that farmland management and the manner in which food is produced in agricultural landscapes influence landscape resilience.

Diverse resource use in agricultural productions is one strategy to tackle shocks (Ashkenazy et al., 2018; International Fund for Agricultural Development, 2021), shocks that in the past could be generated by e.g. crop failure due to pests or adverse weather conditions during the growing season. Being prepared to deal with shocks was an important prerequisite to survive as farmers, most likely not only in Norway but in any farming society. However, Norway's northern location made the ability to tackle shocks especially important due to potentially unstable growing conditions during summer and certain crops being grown on the northern boundary of their potential range.

Thus, within Norwegian agricultural history, access to a variety of resources contributed to a high degree of resilience within the farming system. This resilience was mainly on farm scale and enabled an individual farm or a local farming community to absorb disruptions and shocks. If one resource failed, other resources could be used more intensively. An unintended outcome of this approach was a highly diverse landscape, where a multitude of different types of land could be harvested when needed. This diverse land-use and resource exploration produced very spatially heterogeneous landscapes, again mainly on farm scale. Thus, diversity and heterogeneity seem to be key concepts that can be used to further analyse landscape resilience since they reflect the extent to which agricultural production is based on a diverse resource use.

Heterogeneity and diversity are commonly used in ecology and landscape ecology (Collinge, 2009). When applying these concepts on land and landscape assessment, diversity is used to describe the number of different land cover / land-use types present, and if required also their relative coverage. Heterogeneity adds valuable information about landscape structure (Fjellstad, Dramstad, Strand & Fry, 2001), as the spatial distribution of the differing land cover / land-use types is captured. For both measures, several quantitative indices have been developed based on a desire to enable comparison of landscapes over time (for examples see Andreasen, O'Neill, Noss & Slosser, 2001; Magurran, 2003; McGarigal & Marks, 1995). We will not calculate quantitative indices, however, but intend to discuss the application of the measures from a more theoretical perspective founded in our knowledge about the historical and current state in our three case areas. In the following, we highlight important landscape changes in the case areas and discuss afterwards some consequences of changes in landscape heterogeneity and diversity for the provision of ecosystems services.

FIGURE 3.5 Farming landscape in Rakkestad in 1953 and 1992. White arrows: examples of grasslands that have disappeared in 1992, black arrows: examples of narrow grassy banks that have disappeared in 1992 (Old aerial photo © FotoNor, newer aerial photo © kilden.nibio.no).

The landscape changes in our case studies reflect the three main pathways common for agricultural landscapes throughout the whole of Europe: intensification and upscaling, extensification and land abandonment, and urban and industrial sprawl (Bieling, Plieninger & Trommler, 2011). Intensification and upscaling are clearly visible in Rakkestad (Figure 3.5) and Time (Figure 3.6). Although agricultural production and land use in Rakkestad were already rather specialized in 1953, the changes until 1992 reflect a further upscaling of productions. Fields are merged into larger units, making the majority of the narrow grassy banks disappear by 1992. These narrow grassy banks often separated fields used for different crops, as farming was based on a traditional system of crop rotation. Another important change is diminishing grasslands. While the 1953 aerial photo shows several grasslands, in 1992, their number is strongly reduced.

In Time, similar to Rakkestad, a patchwork of small fields was merged into larger units. The aerial photo from 1953 visualizes a range of different crops and productions, as can be seen from the different shades of grey. This heterogeneity is much reduced in 2019. Fences or stones removed from the fields and located at the edge of the fields were typical field diversions. Some of these narrow diversions are more visible in 2019 than in 1953, as can be seen from the rows of bushes or small trees marking property borders. We assume that bushes and trees have been able to grow since the use of larger machinery does

FIGURE 3.6 Aerial photos of the same agricultural landscape in Time municipality, in 1953 and 2019, respectively (Old aerial photo © FotoNor, newer aerial photo © kilden.nibio.no).

not allow to cut grass close to property borders. Moreover, farmers would try to avoid carrying out any farm operation that could damage elements constituting property borders since 'challenging' the borders would not be considered good conduct. In the 1940s and 1950s when more time was invested in managing the land than in the 2010s, and when how crops developed was more important than at present, trees and bushes would probably have been removed by hand to prevent them from shading the crop and using nutrients from the field.

While merging of smaller fields occurred in Vega as well, Vega is probably the example among our cases where extensification and abandonment of former agricultural land is most common (Figure 3.7). The aerial photos show a number of clearly designated crop fields in 1953 that are hardly visible in 2009. Such a landscape change results in reduced heterogeneity and diversity, as more and more land turns into shrubland. While there could have been a variety of crops and crop varieties grown in 1953, what remains is fields with more or less similar successional stages and very little variation. Finally, the aerial photographs of Time show urban sprawl – the development of housing accompanied by a new road – on what used to be agricultural land.

FIGURE 3.7 Aerial photos showing a landscape from Vega municipality in 1965 and the same landscape in 2009 (Old aerial photo © FotoNor, newer aerial photo © kilden.nibio.no).

Merging of fields and increased average field size as a consequence of small fields being abandoned allow for the use of big machinery and speeding up farm operations (i.e., upscaling of production). However, large fields and an accompanying use of a small selection of crops, and abandonment of marginal land may weaken a farm's availability of resources in times of crises. The dry summer of 2018 with high temperatures and little precipitation during a period of almost five months, resulting in forest fires and impacting agricultural production in the whole of Europe, showed that access to a diversity of resources including mountain resources were important for Norwegian farmers to tackle the climatic shock (Beitnes, Kopainsky & Potthoff, 2022; Skaland et al., 2019). Moreover, on farm scale, the production of a variety of crops made it possible to use grasslands for natural flood control. Grasslands were particularly important in depressions affected by flooding and along the river. When a river raised above its normal level, e.g., during snowmelt, grasslands could function almost like detention ponds. The water would cause limited erosion and little other damage. When a farm only produces grain crops, the adaptation of maintaining grasslands is no longer feasible. In former grasslands – now turned into fields for annual crops – flooding causes severe erosion problems and potentially also crop damage.

The removal of small landscape elements such as grassy banks is not clearly visible in land-use/land cover statistics, since they do not constitute a large proportion of the total in terms of area (Fjellstad & Dramstad, 1999); however, their disappearance may have important effects on a landscape's ability to provide food, as well as the visual appreciation of the landscape by the general public (Stokstad, Krøgli & Dramstad, 2020). Narrow strips across fields are also well documented to be important for biological pest control, e.g., by providing habitat for carabid beetles, key predators of aphids (Altieri, Nicholls & Fritz, 2005;

Dennis, Fry & Andersen, 2000). Further, the use of bigger machinery, necessary to make operations and transport more efficient, is not feasible on small fields. When farm operations were done manually, like haymaking, small fields were no hindrance.

While merging fields and removal of landscape elements may impact on the way food is produced, production is bound to be reduced when the area used for farming is declining as in the example of Vega. What would be needed in terms of investments to bring these areas back into agricultural use again, for example, in times of crises, depends on how far the natural succession has come. An important prerequisite to make use of such areas is the availability of knowledge of and technology for, how to use, for example, marginal resources. Erosion of knowledge may be a bigger challenge than of soil. Housing development in Time is an example of a land-use change that most likely will terminate any possibility to use the land for future agricultural production.

So far, we have argued that a diverse resource use, visible in a high landscape diversity, may enhance resilience. However, the housing development in Time shows that not any kind of increased resource use diversity may result in enhanced resilience, at least not when food production is the goal. The change from farming to housing is a reminder that trade-offs will always occur. Time is located in an economically fast-developing region in which the demand for new housing areas competes with the use of land for agricultural production (Stokka et al., 2018). On the one hand, availability of housing is important for the economic development of the region, on the other hand, we can argue that – taking the small amount of arable land in Norway into consideration – Norway has no agricultural land to lose. This example illustrates that a landscape can never provide all its essential functions within one piece of land, and the provision of one service may exclude the availability of another.

Until now, our discussion of potential consequences of decreasing diversity and heterogeneity for a landscape's ability to produce food has focussed on the farm scale. However, heterogeneity, diversity and related issues of resilience cannot be properly addressed without considering other scales. When policy affects the geographical distribution of different productions on regional scale – as in our case, moving from local to regional scale will increase diversity, since a greater range of productions will be considered. At the national scale, an even larger selection of productions will be included. Thus, at the national scale, the diversity of landscapes and their different productions may in sum still provide a range of products broad enough to ensure a certain degree of resilience. However, considering the national scale only may give an illusion of being resilient, an illusion dependent on many prerequisites. In a country such as Norway, reaching from 58 to 71°N, transport distances are extensive and – as pointed out in the previous section – transport costs are high. Productions taking place in northern Norway cannot contribute to food security in southern Norway without, for example, an efficient transport system. Such a transport system is dependent on a functioning infrastructure, a distribution system, fuel, and durability of products.

Further, farming in Norway is dependent on multiple external input factors, such as import of raw material for concentrated feed production (see previous section). Such a resilience can be called coerced since the maintenance of production levels is dependent on anthropogenic input that in our case even comes from outside the country (Rist et al., 2014). Dependence on external input factors will increase with loss of farmland to other types of land uses and as a result of farmland abandonment.

Reconnection to landscapes' production potentials

In our perspective, the previous sections have highlighted that the current framework in which the agricultural sector operates puts the resilience of local landscapes under pressure. This pressure on resilience is a consequence of reduced opportunities for diverse and flexible use of local resources, visible in a declining heterogeneity and diversity. Climate change may further challenge landscape resilience as the dry summer of 2018 revealed (Beitnes et al., 2022) (see the previous section for more details). Prospects of climate change in all our case municipalities include a probable increase of rain occurring at higher frequency and with greater intensity and – as a consequence – problems with surface water and increased flood events (Hisdal, Vikhamar-Schuler, Førland & Nilsen, 2021). Higher temperatures during summer may result in droughts due to increased evaporation (Hisdal et al., 2021). These changes may all challenge agricultural production, but exactly which production that will suffer the most from any particular event may differ. As long as shocks of any type can be buffered by getting access to resources from other parts of Norway or other parts of the globe; the agricultural sector may continue along the current trajectory for production system development, with its focus on increased efficiency and little attention to the production potential of local landscapes.

However, at the same time as access to resources globally may support a country's resilience, it may make food production dependent on external inputs more vulnerable. Within global interconnected food systems any small-scale event may impact productions at great distances (Nyström et al., 2019). Also food systems in the resource-strong Nordic region can potentially be threatened by the collapse of long supply changes (Nordic Council of Ministers, 2020). The disconnection of where food is produced and where it is consumed increases vulnerability to infrastructure disruptions (Nyström et al., 2019), and disruptions at specific points – 'chokepoints' – such as the Suez Canal may get disproportionally large-scale consequences for food security (Wellesley, Preston, Lehne & Bailey, 2017). In addition, transport contributes to the emission of CO_2, and thus, climate change, which again may have negative impacts on food security. Not at least, relying on external resources can lead to land grabbing; and while providing potential economic opportunities for rural populations, it may also challenge the livelihoods of local people (Santurnino, Hall, Scoones, White & Wolford, 2011).

Thus, a different kind of agriculture would be needed to strengthen landscape resilience. What could such an agriculture be like? We definitely do not mean to romanticize past agriculture with its hardships for farming communities and its restricted choice of products in terms of, for example, vegetables. However, we would argue that we need an agriculture that builds to a larger extent on a production base available in the local landscape. Our concern is that the pendulum now has moved too far towards efficiency, as used in other industries and typically linked to the idea of 'economy of scales', as also discussed by Smiley (1997). In this context, also the land sharing versus land sparing debate is relevant, but as outlined by Grass, Batáry and Tscharntke (2021), multifunctional landscapes should preferably include elements from both approaches.

Key points of our vision for more resilient food productions and thereby resilient landscapes are:

1 A farm size scaled to allow carrying out at least most farm operations during optimal weather conditions. This would contribute to increased production, reducing the existent yield gap, thus decreasing dependence on resources from other places.
2 Field sizes that increase landscape diversity and heterogeneity and give room for 'non-productive' landscape elements such as grassy banks between fields. Accepting and adapting productions also on smaller fields would open up for a greater selection of products on farm scale, while a diversity of different landscape element would help support natural pest control or attenuate the impacts of extreme weather events.
3 Marginal resources that are kept in a productive state. The assessment of the value of resources should consider in which way their use can help reducing need for transport and dependency on global resources, and if their value may increase in times of crisis. However, considerations must be done on a longer timescale, and not be based solely on the situation 'here and now'.
4 A technological development that addresses the needs of a more small-scaled farming. The experiments of a Norwegian agriculture school to sow grass using a drone, or the agriculture robot designed at the Norwegian University of Life Sciences are examples of technological development that according to our understanding would allow the use of smaller fields in an efficient way (Flatås & Alisubh, 2021; Robotikkgruppen, 2021). Such a development may contribute to counteracting the drivers founded in the desire of continuously increased efficiency (i.e., output per unit).

Path to the future – the responsibility of a multiplicity of stakeholders

A large number of stakeholders in addition to farmers and landowners impact directly or indirectly on how food is produced in agricultural landscapes (Meuwissen et al., 2019). This multiplicity of stakeholders implies that the responsibility for remaining or becoming resilient should not rest only on the farmers but be shared among stakeholders (Darnhofer, 2014; Darnhofer, Lamine, Strauss & Navarrete, 2016). As presented in the section 'Norwegian agricultural history in a nutshell', policy makers are key stakeholders of the agricultural sector. The fact that farmers have expressed greater worries about politically induced structural changes than about climate change underlines how strongly policy impacts the Norwegian agricultural sector (Beitnes et al., 2022). In line with the Common Agricultural Policy (CAP) of the EU, the Norwegian subsidy system includes payments to encourage sustainable land use such as area and cultural landscape payments (Bjørkhaug & Rønningen, 2014; Daugstad, Rønningen & Skar, 2006). However, increased efficiency and production are still important policy goals and are encouraged by the subsidy system (Forbord & Vik, 2017). Also, Schiere, Darnhofer and Duru (2012) point out that efficiency and stability –stimulated by European policy – may mean less flexibility and resilience.

Agricultural policy, globally, in an EU and Norwegian context, seems to encourage farming systems to strengthen their robustness (FAO, 2021; Manevska-Tasevska et al., 2021; Meuwissen et al., 2020; Nicholas-Davies, Fowler & Midmore, 2020; Potthoff & Kopainsky, in preparation; Paas et al., 2021; Reidsma et al., 2020). However, robustness, a farming system's ability to tackle stress and shocks (i.e., to persist), is only one resilience capacity (Meuwissen et al., 2019) and buffering shocks by resources external to a local landscape may not be resilient in the long run. Other resilience capacities – adaptability (ability to make changes without changing the structure of the farming system) and transformability (ability to make significant changes) (Meuwissen et al., 2019) – may become more important in the future taking into consideration future challenges for agricultural production, such as climate change.

What kind of agricultural policy could stimulate farming systems to increase their resilience? Basically, such a policy and its related subsidy system would need to give more room for a flexible and diverse resource use. More flexibility could, for example, encourage farmers to keep marginal areas in use that can become important in times of crisis and to sustain a broader selection of productions. Moreover, 'hidden' costs such as transport costs should be considered carefully when evaluating cost efficiency within productions. Not at least, trade-offs between increasing resilience through, for example, diversification and efficiency would have to be addressed (FAO, 2021). Encouraging technological development and innovation could reduce the costs of declining efficiency by making the management of small fields less time consuming and by identifying values and usages of seemingly valueless or marginal resources. Examples of the latter are projects about using 'worthless' sheep wool as fertilizer (McKinnon, 2021) and identifying values of outfield resources (Strand et al., 2021).

Among the multiplicity of stakeholders impacting on food production we would like – besides the role of policy makers – to stress the importance of consumers. Through our choices as consumers of food, we have an impact also on the agricultural landscape and its resilience. Buying locally and regionally produced and processed food can support farmers in offering a larger diversity of products and in taking into use a broader selection of their farm resources. Consumer choices and the willingness to pay a higher price for food have thereby also an important impact on the degree to which knowledge to harvest local resources is preserved and farmers can benefit from natural pest control. The possibility to make informed choices depends, of course, on whether information is available to us. In this context, media and the processing industry can play a role as stakeholders. Media can ensure, through informing the public and policymakers, that local food production and product diversity is on the agenda. The processing industry can make sure that information about origin of products is easily available. Increased interest in locally produced food bought in shops and directly from the producer during the COVID-19 pandemic in Norway (September 2020 to August 2021) indicate a trend of a higher consumer awareness for whatever reason (Johnsen, 2021).

We believe that through our vision for agriculture – keeping land in production, adapting production to the landscape, and lessening the drivers for increased efficiency – we can increase agricultural resilience. This vision, we would argue, should be a key to future agricultural development, in Norway as in other countries, even though the way forward may be different in different parts of the world and different production and policy systems.

Acknowledgements

Dramstad's work at NIBIO was financed by the Norwegian Ministry of Agriculture and Food, through the Research Council of Norway (grant No. 194051). We thank Frode Bentzen (NIBIO) for assistance with the aerial photos and Ulrike Bayr (NIBIO) for making the map.

References

Acksel, A., Giani, L., Stasch, C., Kühn, P., Eiter, S., Potthoff, K., Regier, T., & Leinweber, P. (2019). Humus-rich topsoils in SW Norway – Molecular and isotopic signatures of soil organic matter as indicators for anthropo-pedogenesis. *Catena, 172*, 831–845.

Almås, R. (2002). *Norges Landbrukshistorie 1920–2000. Frå bondesamfunn til bioindustri* (Vol. IV). Oslo: Det Norske Samlaget.

Almås, R. (Ed.) (2004). *Norwegian Agricultural History*. Trondheim: Tapir Akademisk Forlag.

Altieri, M. A., Nicholls, C. I., & Fritz, M. A. (2005). *Manage Insects on Your Farm. A Guide to Ecological Strategies*. College Park: Sustainabile Agricultural Research and Education.

Andreasen, J. K., O'Neill, R. V., Noss, R., & Slosser, N. C. (2001). Considerations for the development of a terrestrial index of ecological integrity. *Ecological Indicators, 1*(1), 21–35.

Animalia. (2020). Ullstasjoner i Norge. Retrieved from https://www.animalia.no/no/Dyr/ull-og-ullklassifisering/ullstasjoner-i-norge/

Ashkenazy, A., Chebach, T. C., Knickel, K., Peter, S., Horowitz, B., & Offenbach, R. (2018). Operationalising resilience in farms and rural regions – Findings from fourteen case studies. *Journal of Rural Studies, 59*, 211–221.

Beitnes, S. S., Kopainsky, B., & Potthoff, K. (2022). Climate change adaptation processes seen through a resilience lens: Norwegian farmers' handling of the dry summer of 2018. *Environmental Science & Policy, 133*, 146–154.

Bieling, C., Plieninger, T., & Trommler, K. (2011). Cross the border – Close the gap: Resilience-based analysis of landscape change (Editorial). *European Countryside, 2*, 83–92.

Bjartnes, A. (2018). Utslipp fra jord til bord. In A. Bjartnes (Ed.), *Matsystemet under Press* (Vol. Rapport nr. 02/2018, pp. 6–7). Bergen: Norwegian Climate Foundation.

Bjørkhaug, H., & Rønningen, K. (2014). Crisis? What crisis? Marginal farming, rural communities and climate robustness: The case of Northern Norway. *International Journal of Sociology of Agriculture and Food, 21*(1), 51–69.

Burton, R. J. F., & Farstad, M. (2020). Cultural lock-in and mitigating greenhouse gas emissions: The case of dairy/beef farmers in Norway. *Sociologia Ruralis, 60*(1), 20–39.

Collinge, S. K. (2009). *Ecology of Fragmented Landscapes*. Baltimore: John Hopkins University Press.

Cumming, G. S. (2011). Spatial resilience: Integrating landscape ecology, resilience and sustainability. *Landscape Ecology, 28*, 899–909. doi:10.1007/s10980-011-9623-1

Darnhofer, I. (2014). Resilience and why it matters for farm management. *European Review of Agricultural Economics, 41*(3), 461–484.

Darnhofer, I., Lamine, C., Strauss, A., & Navarrete, M. (2016). The resilience of family farms: Towards a relational approach. *Journal of Rural Studies, 44*, 111–122.

Daugstad, K., Rønningen, K., & Skar, B. (2006). Agriculture as an upholder of cultural heritage? Conceptualizations and value judgements—A Norwegian perspective in international context. *Journal of Rural Studies, 22*, 67–81.

Dennis, P., Fry, G. L. A., & Andersen, A. (2000). The impact of field boundary habitats on the diversity and abundance of natural enemies in cereals. In B. Ekborn, M. E. Irwin & Y. Robert (Eds.), *Interchanges of Insects between Agricultural and Surrounding Landscapes* (pp. 195–214). Dordrecht: Springer.

Eurostat. (2021). Farms and farmland in the European Union - statistics. Evolution of farms and farmland from 2005 to 2016. Retrieved from https://ec.europa.eu/eurostat/statistics-explained/index.php?title=Farms_and_farmland_in_the_European_Union_-_statistics#Farms_in_2016

FAO. (2021). *In Brief. The State of Food and Agriculture 2021. Making Agrifood Systems more Resilient to Shocks and Stresses*. Rome: FAO.

Fjellstad, W., Dramstad, W. E., Strand, G.-H., & Fry, G. L. A. (2001). Heterogeneity as a measure of spatial pattern for monitoring agricultural landscapes. *Norsk Geografisk Tidsskrift – Norwegain Journal of Geography, 55*(2), 71–76. doi:10.1080/00291950119811

Fjellstad, W. J., & Dramstad, W. E. (1999). Patterns of change in two contrasting Norwegian agricultural landscapes. *Landscape and Urban Planning, 45*(4), 177–191.

Flatås, S., & Alisubh, T. (2021). I denne åkeren er gresset sådd med drone – kaller prøveprosjektet en suksess. Retrieved from https://www.nrk.no/trondelag/naspirer-gressfroene-i-norges-forste-dronesadde-aker-1.15523604

Folke, C., Carpenter, S. R., Walker, B., Scheffer, M., Chapin, T., & Rockström, J. (2010). Resilience thinking: Integrating reslilience, adaptability and transformability. *Ecology and Society, 15*(4), 1–9.

Forbord, M., & Vik, J. (2017). Food, farmers and the future: Investigating prospects of increased food production withing a national context. *Land Use Policy, 67*, 546–557.

Gjerdåker, B. (2004). Continuity and modernity 1815–1920. In R. Almås (Ed.), *Norwegian Agricultural History* (pp. 235–293). Trondheim: Tapir Academic Press.

Grass, I., Batáry, P., & Tscharntke, T. (2021). Combining land-sparing and land-sharing in European landscapes. *Advances in Ecological Research, 64*, 251–303. doi:10.1016/bs.aecr.2020.09.002

Hillestad, M. E. (2014). *Transport i landbruket. Fylkesveier som like viktige stamveier* (Vol. 2). Oslo: AGRI Analyse.

Hillestad, M. E., & Bunger, A. (2019). *Kornhøsting i våtere klima*. Oslo: AgriAnalyse.

Hisdal, H., Vikhamar-Schuler, D., Førland, E. J., & Nilsen, I. B. (Eds.). (2021). *Klimaprofiler for fylker. Et kunnskapsgrunnlag for klimatilpasning*: Norsk klimaservicesenter.

International Fund for Agricultural Development. (2021). *Transforming food systems for rural prosperity*. Rome: International fund for Agricultural Development.

Johnsen, A. M. (2021). Eventyrlig vekst i salget av lokalmat i butikk, nedgang i storhusholdning. Retrieved from https://stiftelsennorskmat.no/no/aktuelt/salgstall-lokalmat-2020-2021

Jones, M., & Rønningen, K. (2007). Jordskifte og kulturlandskap. In Ø. Ravna (Ed.), *Areal og eiendomsrett* (pp. 371–390). Oslo: Universitetsforlaget.

Landbruksdirektoratet. (2022). Kraftfôrstatistikk. Retrieved from https://www.landbruksdirektoratet.no/nb/statistikk-og-utviklingstrekk/utvikling-i-jordbruket/kraftforstatistikk

Lunden, K. (2004). Recession and new Expansion 1350–1814. In R. Almås (Ed.), *Norwegian Agricultural History* (pp. 144–158). Trondheim: Tapir Akademisk Forlag.

Magurran, A. E. (2003). *Measuring Biological Diversity*. Oxford: John Wiley & Sons.

Manevska-Tasevska, G., Petitt, A., Larsson, S., Bimbilovski, I., Meuwissen, M. P. M., Feindt, P. H., & Urquhart, J. (2021). Adaptive governance and resilience capacity of farms: The fit between farmers' decisions and agricultural policies. *Frontiers in Environmental Science, 9*, 1–13. doi:10.3389/fenvs.2021.668836

McGarigal, K., & Marks, B. (1995). *FRAGSTATS: Spatial pattern analysis program for quantifying landscape structure* (Vol. 351). Portland: U.S. Department of Agriculture, Forest Service, Pacific Northwest Research Station.

McKinnon, K. (2021). Lavverdiull til planteproduksjon – utprøving av ullpellets som gjødsel. Retrieved from https://www.norsok.no/prosjekter/2021/lavverdiull-til-planteproduksjon-utproving-av-ullpellets-som-gjodsel

Meld. St. 9 (2011–2012). *Landbruks- og matpolitikken. Velkommen til bords*. Oslo: Landbruks- og matdepartementet.

Meuwissen, M. P. M., Feindt, P. H., Midmore, P., Wauters, E., Finger, R., Appel, F., Spiegel, A., Mathijs, E., Termeer, K. J. A. M., Balmann, A., de Mey, Y., & Reidsma, P. (2020). The struggle of farming systems in Europe: Looking for explanations throught the lens of resilience. *EuroChoices, 19*(2), 4–11. doi:10.1111/1746-692X.12278

Meuwissen, M. P. M., Feindt, P. H., Spiegel, A., Termeer, C. J. A. M., Mathijs, E., de Mey, Y., Finger, R., Balmann, A., Wauters, E., Urquhart, J., Vigani, M., Zawalińska, K., Herrera, H., Nicholas-Davies, P., Hansson, H., Paas, W., Slijper, T., Coopmans, I., Vroege, W., Ciechomska, A., Accatino, F., Kopainsky, B., Poortvliet, P. M., Candel, J. J. L., Maye, D., Severini, S., Senni, S., Soriano, B., Lagerkvist, C.-J., Peneva, M., Gavrilescu, C., & Reidsma, P. (2019). A framework to assess the resilience of farming systems. *Agricultural Systems, 176*, 1–10.

Neset, T.-S., Wiréhn, L., Klein, N., & Käyhkö, J. (2019). Maladaptation in Nordic agriculture. *Climate Risk Management, 23*, 78–87.

Nicholas-Davies, P., Fowler, S., & Midmore, P. (2020). Telling stories – farmers offer new insights into farming reslience. *EuroChoices, 19*(2), 12–17. doi:10.1111/1746-692X.12279

Nordic Council of Ministers. (2020). *What Can the COVID-19 Pandemic Teach Us about Resilient Nordic Food Systems?* Copenhagen: Nordic Council of Ministers.

Norges Offisielle Statistikk. (1950). *Jordbrukstellingen i Norge 20. juni 1949. Første Hefte. Areal, Husdyrholdet M.V. Herrredvise oppgaver.* Oslo: Statistisk sentralbyrå.

Nysted, T. E., Uldal, S. H., & Vakse, I. (2020). Hva spiser norske husdyr – egentlig. In A.-K. Kjos, O. Nafstad, H. Odden, T. A. Ruud, T. Saltnes & M. Ytterdahl (Eds.), *Kjøttets tilstand 2020. Status i norsk kjøtt- og eggproduksjon* (pp. 6–11). Oslo: ANIMALIA.

Nyström, M., Jouffray, J. B., Norström, A. V., Crona, B., Jørgensen, P. S., Carpenter, S. R., Bodin, Ö., Galaz, V., & Folke, C. (2019). Anatomy and resilience of the global production system. *Nature, 575*(5), 98–108.

OECD. (2021). Policies for the future of farming and food in Norway. Retrieved from https://www.oecd-ilibrary.org/sites/177645e6-en/index.html?itemId=/content/component/177645e6-en

Olsson, E. G. A., Austrheim, G., & Grenne, S. N. (2000). Landscape change patterns in mountains, land use and environmental diversity, Mid-Norway 1960–1993. *Landscape Ecology, 15*, 155–170.

Potthoff, K. (2020). Endringer i norske jordbrukslandskap. In J. Aasetre & J. Cruickshank (Eds.), *Innføring i miljø- og ressursgeografi* (pp. 199–223). Bergen: Fagbokforlaget.

Potthoff, K., & Kopainsky, B. (in preparation). Norwegian low-intensity ovine and bovine farming systems – a reslience perspective.

Paas, W., Coopmans, I., Severini, S., van Ittersum, M. K., Meuwissen, M. P. M., & Reidsma, P. (2021). Participatory assessment if sustainability and resilience of three specialized farming systems. *Ecology and Society, 26*(2). doi:10.5751/ES-12200-260202

Reidsma, P., Meuwissen, M., Accatino, F., Appel, F., Bardaji, I., Coopmans, I., Gavrilescu, C., Heinrich, F., Krupin, V., Manevska-Tasevska, G., Peneva, M., Rommel, J., Severini, S., Soriano, B., Urquhart, J., Zawalińska, K., & Paas, W. (2020). How do stakeholders perceive the sustainability and resilience of EU farming systems? *EuroChoices, 19*(2), 18–27.

Rist, L., Felton, A., Nyström, M., Troell, M., Sponseller, R. A., Bengtsson, J., Österblom, H., Lindborg, R., Tidåker, P., Angeler, D. G., Milestad, R., & Moen, J. (2014). Applying resilience thinking to production ecosystems. *Ecosphere, 5*(6), 1–11.

Robotikkgruppen. (2021). Landbruksroboten Thorvald. Retrieved from https://www.nmbu.no/fakultet/realtek/forskning/forskergrupper/robotikk/prosjekter/thorvald

Rustad, L. J. (2020). Ferske tal om norsk sjølvforsyning. Retrieved from https://www.nibio.no/nyheter/ferske-tal-om-norsk-sjolvforsyning

Rønningen, K., Fugestad, E. M., & Burton, R. (2021). Path dependencies in Norwegian dairy and beef farming communities: Implications for climate mitigation. *Norsk Geografisk Tidsskrift – Norwegain Journal of Geography, 75*(2), 65–78.

Santurnino, M. B. J., Hall, R., Scoones, I., White, B., & Wolford, W. (2011). Towards a better understanding of global land grabbing: An editorial introduction. *The Journal of Peasant Studies, 38*(2), 209–216.

Schiere, J. B., Darnhofer, I., & Duru, M. (2012). Dynamics in farming systems: Of changes and choices. In I. Darnhofer, D. Gibbon & B. Dedieu (Eds.), *Farming Systems Research into the 21st Century: The New Dynamics* (pp. 337–363). Dordrecht: Springer.

Selman, P. (2012). Landscapes as integrating frameworks for human, environmental and policy processes. In T. Plieninger & C. Bieling (Eds.), *Resilience and the Cultural Landscape. Understanding and Managing Change in Human-Shaped Environments* (pp. 27–48). Cambridge: Cambridge University Press.

Skaland, R. G., Colleuille, H., Andersen, A. S. H., Mamen, J., Grinde, L., Tajet, H. T. T., Lundstad, E., Sidselrud, L. F., Tunheim, K., Hanssen-Bauer, I., Benestad, R., Heiberg, H., & Hygen, H. O. (2019). *METinfo. Tørkesommeren 2018* (Vol. 14/2019). Olso: Meteorologisk institutt.

Smiley, J. (1997). Farming and the landscape. In J. I. Nassauer (Ed.), *Placing Nature: Culture and Landscape Ecology* (pp. 33–44). Washington, DC: Island Press.

Statistics Norway. (2021a). Arealbruk og Arealressurser. Tabell 09594: Arealbruk og arealressurser, etter arealklasser (km^2) (B) 2011–2021. Retrieved from https://www.ssb.no/statbank/table/09594

Statistics Norway. (2021b). Gardsbruk, jordbruksareal og husdyr. Tabell 03312: Jorbruksbedrifter (F) 1969–2020. Retrieved from https://www.ssb.no/statbank/table/03312

Statistics Norway. (2021c). Gardsbruk, jordbruksareal og husdyr. Tabell 06447: Husdyr, etter husdyrslag (K) 1969–2020. Retrieved from https://www.ssb.no/statbank/table/06447/

Statistics Norway. (2021d). Gardsbruk, jordbruksareal og husdyr. Tabell 06462: Jordbruksareal, etter bruken (dekar) (K) 1969–2020. Retrieved from https://www.ssb.no/statbank/table/06462/

Statistics Norway. (2021e). Gardsbruk, jordbruksareal og husdyr. Tabell 12658: Jordleie (K) 1969–2019. Retrieved from https://www.ssb.no/statbank/table/12658/

Stokka, T., Dramstad, W. E., & Potthoff, K. (2018). The use of rented farmland in an area of intensive agricultural production in Norway. *International Journal of Agricultural Sustainability, 16*(3), 243–254. doi:10.1080/14735903.2018.1471262

Stokstad, G., & Krøgli, S. O. (2012). *Størrelsen på jordstykker.* Ås: Skog og landskap.

Stokstad, G., Krøgli, S. O., & Dramstad, W. (2020). The look of agricultural landscapes – How do non-crop landscape elements contribute to visual preferences in a large-scale agricultural landscape? *Norsk Geografisk Tidsskrift–Norwegian Journal of Geography, 74*, 111–122.

Strand, G.-H., Svensson, A., Rekdal, Y., Stokstad, G., Mathiesen, H., & Bryn, A. (2021). *Verdiskaping i utmark. Status og muligheter* (Vol. Vol. 7, Nr. 175). Ås: NIBIO.

Svin. (2018). Her er Norges 23 griseslakterierer. Retrieved from https://svineportalen.no/her-er-norges-23-griseslakterier/

Uleberg, E., Hanssen-Bauer, I., van Oort, B., & Dalmannsdottir, S. (2014). Impact of climate change on agriculture in Northern Norway and potential strategies for adaptation. *Climatic Change, 122*, 27–39.

Vik, J., & Flø, B. E. (2017). *Scenarioer for norsk landbruksproduksjon. En snål rapport, et tverrfaglig eksperiment, og et diskusjonsgrunnlag* (Vol. 3). Ås: NIBIO.

Wellesley, L., Preston, F., Lehne, J., & Bailey, R. (2017). Chokepoints in global food trade: Assessing the risk. *Research in Transport Business & Management, 25*, 15–28.

Wiréhn, L. (2018). Nordic agriculture under climate change: A systematic review of challenges, opportunities and adaption strategies for crop production. *Land Use Policy, 77*, 63–74.

4

REDEVELOPING RELATIONSHIPS WITH LANDSCAPES FOR FOOD, WATER, AND ENERGY SELF-SUFFICIENCY IN SOUTHEASTERN SOUTH DAKOTA, USA

Meghann Jarchow

Author positioning

I am a white woman of European ancestry who was raised in a rural area of southeastern South Dakota in the United States. I was educated in universities in the Midwest of the United States, studying within the dominant Western scientific paradigm. I received my bachelor of arts degree in biology, a master of science degree in biology, and a doctor of philosophy in sustainable agriculture and ecology and evolutionary biology – focusing on prairie and wetland plant ecology. I started at the University of South Dakota as an assistant professor in July 2012 to coordinate and grow the Sustainability Program, which at that point was an undergraduate minor and major and has since grown to include master of science and doctor of philosophy degrees. Through this work, I have greatly expanded my knowledge of the social aspects of sustainability and of sustainability education. My research tends to be place based and generally focused on the Midwest and Northern Great Plains. Because of my connection to this place, I use "our" when writing about this region in the chapter.

Vision from the future

This section describes a sustainability vision of southeastern South Dakota written from the perspective of the year 2070.

We now live in relation(ship) with our landscapes in southeastern South Dakota (Figure 4.1). The concepts of general resilience and transformability guided the reorganization of this region and the development of these relationships. General resilience is the property of systems to change, innovate, and reorganize while still maintaining their overall structure and functioning in the face of uncertainty and shocks (Folke et al. 2010). This differs from optimizing

DOI: 10.4324/9781003266440-4

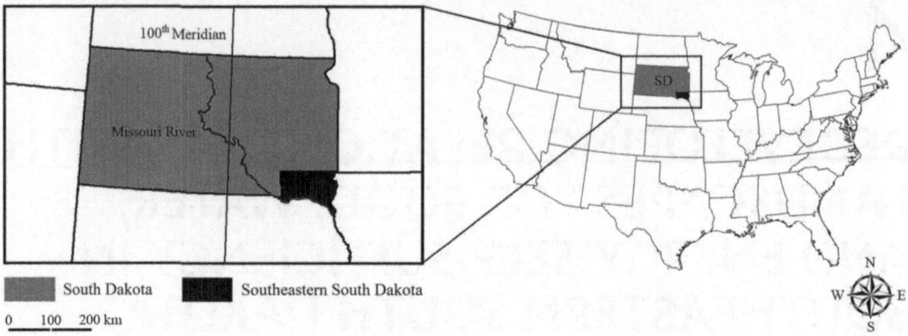

FIGURE 4.1 Map of southeastern South Dakota in relation to the United States, 100th meridian longitudinal line, and Missouri River. The extent of the study area and states was preliminarily mapped based on USDA NASS (2022).

systems to withstand or recover from likely or expected system shocks in that maintaining constancy is not the goal of general resilience (Folke et al. 2004). Transformability is "the capacity to create untried beginnings from which to evolve a fundamentally new way of living when existing ecological, economic, and social conditions make the existing system untenable" (Folke et al. 2004, p. 575). In this sense, having resilient landscapes in southeastern South Dakota is not a form of intensive management; it is a way embracing change and celebrating innovation while maintaining high levels of functioning over the long term.

The landscapes of the region now support high levels of human and more-than-human biodiversity.[1] The plant biodiversity includes prairie, wetland, riverine, forest, and cropped plants – generally growing in communities with high structural and functional-group diversity. The communities are spatially arranged in mosaics with shapes that align with landscape features. Human activities influence the structure and diversity of almost all of the plant communities in a manner that is similar to Indigenous People's traditional relationships with the plant nations (Ellis et al. 2021, Simpson 2017, Wall Kimmerer 2013) except that now the human population is much higher than it was when only Indigenous People inhabited the region.

The species compositions of the plant communities differ from the historical communities to include species from more southerly historical ranges, species native to other continents, and a wide variety of domesticated crops. The overall functioning of the landscape is distinct from the less densely populated times before Euro-American colonization and the intensively managed period when the region was part of the "Corn Belt."[2] The region has the high functional redundancy common before colonization and also the high productivity of human-usable goods characteristic of when the land was nearly exclusively managed for annual row-crop production. The high diversity and productivity are possible because people now invest significant amounts of time into carefully interacting with the landscape including having deep context-specific knowledge of the different plant communities.

This diversity and redundancy expand beyond the plant communities to all the living and non-living world. Wild and domesticated animals are abundant and have their physical and social needs met. The soil is understood to be a complex, living world in and of itself that is foundational to nurturing wellbeing. Water is highly valued, and its use, storage, and transport across the landscape are carefully observed and directly inform human activities. The sun and wind and heating and cooling from the earth are the primary sources of energy for humans.

As the effects of climate change increasingly negatively impacted some of the mostly densely populated regions of the United States and world, people began to move to regions such as ours that were sparsely populated and had abundant gifts from the more-than-human world. Additionally, some of the changes to our climate, including warmer winter temperatures and more precipitation, increased the overall productivity of the region. The initial influxes of people to the region created chaos because they coincided with other transitions in the region such as the significant deterioration of our social, economic, and environmental systems (see "Current and Future Stressors" section). This chaos created an opportunity for fundamental changes in people's relationships with each other and the more-than-human world: there was broad acceptance of nondualism and a willingness to let go of our desire for control. Acceptance of nondualism created a shared belief in the fundamental interconnectedness among everything on the earth. By lessening our desire for control, people are able to live more mindfully, in the present moment, and in tune with what is occurring around them.

These fundamental changes in beliefs had significant effects on people's presence and impacts on the landscape. People have shifted to a "gift and responsibility" mindset (Wall Kimmerer 2013); we do not see the earth as a set of resources to be used but rather as a gift from the earth, who is viewed as our mother. We have come to have a deep gratitude for the more-than-human lives that must be given to support our own lives (Wall Kimmerer 2013), which has reduced our desire to consume beyond our needs. We have learned how to use "two-eyed seeing," which is the use of both Indigenous and Western ways of knowing, in our interactions with the more-than-human world (Marshall as cited in Bartlett 2006). Rather than previous beliefs that advocated for the creation of wilderness areas (i.e., areas without humans), these fundamental changes in our beliefs have brought us in more – and more intimate – contact with the earth and have resulted in us as not seeing ourselves as separate from our landscapes.

History and current context

Human's role in and relationship with what is now southeastern South Dakota has changed multiple times over the past 12,000 years. From about 115,000 to 12,000 years ago, this region was on the southwestern end of an expanding and retracting ice sheet, sometimes more than 3 km thick, that scraped the existing bedrock and created a relatively flat landscape that contains many "pothole," also called geographically isolated, wetlands. Approximately 80% of the land in eastern South Dakota drained into wetlands, rather than rivers, before Euro-American colonization (Johnson et al. 1997). This gave the landscape the capacity to retain water even through the periodic droughts that occur in the region. Following the retreat of the ice sheet, the region underwent a period of transformation where the tallgrass prairie and associated pothole wetlands developed (Samson and Knopf 1996). Maintenance of the tallgrass prairie, which requires disturbance to prevent conversion to forest, occurred through ongoing grazing and fire (Howe 1994). Bison and elk were likely the dominant grazers, and lightning and Indigenous People were likely the dominant sources of fire (Samson and Knopf 1996).

The Oceti Sakowin (Sioux), most recently the Inhanktonwan (Yankton Sioux), and other Indigenous People have lived in this region for thousands of years. Enabled by the Doctrine of Discovery from the 13th century, European colonization of the region began in the 17th century when it was claimed as a colony of France, but European presence in the region was light and primarily used for harvesting animals for the fur trade. More dramatic changes to the region began after the United States became a country in 1776, and Manifest Destiny, an expansion of the Doctrine of Discovery, provided a moral basis for continued westward colonization, which resulted in the Indigenous Holocaust in the Americas (Thornton 1987, Stannard 1992). This also began the shift from landscapes with high general resilience to ones with lower general resilience.

In 1803, this region was purchased from France as part of the Louisiana Purchase. It was part of the Wisconsin Territory from 1836 to 1838, was part of the Minnesota Territory until 1849, and then became part of the Dakota Territory until 1861. The region, which was part of an area known as the Yankton Triangle, was ceded by the Inhanktonwan Nation in the 1858 Treaty of Washington. Colonization within the Dakota Territory was initially concentrated in the Yankton Triangle, and the land was said to have approximately 4.5 million hectares of land that was well-suited for agriculture (Brooks and Jacon 1994).

Colonization of the region occurred most intensely from 1878 to 1887, which was known as the Great Dakota Boom (Schell 2004). This colonization occurred more than a decade after the passage of the Homestead Act (1862)[3] and ceding of the land by the Inhanktonwan because of Oceti Sakowin resistance (referred to as "Indian problems"), heavy involvement of Euro-American men in the U.S. Civil War, multiple years of drought, and the lack of railroads to and within the region (Hamburg 1975). During the Great Dakota Boom, however, thousands of miles of railroad track were laid in eastern South Dakota and the population increased from 11,766 in 1870 to 328,808 in 1890, and the number of farms increased from 1,700 to 50,158 over the same period (Hamburg 1975). South Dakota became a state in 1889, and by the end of 19th century, eastern South Dakota was "fully settled" (Brooks and Jacon 1994).

Euro-American colonization resulted in large declines in biodiversity and increases in landscape simplification. Because genocide was part of the U.S. government's plans for addressing the "Indian problems" in this region, the elimination of the bison was an important part of the strategy. U.S. Army General Philip Sheridan stated about the bison hunters in 1875:

> These men have done in the last two years and will do more in the next year to settle the vexed Indian question, than the entire regular army has done in the last thirty years. They are destroying the Indian's commissary, and it is a well-known fact that an army losing its base of supplies is placed at a great disadvantage. Send them powder and lead, if you will; for the sake of lasting peace, let them kill, skin and sell until the [bison] are exterminated. Then your prairies can be covered with speckled cattle and the festive cowboy.
>
> *(cited in Geist 1996, p. 91)*

Some have estimated that 26 bison, from more than 50 million, survived in the United States by 1900 (LaDuke 1999). The farming in southeastern South Dakota at the end of the 19th century was described as "diversified or mixed farming practices" that included growing corn, oats, and wheat; raising cows for milking; and raising pigs for meat (Schell 2004).

The trends in landscape change in this region that began with colonization have continued to the present day. Row-crop agriculture is the dominant land use in eastern South Dakota, and southeastern South Dakota has had among the greatest direct land-use conversion within the state (SD GF&P 2014), with nearly two-thirds of land in the region planted to corn and soybeans (NASS 2021, Figure 4.2). Since the 1930s, the total number of farmers in the state has been decreasing while the total farm size has been increasing – more than tripling from 180 hectares in 1935 to 590 hectares in 2020 (Diersen et al. 2000, NASS 2020). Population density in southeastern South Dakota remains low at 13.7 people per square kilometer in 2020 (US Census Bureau 2022).

There has been significant loss of, or damage to, the major ecosystems in the region. The tallgrass prairie ecosystem has been almost extirpated from this region, and overall, less than 2.5% of the northern tallgrass prairie remains (Samson et al. 2004). More than 77% of the wetlands in the region have been drained through the instillation of sub-surface tile drainage and urbanization (SD GF&P 2014). Most of the water bodies in the state – 91% of the lakes and reservoirs and 78% of rivers and streams that were surveyed – are impaired in some way (SD DENR 2020). The extent of riverine cottonwood forest, the most common riverine ecosystem in the region, is declining and the likelihood of large-scale restoration of this ecosystem has been described as "doubtful" because of significant modifications, especially damming and shoreline armoring, to our river systems (Johnson et al. 2012).

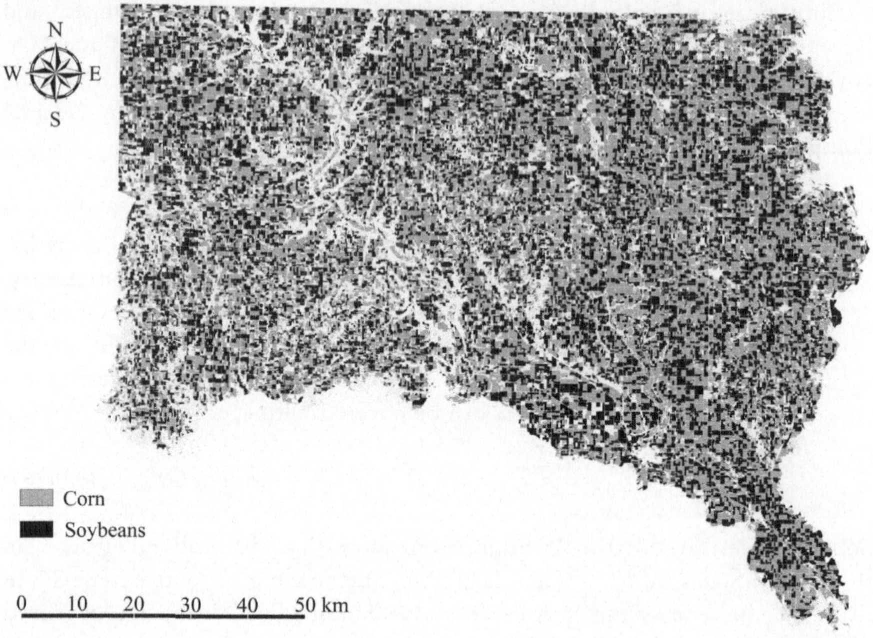

FIGURE 4.2 Land-use map of southeastern South Dakota from 2021 showing the distribution of corn and soybean acreage across the region (USDA NASS 2022). White areas indicate a land use other than corn or soybeans.

South Dakota produces relatively large amounts of renewable energy, although most of the renewable energy is produced outside of southeastern South Dakota. Eighty-three percent of the electricity produced in the state in 2020 was renewable – 50% from hydropower and 33% from wind turbines (EIA 2021). In 2018, South Dakota produced 7% of total fuel ethanol in the United States, and all of the ethanol plants within South Dakota used corn as the feedstock (EIA 2021). South Dakota has limited solar photovoltaic electricity production and has no significant crude oil, coal, or natural gas reserves (EIA 2021).

Current and future stressors

Many of the region's current and future stressors are related to the extreme land-use changes that have occurred in this region over the past 150 years. We have gone from a region with tremendous biodiversity – the North American grasslands contain about 7,500 plant species (Samson and Knopf 1996) – to one where most of the landscape is actively managed for monocultures of two annual plant species. Because of this, multiple scientists have referred to the Corn Belt as a mono-functional landscape, in contrast to multifunctional landscapes that provide a wide array of ecosystem services (Foley et al. 2005; Fischer et al. 2017).

With the loss of the prairie plants comes the associated loss of animal and other taxonomic diversity. The loss of continuous living plant cover contributes to soil erosion and nutrient pollution into water bodies.

Although they produce large amounts of saleable products, corn and soybean cropping systems are brittle systems – they thrive under a relatively narrow range of environmental conditions. The corn and soybean systems in this area require spring weather where the soil is dry enough to allow large equipment to access the fields yet wet enough to support seedling germination and growth. The installation of sub-surface tile drainage reduces ponding and saturated soil conditions in the spring because it lowers the water table and moves water, including any nutrients and biocides dissolved in the water, more quickly off the crop fields into nearby water bodies or drainage ditches. This results in more "flashy" systems during deluges because the landscape has a reduced capacity to hold water and has a mechanism for moving the water more quickly out of the soil.

Corn especially has been described as a "leaky" system with regard to nutrients because it requires relatively large inputs of nutrients for optimal growth but is unable to keep those nutrients in the plant-soil system because of its small size in the spring when growing from seed and senescence in the fall (Heggenstaller et al. 2008). This is in contrast to prairie plants which have much smaller and "tighter" nutrient cycles (Jarchow et al. 2015). The economics of our agricultural system are leaky beyond just the nutrients. Most of our agricultural products are sold as commodities that leave the region thereby having the region function as an extraction area (Sieverding et al. 2016). In spite of the brittleness of our farming systems, they have become entrenched due to federal farm policy, heavy reliance on relatively few agribusiness corporations for inputs and as purchasers of the grain, and use of increasingly large and expensive equipment.

We have also changed to a region where few people directly live off of the land. The high cost of entry into row-crop farming because of the tight profit margins and need for large equipment prohibits most people from entering into farming. As people move away from land-centered lifestyles and as the dominant culture continues to suppress the deep wisdom that the Indigenous People have of this land, we are becoming less connected to and knowledgeable about our land (Louv 2008; Wall Kimmerer 2013).

Climate change is both an opportunity and a stressor in our region – now and increasingly into the future. Climate change is increasing the length of our growing season and is making our winters warmer (Todey 2021), both of which have positive implications for plant growth in the region. Climate change has and is continuing to shift the "100th meridian" east. The 100th meridian longitudinal line, which is slightly west of southeastern South Dakota (Figure 4.1), roughly aligns with a sharp climatic shift within North America that divides the United States into the humid eastern half and the arid western half (Powell 1879). The eastward shift in this zone has resulted in southeastern South Dakota now being to the west of the "100th meridian" (Seager et al. 2018). Climate change is contributing to current and modeled future increases in spring and fall precipitation as well as increases in the frequency of extreme precipitation events (Todey 2021). In March 2019, for example, a bomb cyclone[4] followed by continuing heavy rain resulted in spring flooding that caused South Dakota to lead the United States in the number of unplanted cropland acres (USGS 2019). It also resulted in extensive urban flooding.

I believe that climate change will also be a driver of significant human migration into this region. In this region, many of the effects of climate change could be positive with appropriate adaptation. Many of the more populated parts of the United States will experience significant negative impacts of climate change where adaptation could be more difficult including increased frequency and intensity of wildfires, increased flooding due to sea level rise and ocean storms, reduced water availability, increased frequency of extreme heat, increased frequency and severity of drought, and reduced suitability of lands for agriculture (IPCC 2012).

Path to the future

Whereas the Great Dakota Boom set this region on a path of biodiversity loss and landscape simplification, a "Resilience Reorganization" could ameliorate past damages and enhance landscape resilience. It seems possible that we are nearing a period of "release" (*sensu* Gunderson and Holling 2002; Figure 4.3) in the Corn Belt where some of the stressors described previously could cause cascading disturbances (Sundstrom and Allen 2019). Such a release would provide an opportunity for transformability within southeastern South Dakota.

We could make changes to the underlying objectives of the system – moving from one of brittleness, extraction, and export to one of general resilience and circularity. The path to creating a more resilient and circular future in southeastern South Dakota could include a shift toward more localization (Norberg-Hodge 2019). Norberg-Hodge (2019) defines economic localization as "the removal of fiscal and other supports that currently favor giant transnational corporations and banks" and "reducing dependence on export markets in favor of production for local needs" (pg. 45). Southeastern South Dakota has the capacity to produce significant amounts of food and energy and has access to plentiful water, especially if water is allowed to remain in the soil and wetlands longer. Our tallgrass prairie ecosystem is one that could be mostly restored within decades.

One step in the path to creating more resilient landscapes includes creating a *shared* sustainability vision for the region and identifying and nurturing the "seeds" of that vision (*sensu* Meadows 1994) that are already occurring in our region and elsewhere. We could use diverse knowledges and be open to radical innovations to enable southeastern South Dakota to become mostly self-sufficient in food, energy, water, and material needs. This would be supportive of our current and likely growing future human population, while also meeting the needs of the more-than-human world.

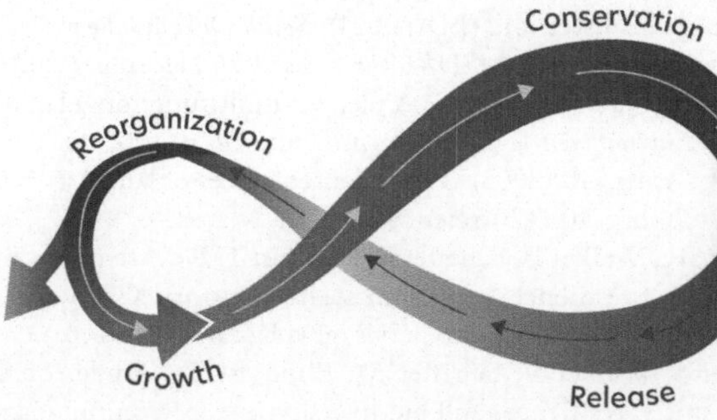

FIGURE 4.3 Adaptive cycle of systems. Adapted from Gunderson and Holling (2002).

Acknowledgment

This work was supported by the National Science Foundation under the EPSCoR Track II cooperative agreement number OIA-1632810. Thank you to Amin Rastandeh for creating Figures 4.1 and 4.2 and Mikayla Meyer for creating Figure 4.3.

Notes

1 I use the term "more-than-human" rather than "non-human" to encourage more ecocentric thinking.
2 The region is called the Corn Belt because maize, which is called "corn" in the United States, is the dominant crop.
3 The Homestead Act provided 160 acres (65 ha) of land to heads of household "for the purpose of actual settlement and cultivation" for a small fee if the person lived on and cultivated the land for at least five years (United States Government 1862).
4 A bomb cyclone is a "large, intense midlatitude storm that has low pressure at its center, weather fronts, and an array of associated weather, from blizzards to severe thunderstorms to heavy precipitation. It becomes a bomb when its central pressure decreases very quickly by at least 24 millibars in 24 hours" (Mullens 2022).

References

Bartlett CM (2006) Knowledge inclusivity: "Two-eyed seeing" for science for the 21st century. In Wilber M, Kearney J (eds) *Proceedings of the Workshop on Learning Communities as a Tool for Resource Management*, 4–5 November 2005, Halifax, NS.

Brooks A, Jacon S (1994) *Homesteading and agricultural development* context, South Dakota State Historical Preservation Center, Vermillion. https://history.sd.gov/preservation/docs/HomesteadAgDevelop.pdf. Accessed 26 December 2021.

Diersen MA, Janssen L, Loewe P (2000) *The structure of South Dakota agriculture: Changes and projections*. https://openprairie.sdstate.edu/econ_research/64/. Accessed 26 December 2021.

Ellis EC, Gauthier N, Klein Goldewijk K, Bliege Bird R, Boivin N, Díaz S, et al. (2021) People have shaped most of terrestrial nature for at least 12,000 years. *Proceedings of the National Academy of Sciences* 118(17): e2023483118. https://doi.org/10.1073/pnas.2023483118.

Energy Information Administration. (EIA; 2021). *South Dakota state energy profile*. https://www.eia.gov/state/print.php?sid=SD#61. Accessed 26 December 2021.

Fischer J, Meacham M, Queiroz C (2017) A plea for multifunctional landscapes. *Frontiers in Ecology and the Environment* 15(2): 59. https://doi.org/10.1002/fee.1464.

Foley JA, DeFries R, Asner GP (2005) Global consequences of land use. *Science* 309(5734): 570–557. https://doi.org/10.1126/science.1111772.

Folke C, Carpenter SR, Walker B, Scheffer M, Chapin T, Rockström J (2010) Resilience thinking: Integrating resilience, adaptability and transformability. *Ecology and Society* 15(4): 20. https://www.ecologyandsociety.org/vol15/iss4/art20/.

Folke C, Carpenter S, Walker B, Scheffer M, Elmqvist T, Gunderson L, Holling CS (2004) Regime shifts, resilience, and biodiversity in ecosystem management. *Annual Reviews of Ecology, Evolution, and Systematics* 35: 557–581. https://doi.org/10.1146/annualrev.ecolsys.35.021103.105711.

Geist V (1996) *Buffalo Nation: History and Legend of the North American Bison*. Voyageur Press, Stillwater, MN.

Gunderson LH, Holling CS (2002) *Panarchy: Understanding Transformations in Human and Natural Systems*. Island Press, Washington, DC.

Hamburg JF (1975) *Railroads and the settlement of South Dakota during the Great Dakota Boom, 1878–1887*. South Dakota State Historical Society, Pierre, SD. https://www.sdhspress.com/journal/south-dakota-history-5-2/railroads-and-the-settlement-of-south-dakota-during-the-great-dakota-boom-1878-1887/vol-05-no-2-railroads-and-the-settlement-of-south-dakota-during-the-great-dakota-boom-1878-1887.pdf. Accessed 26 December 2021.

Heggenstaller AH, Anex RP, Liebman M, Sundberg DN, Gibson LR (2008) Productivity and nutrient dynamics in bioenergy double-cropping systems. *Agronomy Journal* 100(6): 1740–1748. https://doi.org/10.2134/agronj2008.0087.

Howe HF (1994) Managing species diversity in tallgrass prairie: Assumptions and implications. *Conservation Biology* 8(3): 691–704. https://doi.org/10.1046/j.1523-1739.1994.08030691.x.

Intergovernmental Panel on Climate Change (IPCC; 2012) *Managing the Risks of Extreme Events and Disasters to Advance Climate Change Adaptation. A Special Report of Working Groups I and II of the Intergovernmental Panel on Climate Change* [Field, CB, Barros V, Stocker TF, Qin D, Dokken DJ, Ebi LK, Mastrandrea MD, Mach KJ, Plattner GK, Allen SK, Tignor M, Midgley PM (eds)] Cambridge University Press, Cambridge, UK, and New York, NY.

Jarchow M, Swanson D, Kerby J (2020) North American Grasslands as multifunctional landscapes. In Leal Filho W, Azul A, Brandli L, Lange Salvia A, Wall T (eds) *Life on Land, Encyclopedia of the UN Sustainable Development Goals*. https://doi.org/10.1007/978-3-319-71065-5_123-1.

Johnson RR, Higgins KF, Kjellsen ML, Elliott CR (1997) *Eastern South Dakota wetlands*. South Dakota State University, Brookings, SD. https://www.fws.gov/wetlands/Documents/Eastern-South-Dakota-Wetlands.pdf. Accessed 26 December 2021.

Johnson WC, Dixon MD, Scott ML, Rabbe L, Larson G, Volke M, Werner B (2012) Forty years of vegetation change on the Missouri River floodplain. *BioScience* 62(2): 123–135. https://doi.org/10.1525/bio.2012.62.2.6.

LaDuke W (1999) *All Our Relations: Native Struggles for Land and Life*. South End Press, Cambridge, MA.

Louv R (2008) *Last Child in the Woods: Saving Our Children from Nature-deficit Disorder*. Algonquin Books of Chapel Hill, Chapel Hill, NC.

Meadows D (1994) *Down to Earth*. Third Biennial Meeting of the International Society for Ecological Economics. 24–28 October 1994, San Jose, Costa Rica. https://donellameadows.org/archives/envisioning-a-sustainable-world-video/. Accessed 26 December 2021.

Mullens E (2022) *What is a bomb cyclone: An atmospheric scientist explains*. https://www.scientificamerican.com/article/what-is-a-bomb-cyclone/. Accessed 3 May 2022.

National Agricultural Statistics Service (NASS; 2020) *2020 state agriculture overview*. https://www.nass.usda.gov/Quick_Stats/Ag_Overview/stateOverview.php?state=SOUTH%20DAKOTA. Accessed 26 December 2021.

National Agricultural Statistics Service (NASS; 2021) *USDA's National Agricultural Statistics Service South Dakota Field Office*. https://www.nass.usda.gov/Statistics_by_State/South_Dakota/Publications/County_Estimates/index.php. Accessed 26 December 2021.

Norberg-Hodge H (2019) *Local is Our Future: Steps to an Economics of Happiness*. Local Futures, East Hardwick, VT.

Powell JW (1879) *Report on the Lands of the Arid Region of the United States, with a More Detailed Account of the Lands of Utah, with Maps*. Government Printing Office, Washington, DC. https://pubs.usgs.gov/unnumbered/70039240/report.pdf. Accessed 26 December 2021.

Samson FB, Knopf FL (1996) *Prairie Conservation: Preserving North America's Most Endangered Ecosystem*. Island Press, Washington, DC.

Samson FB, Knopf FL, Ostlie WR (2004) Great Plains ecosystems: Past, present, and future. *Wildlife Society Bulletin* 32(1): 6–15. https://doi.org/10.2193/0091-7648(2004)32[6:GPEPPA]2.0.CO;2.

Schell HS (2004) *History of South Dakota*, 4th Edition, Revised. South Dakota State Historical Society Press, Pierre, SD.

Seager R, Lis N, Feldman J, Ting M, Williams AP, Nakamura J, Liu H, Henderson N (2018) Whither the 100th meridian? The once and future physical and human geography of America's arid-humid divide. Part I: The story so far. *Earth Interactions* 22(5): 1–22. https://doi.org/10.1175/EI-D-17-0011.1.

Sieverding HL, Clay DE, Khan E, Sivaguru J, Pattabiraman M, Koodali RT, Ndiva-Mongoh M, Stone J (2016) A sustainable rural food-energy-water nexus framework for the Northern Great Plains. *Agricultural & Environmental Letters* 1(1): 160008. https://doi.org/10.2134/ael2016.02.0008.

Simpson LB (2017) *As We Have Always Done: Indigenous Freedom Through Radical Resistance*. University of Minnesota Press, Minneapolis, MN.

South Dakota Department of Environment and Natural Resources (SD DENR; 2020) *The 2020 South Dakota Integrated Report for Surface Water Quality Assessment*. https://danr.sd.gov/OfficeOfWater/SurfaceWaterQuality/docs/DANR_2020_IR_final.pdf. Accessed 26 December 2021.

South Dakota Game, Fish & Parks (SD GF&P; 2014) *South Dakota Wildlife Action Plan*. https://gfp.sd.gov/UserDocs/nav/SD_Wildlife_Action_Plan_Revision_Final.pdf. Accessed 26 December 2021.

Stannard DE (1992) *American Holocaust: The Conquest of the New World*. Oxford University Press, New York, NY.

Sundstrom SM, Allen CG (2019) The adaptive cycle: More than a metaphor. *Ecological Complexity* 39: 100767. https://doi.org/10.1016/j.ecocom.2019.100767.

Thornton R (1987) *American Indian Holocaust and Survival: A Population History since 1492*. University of Oklahoma Press, Norman, OK.

Todey D (2021) Understanding and managing changing climate conditions in Midwest/Plains agriculture: Information and tools. *SD Soil and Water Conservation Society Annual Conference*. 9 December 2021.

United States Census Bureau (US Census Bureau; 2022) *South Dakota, State in United States*. https://data.census.gov/cedsci/profile?g=0400000US46. Accessed 3 May 2022.

United States Department of Agriculture National Agricultural Statistics Service Cropland Data Layer (USDA NASS; 2022) *CropScape – Cropland Data Layer*. https://nassgeodata.gmu.edu/CropScape/. Accessed 23 May 2022. USDA-NASS, Washington, DC.

United States Geological Survey (USGS; 2019) *Image of the week: Unplanted acres in South Dakota*. https://eros.usgs.gov/image-gallery/image-of-the-week/south-dakotas-unplanted-acres-of-2019. Accessed 26 December 2021.

United States Government (1862) Act of May 20, 1862 (Homestead Act), Public Law 37–64 (12 STAT 392), Enrolled Acts and Resolutions of Congress 1789–2011, General Records of the United States Government, Record Group 11. National Archives Building, Washington, DC.

Wall Kimmerer R (2013) *Braiding sweetgrass: Indigenous wisdom, scientific knowledge, and the teachings of plants*. Milkweed Editions, Minneapolis, MN.

5

A SOCIAL PERENNIAL VISION FOR THE NORTH AMERICAN GREAT PLAINS ROOTED IN THE RESILIENCE OF A NATURAL SYSTEM-INSPIRED AGRICULTURE

Aubrey Streit Krug, Timothy E. Crews and Thomas P. McKenna

Vision

The North American Great Plains is often pictured as a landscape defined by a horizon line, with skies above and plants below. The skies may be broad and blue, or a tumult of thunderous clouds. The plants may be a lush sea of grass, spotted with grazing animals, or a monochromatic stand of wheat, waving in the wind before harvest. For people in other places, these pictures may depict the Great Plains as flyover country, a landscape to be crossed by plane or car.

For those of us who have come to live here, at the crossroads, these pictures may be simply incomplete. What could be added? The belowground world of roots and soil with its habitat for microbes, the rocks laid down and the sediments blown in, the bison and cattle grazing and also the grasshoppers, the mesh of rivers and aquifers, the tracks of migration as well as survey lines and railroads, the clustered trees and brush with fruits and berries, the gardens and the machines, the pipelines and the barbed wire fences, the lively insects and birds and mammals, and all of us humans in villages and cities and dynamic communities who have participated in shaping the landscape, taking and sometimes giving back.

The Great Plains stretches across the United States from Texas to the prairie provinces of Canada, between the Rocky Mountains and the Mississippi River in the middle of the North American continent (Figure 5.1). The Great Plains tends not to be mapped primarily as an economic or political region. It is instead delineated in terms of a semi-arid climate and flat, sloping, rolling topography, physiogeographic factors shaped by the geologic forces of marine sedimentation, mountain uplift, and erosion, which manifest in ecoregional plant communities and which, in tandem with human sociocultural and economic activities, help determine the region's agricultural vegetation. At the risk of oversimplifying: the "plains" shape and support the "grassland" of tallgrass, mixed-grass, and shortgrass "prairies."

DOI: 10.4324/9781003266440-5

In the late 1800s, US federal government incentives to encourage settlement westward, such as the Homestead Act, brought many settlers into the Great Plains. With this migration, settlers began converting prairie landscapes into agricultural production through farming and ranching to produce annual grains and livestock (Cunfer & Krausmann, 2015). According to the 2011 *Atlas of the Great Plains*,

> Throughout the region, more than 80% of the land is used for agricultural activities, including ranching. Only on the margins of the Plains is less than half of the land devoted to agriculture. This pattern emerged in the early 20th century and has continued ever since.
>
> *(Lavin et al., 2011, p. 131)*

The reality of the agricultural landscape that has been built and maintained through settler colonial systems in Indigenous homelands is not the totality of the Great Plains, but given its current dominance, we take it as one important starting point for grappling with questions of resilience. The farming part of this agricultural reality has led to the region being both recognized and mythologized as the "breadbasket," ecologically vulnerable to drought as during the Dust Bowl[1] but nevertheless stocking the world's granaries with staple grains, with wheat in particular continuing to reign "both on the ground and in the region's psyche" (Morris, 2011).

Contemporary agricultural fields of wheat, sorghum, corn, and other cereal grains, oilseeds, and legumes are populated with plants humans have selected and invested in breeding for yield, vigor, disease and pest resistance, and other traits that enhance their value as sources of calories for humans and cattle as well as for biofuels. Most of these fields grow in dependence with energy-intensive inputs: tillage and chemicals that eliminate and reduce competition from other plants, fertilizers containing nutrients that feed plants, other chemicals that kill insects and pathogens that flourish in monocrop stands, and sometimes irrigated water that buffers or alters the reality of climatic variability and limited precipitation.

Yet these fields also grow in the context of, and in dependence on, natural systems: the highly variable semi-arid continental climate, the grasslands that persist, fertile Mollisol soils that perennial and diverse prairie plant communities have helped build and hold, and dynamic cycles of climate, water, carbon, nitrogen, and more. Natural systems have provided for an adapting array of ongoing Indigenous communities and food systems, and peoples of the Great Plains have engaged these systems and helped shape landscapes through the use of fire as well as gathering, hunting, and agricultural activities. On the whole, these natural systems of the Great Plains (which are connected with the natural systems of the continent and planet, and which become coupled with human systems) provide a key counterpoint to the breadbasket in our effort to assess and envision resilience.

Though it is much debated in the research literature, we find it helpful to remember that the concept of resilience exists for a reason: because humans are interested in understanding how systems change, as well as what properties systems can maintain during change. In the case of Great Plains landscapes, we are interested in what properties *can* be maintained, or persist, because we further think some properties *should* persist and are interested in identifying, encouraging, and collectively pursuing actions that could increase the possibility of their persistence. We are interested in applying the concept of resilience as we envision pathways to creating more resilient landscapes, especially in the face of climate change.

Our approach is informed by Thóren and Olsson (2018), who argue that applying the concept of resilience "relies on the implicit or explicit presence of a normative or ontological framework that answers questions about what is important or central about the systems to which the concept is applied" (p. 114). They define resilience as "the ability of a system ***S*** to absorb some disturbance ***D*** whilst maintaining property ***I***" (p. 4). Applying the resilience concept to a system, especially with regard to social dimensions of a system, requires specifying the *I* for that system, or giving the "persistence criteria" (p. 114–115).

The landscapes of the Great Plains include many biological communities, including human communities, and as such are social-ecological systems. In the sections below, we discuss disturbances to these systems, ranging from historical to current to future stressors, with a particular focus on climate change. We are interested in sustaining the soils of the Great Plains, and specifically their ability to provide staple foods which we envision equitably and sustainably nourishing human communities. So the property we aim to maintain is the production of staple foods.

While the current system is indeed producing large quantities of annual grains, it is precarious in the face of long-term disturbance, and it features social dimensions of human inequity and injustice that we are keen *not* to maintain. Therefore, in applying the resilience concept, we find ourselves envisioning more resilient Great Plains landscapes through a process of backward design, in which maintaining the production of staple foods necessitates grappling with both positive and negative disruptions, which, in turn, leads to an understanding that (a) the currently dominant agricultural system must be reinvented and that (b) the natural systems, the context which the currently dominant system depends upon and depletes, must be restored. The result is a social perennial vision for the Great Plains rooted in a resilient, natural system-inspired agriculture.

First, the roots in a natural system-inspired agriculture: we envision the re-perennialization and diversification of the agricultural landscapes of the Great Plains, grounded in the widespread restoration and conservation of grasslands that support grazing animals, and featuring perennial grain staple foods that support humans and are grown in ecologically intensified agroecosystems—new kinds of grasslands—along with tree crops, vegetables, and ancestral foods.

Second, the social perennial vision: we envision that these reinvented and restored landscapes (and the turn from the currently dominant agriculture system that their existence requires) are made possible by and reciprocally sustain human communities that live within the ecological limits of their environments. We envision that these communities have learned, and are still persisting in learning, how to fairly and justly relate to each other and the more-than-human world (especially perennial polycultures) through relationships based on qualities of sufficiency, responsibility, and co-creativity (Streit Krug & Tesdell, 2020).

We believe this vision sets important parameters, while being roomy enough to house plural possibilities and both long-standing and emergent visions from other intellectual, scientific, and cultural lineages. Our vision emerges in part through the life of an evolving and increasingly global perennial grain agriculture research community that spans generations and places. We acknowledge our positions in Kansas in the homelands of the Kaw, Pawnee, Osage, and other Indigenous nations, and our current professional roles and relationships to this perennial grains research community, both in gratitude and in service of reckoning with pathways forward now, when climate justice is so urgently needed.

In *New Roots for Agriculture*, first published in 1980, co-founder of The Land Institute Wes Jackson wrote a vision set in the year 2030, in which humans have made the transition from fossil fuels into a solar era and are living more sustainably across a range of regions in a pluralistic society that abides by common limits. While he has an agrarian focus on "one utopian farm" in Kansas and its "nearby solar village," Jackson's overall vision for the region has many familiar attributes: human communities that practice intergenerational learning and are embedded in a landscape that features perennial polycultures, including perennial grains, the re-integration of livestock and animals into native ecosystems, and annual crops in bottomlands. In this vision, "soil loss has been reduced to replacement levels" (p. 129).

Four years later, Jackson wrote the afterword to a new edition of *New Roots* in which he concluded:

> The task before us, therefore, is to build an agriculture that is resilient to human folly, an agriculture that rewards wisdom and patience, an agriculture in which the land remains resilient but not silent during those excursions toward some dangerous unknown, dangerous because we have become too enamored with our own cleverness and enterprise.
>
> *(p. 148).*

Less than a decade away from the 2030 he envisioned, we find ourselves confronted with the long view and the growing dangers at hand in order to discern possible paths forward.

History and current context

Let's begin with the long view. The geological context of what we call the Great Plains (Figure 5.1) dates back to the formation of the North American continent more than a billion years, followed by the inland seas that laid down marine

FIGURE 5.1 The North American Great Plains spanning the United States and Canada. This map was created by the editors of the *Encyclopedia of the Great Plains* and permission for its use was granted by the University of Nebraska Center for Great Plains Studies.

sediments half a billion years later, and then the relatively recent rising up of the Rocky Mountains in the Laramide Orogeny around 65–70 million years ago (Swineheart, 2011; Kaye, 2011, p. 18). The rain shadow of these mountains to the west exerts a biophysical force that has come to shape the semi-arid climate of the Great Plains, and mineral particles eroded from these mountains, carried by rivers, has contributed to the region's lands and soils. So too has the wind deposition of sediment, sand and silt, over past millions of years, and the till deposited by glaciers. And the Great Plains has itself been shaped by wind and water: "most of the present physiographic regions of the Great Plains are a result of erosion in the last five million years" (Swineheart, 2011).

The grassland ecosystems that have emerged and reigned on the Great Plains, including since the Holocene glacial retreat beginning around 12,000 years ago, feature diverse mixtures of mostly herbaceous, perennial plants that live for more than two years, dying back above ground but persisting through living roots. Great Plains ecoregions are "grassful" because many grasses evolved the ability to survive and thrive with normalized disturbances of fire (both from lightning and cultural, in the form of Indigenous practices, see Roos et al., 2018), animal grazing (see O'Brien, 2017), and because of the fluctuating Holocene climate (Commerfeld et al., 2018). Grasslands are not resilient to these disturbances, but are resilient because of these disturbances. Changes to or repression of these disturbance regimes can result in loss of grassland biodiversity or conversion to woodland ecosystems (Helzer, 2010, p. 1–2).

Deeply rooted grassland plants built and held Mollisol soils, persisted through drought and other climatic variability, and sufficiently provided for an adapting array of Indigenous communities and cultures across many millennia, whose food systems have been built on gathering, hunting, and farming, which began in the region's river valleys (Wishart, 2016, p. 1–15). Contemplating the creatures and inhabitants who have co-evolved lifeways with the land leads to a recognition that the Great Plains can be landscapes of sufficiency. In her book *Goodlands*, Fran Kaye gives first priority to the land, and describes the general response of Native Americans and First Nations to the Great Plains over time as being "to use it, appreciate it, learn it, and manipulate it, but not to replace it or make drastic changes" (2011, p. 4). Kaye demonstrates the durability and resilience of this perspective, in contrast to the rapid degradation of the land by cultures who in an effort to "fix" what was "broken" failed to see what was actually present and working. Kaye explains how in a perspective of sufficiency, the Great Plains is better named as *grassful* than *treeless*.

But instead a perspective of deficiency has dominated in the Great Plains over the past several centuries, with the imposition of settler colonial social, economic, and legal systems that deem the peoples and places of the Great Plains as lacking, in need of improvement, and thus open for exploitation. In the 19th century, the US government and military actions, capitalist ventures, migrants and immigrants, and agricultural technologies, including seeds and barbed wire, all marked Euro-American colonization and industrialization of the landscape. Great Plains landscapes were rapidly converted to cattle ranching and farming, based on widespread plowing of soils and planting of annual grain systems.

Tillage of prairies not only kills perennial plants but also breaks open soil aggregates that stabilize and protect organic matter, allowing microbial activity to increase and Great Plains prairies to lose organic matter and emit carbon dioxide (DeLuca and Zabinski, 2011). Carbon sequestration, nutrient retention, and soil protection from erosion are just a few of the essential ecosystem services provided by perennial, high-diversity grasslands that become converted into "ecosystem dis-services" following grassland conversion into annual, low-diversity agricultural systems (Crews, 2017; Crews et al., 2018; See Figure 5.2).

The industrial growth of agriculture, including the use of grain to feed livestock, on the Great Plains has been made possible by the mining of soil, water, and other resources to produce grain and meat commodities for national and global markets (Cronon, 1991). These agricultural products are made possible by an unprecedented growth in the dependence on fossil-fueled inputs, particularly agrochemicals. The workers and laborers involved in their production, and the arrival and mobility of settlers and immigrants and distribution of people across Great Plains landscapes, have also been shaped by fossil fueled-infrastructure development. Fossil fuels provided the majority of "work" energy to grow crops, effectively replacing humans and draft animals that contributed to the de-population of rural communities. The remaining jobs were to be found in food processing such as slaughterhouses that concentrated people in more urban settings.

Railroads, highways, planes, and the internet have in recent centuries made it possible for communities to exist on and move across landscapes in ways that over-ride the ecological cycles and limits that govern natural prairie systems (as well as the ecosphere more generally). For example, the locations of large population centers of humans and bovines in the Great Plains have little to do with the fertility of the land that surrounds them. Since 1870, when the first US Census data is available, "Depopulation in the Plains has been a depopulation of those who live on the land rather than a loss in total numbers" (Lavin et al., 2011, p. 6). While a dynamic tradition of environmental justice activism persists on the Great Plains, for instance, in the vital work of Indigenous water protectors, Great Plains landscapes in the 21st century lack resilience, both social and ecological.

Current and future stressors

Conservation of the remaining perennial, diverse grasslands of the Great Plains is threatened by land-use change and climate change. Grasslands continue to be converted into annual row-crop agriculture. The most recent 2021 *Plowprint Report* (World Wildlife Fund), which looks at data from 2018 to 2019, estimated that 2.6 million acres of grassland were plowed during that period. The primary reason for plowing is for planting into the annual grain crops of corn, soy, and wheat. Past *Plowprint Report* data from 2013 to 2017 have also been recently re-analyzed using updated methods, and the findings show that "the overall trend observed in annual rates of grassland conversion remained consistent" (Olimb & Lendrum, 2021, p. 114). Increases in commodity prices can dramatically alter the landscape by incentivizing producers to convert prairies and grasslands, including those in the Conservation Reserve Program, to annual row crops (Hendricks & Er, 2018). Producers try to capitalize before a market shift.

In contrast, economic and policy incentives for retaining grasslands and preserving prairie have room for improvement (Lark, 2020), and existing grasslands require proactive and increasingly skillful management by humans in the face of climate change. Indigenous Peoples hold management experience and expertise in their prairie homelands, but their knowledge and sovereignty has been threatened, and colonization of the Great Plains has largely involved the suppression of natural and cultural fire regimes (Roos et al., 2018). Grasslands require some disturbance to remain grasslands. Fire suppresses woody species and allows desirable and native species to thrive, while the presence of grazers and browsers can stimulate herbaceous plant growth and reduce the vigor of invasive species and woody vegetation.

However, it is possible for disturbance to be too intensive in frequency or rate. Too frequent of fires or too high of a stocking rate for grazing can be detrimental. (Widespread tillage, meanwhile, is so intensive it is perhaps more of a catastrophe than a disturbance in a grassland context.) Mimicking the natural and historical fire regime and grazing patchwork has the potential to increase biodiversity and support important ecosystem processes (Fuhlendorf et al., 2012). Yet stocking and managing cattle and designing and implementing burn plans require time, money, and knowledge, and often some social capital. Restoring species diversity and removing woody species similarly involve costs. If grasslands and prairies are not well-managed and begin to lose biodiversity or convert to woodlands, the economic value of the land may decrease.

The ability of grasslands to persist during climate change may also decrease if they are not well-managed. While many impacts are not known, the atmospheric enrichment of CO_2, increased temperatures, and changes in precipitation patterns already occurring with climate change may favor invasive, often woody plant species in some regions of the Great Plains (Archer et al., 2017; Gaskin et al., 2021). This emphasizes the importance of maintaining biodiversity in perennial landscapes.

When grasslands are converted to annual grain agriculture systems, social-ecological stressors are intensified rather than alleviated. Modern grain agriculture evolved with ever increasing reliance on fossil fuels to relax or eliminate the major ecological limiting factors of crop growth, whether through machinery or chemical inputs. These practices exacerbate pollution of watersheds, soil loss and degradation, and loss of deep-rooted carbon storage. The bevy of ecosystem dis-services, listed in Figure 5.2 (Crews et al., 2018), have both near- and long-term effects. Given the slow rates at which soils are formed, and the limits of aquifers like the Ogallala, erosion and depletion of land and water now have the potential to diminish ecological capacity for human and more-than-human communities for many years to come. The near-term impacts of other ecosystem dis-services require the purchase of technological solutions that bring farmers on to a self-reinforcing "agricultural treadmill" that is difficult to escape and involves significant social and economic stressors (Crews et al., 2018; Cochrane, 1958). For instance, buying and applying fertilizer to increase grain yields despite reduced soil organic matter and nutrient losses becomes a regular, and regularly increasing, cost. The increased efficiency and output may lead to declining prices, increasing farm size, and thus more technological solutions.

Nationally, including in the Great Plains, structures of land dispossession and land access and the costs of renting or purchasing property shape agricultural landscapes, and what people and practices are on the land. Significant barriers in terms of "access, assets, and assistance" exist for new and beginning producers who would like to pursue sustainable agriculture, and support is needed for appropriate policy and funding responses (Carlisle et al., 2019). Because of exclusionary and racist systems, such barriers to opportunity are especially acute for socially disadvantaged and minoritized farmers and farmworkers (Union of Concerned Scientists & HEAL Food Alliance, 2020).

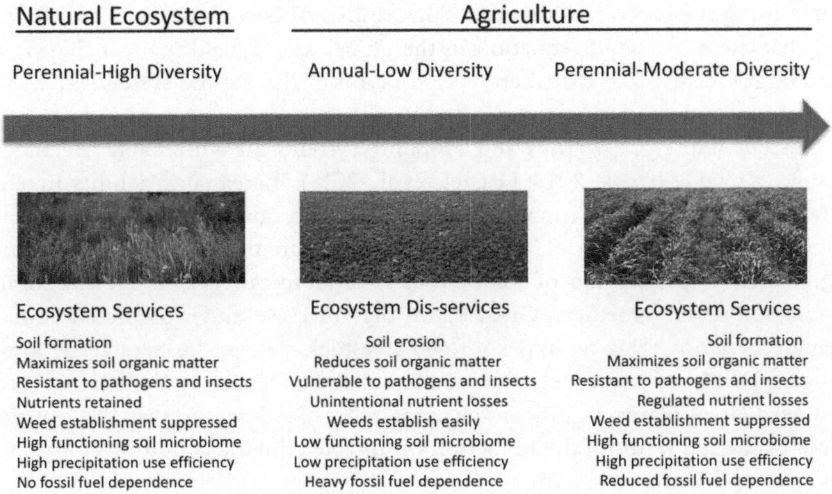

FIGURE 5.2 The evolution of ecosystem services and disservices in agriculture. This figure is reproduced from Crews, Carton, and Olsson's 2018 article "Is the Future of Agriculture Perennial?" available open-access at https://doi.org/10.1017/sus.2018.11.

There are also socioeconomic stressors to human communities, and everyone who eats, associated with the currently dominant agricultural system in the Great Plains. Rural communities have experienced political de-skilling (Epp, 2001) and depopulation, and settler and Indigenous communities have been displaced from potentially coherent place-based cultures, in the sense of norms, patterns, and expectations that social groups develop to live within the ecological realities of a local and regional place. Across the Great Plains, people's norms, patterns, and expectations are more likely to be shaped by the national media and global internet than they are by the lived experience of interacting with their neighbors or phosphorus weathering rates of the landscape they inhabit. The socioeconomic drivers of, and outputs from, the current agricultural system have also not resulted in food justice, relief from food insecurity, affordable access to healthy foods, and sustainable mainstream food cultures in the Great Plains.

The social-ecological stressors described so far might be understood as negative disturbances that can and should be challenged, not absorbed, in the effort to create more resilient Great Plains landscapes. The vision we describe is an attempt to challenge these stressors by reinventing the currently dominant agricultural system and maintaining the persistence criterion of producing staple foods to equitably feed people.

Climate change is an overarching current and future social-ecological stressor, and negative disturbance, that can and should be mitigated. But realistically, we recognize that the near- and long-term reality of climate change is the very disturbance we envision Great Plains landscapes as needing resilience to while maintaining staple food production. Such resilience becomes possible to the extent that there is a rapid reduction in the global use of fossil fuels, including to power agriculture, and that there is restoration of the natural systems in which agricultural systems are rooted.

Current staple grain crops and cropping systems are vulnerable to climate change (Kukal & Irmak, 2018; Hatfield et al., 2018). Increased variability in temperature and precipitation may negatively impact productivity of economically important crops, which could cause dramatic shifts in management practices and regions where production of some crops is economically viable. While forage production in the Northern Great Plains may increase under predicted climate change scenarios, grazing and confined livestock will be vulnerable to lower forage quality and more extreme weather events (Derner et al., 2018). In the Southern Great Plains, significant changes in grassland community composition could occur as a result of the reduction in suitable climatic conditions for prevalent grasses (Manier et al., 2019).

Across the Great Plains, the location of the 100th meridian has effectively marked the divide between more arid western climate and more humid eastern climate. Differences realized in soil moisture shape the natural systems gradient from shortgrass to tallgrass prairie and in the currently dominant grain agricultural systems, with wheat in the west (along with ranching) and corn in the east. There are corresponding differences in human settlement density and farm size, with larger and fewer farms and less dense settlement in the west (Seager et al., 2018a). But under models of climate change that one research team corrected for bias, aridity is expected to increase with rising temperatures, and the "effective" 100th meridian is predicted to move east over the 21st century. Subsequent shifts in the agricultural economy may be changes in crops grown, increase in grassland ranching, and increase in farm size by consolidation (Seager et al., 2018b). Such shifts, if they happen, may keep remaining producers on the "agricultural treadmill" described above. Or other adaptations, for better or worse, may ensue based on "changes in crop technology, farming practice, farm policy, and the wider economy" (Seager et al., 2018b, p. 21). In an era of increased climatic variability and uncertainty, there are still choices and actions available.

Human communities, and the agricultures by which they are inextricably connected to Great Plains landscapes, are vulnerable to food insecurity and injustice due to climate change, soil erosion and degradation, biodiversity loss, and more. We envision in response, echoing Jackson's words from nearly 40 years ago, a just and perennial food future thanks to "an agriculture in which the land remains resilient but not silent." How might we collectively work toward that future?

Path to the future

A more resilient future that maintains staple food production requires urgent re-perennialization and diversification of landscapes, through grassland restoration and conservation (Lark 2020), as well as a reinvention of agriculture that is rooted in features of natural systems such as perenniality and diversity (Crews et al. 2018). By engaging communities, researchers, and a broad array of partners, it may be possible to build both the scientific knowledge and social movement necessary to accomplish a social perennial vision for the North American Great Plains.

We see three interconnected components of the path forward.

Science: rigorous study and analysis based on evidence, observation, experimentation, and experience is needed to accelerate and grow the development of perennial grain staple crops in diverse agroecosystems; to advance the conservation and adaptive management of grasslands; to evaluate agricultural decisions and stewardship options in the context of the nested natural systems in which they are rooted; and to challenge humans to apply our curiosity toward seeking knowledge that endures.

Community: transdisciplinary and participatory research methodologies that ethically engage and are responsibly co-created with communities are needed to build learning and legitimacy for reinvented agricultures and resilient landscapes in the Great Plains. Communities, including human social groups and the broader land communities in which we live, are made and sustained through common senses of purpose and practices of care work. Inclusive, skillful communities are needed to advocate for and enact equity and justice in food systems.

Story: shared stories that connect human memory, imagination, and emotion—from love to grief and back again—and that open people to hear the stories of the more-than-human world are needed to drive cultural change. New, and old, stories of sufficiency can take root through social learning. Great Plains cultures and scholars have long pointed out how the revision and re-creation of stories about this region has material consequences; social, historical, artistic, and literary efforts have ecological impacts, however indirect or complex they may be to trace. Stories of soil, perennial staple foods, and people coming together to learn to make change all matter.

Science, community, and story must share a relational context in the landscapes of the Great Plains. Great Plains cultures and scholars have also long pointed out the power of the Great Plains, how the land requires and constrains certain human responses to it over various scales and settings. The land is not silent. If we are listening, we have work to do together: researchers and scientists, farmers and ranchers and eaters, teachers and students, youth and elder activists, Indigenous communities, restoration and conservation groups, community members of all kinds. Reinvention of currently dominant agricultural systems, restoration of natural systems, resistance to injustice and exploitation, and resilience to climate change are all feasible outcomes of focusing on the persistence criteria of maintaining staple food production on the Great Plains and pursuing a social perennial vision rooted in a natural system-inspired agriculture.

As burning fossil fuels to power agriculture (and society as a whole) necessarily diminishes, farmers and eaters here and from around the world will benefit from agroecosystems modeled after the prairie, forest, desert, or whatever natural ecosystem grows with a minimum of human intervention, and thus requires a minimum amount of human labor. Where are the perennials, materially? Who are the perennials, metaphorically? Investigating these questions in diverse landscapes may help us all.

Acknowledgment

We gratefully acknowledge the Perennial Agriculture Project, a joint project between The Land Institute and the Malone Family Land Preservation Foundation, for financial support of McKenna's position at the University of Kansas.

Note

1 In the 1930s, the combination of low rainfall, high winds, and ploughed barren crop fields led to severe wind erosion in the Great Plains. Frequent dust storms led to naming this period the Dust Bowl (Worster, 2004).

References

Archer, S. R., Andersen, E. M., Predick, K. I., Schwinning, S., Steidl, R. J., and Woods, R. (2017). Woody plant encroachment: Causes and consequences. In D. D. Briske (Ed.), *Rangeland Systems, Processes, Management and Challenges* (pp. 25–84). Springer Open.

Carlisle, L., Montenegro de Wit, M., DeLonge, M. S., Calo, A., Getz, C., Ory, J., Munden-Dixon, K., Galt, R., Melone, B., Knox, R., Iles, A., and Pres, D. (2019). Securing the future of US agriculture: The case for investing in new entry sustainable farmers. *Elementa: Science of the Anthropocene* 7(17). https://doi.org/10.1525/elementa.356

Cochrane, W. (1958). *Farm Prices: Myth and Reality*. University of Minnesota Press.

Commerford, J. L., Grimm, E. C., Morris, C. J., Nurse, A., Stefanova, I., and McLauchlan, K. K. (2018). Regional variation in Holocene climate quantified from pollen in the Great Plains of North America. *International Journal of Climatology*, 38(4), 1794–1807. https://doi.org/10.1002/joc.5296

Crews, T. E. (2017). Closing the gap between grasslands and grain agriculture. *Kansas Journal of Law and Public Policy* (26)3, 274–296. https://lawjournal.ku.edu/wp-content/uploads/2020/08/Crews-V26I3.pdf

Crews, T. E., Carton, W., and Olsson, L. (2018). Is the future of agriculture perennial? Imperatives and opportunities to reinvent agriculture by shifting from annual monocultures to perennial polycultures. *Global Sustainability* 1, 1–18. https://doi.org/10.1017/sus.2018.11

Cronon, W. (1991). *Nature's Metropolis: Chicago and the Great West*. W. W. Norton & Company.

Cunfer, G., and Krausmann, F. (2015). Adaptation on an agricultural frontier: Socio-ecological profiles of Great Plains settlement, 1870–1940. *Journal of Interdisciplinary History* 46(3), 355–392. https://doi.org/10.1162/JINH_a_00868

DeLuca, T. H., and Zabinski, C. (2011). Prairie ecosystems and the carbon problem. *Frontiers in Ecology and the Environment* 9(7), 407–413. https://doi.org/10.1890/100063

Derner, J., Briske, D., Reeves, M., Brown-Brandl, T., Meehan, M., Blumenthal, D., Travis, W., Augustine, D., Wilmer, H., Scasta, D., Hendrickson, J., Volesky, J.,

Edwards, L., Peck, Dannele. (2018). Vulnerability of grazing and confined livestock in the Northern Great Plains to projected mid- and late-twenty-first century climate. *Climatic Change* 146, 19–32. https://doi.org/10.1007/s10584-017-2029-6

Epp, R. (2001). The political de-skilling of rural communities. In R. Epp and D. Whitson (Eds.), *Writing Off the Rural West: Globalization, Governments, and the Transformation of Rural Communities* (pp. 301–324). The University of Alberta Press and Parkland Institute.

Fuhlendorf, S. D., Engle, D. M., Elmore, D., Limbs, R. F., and Bidwell, T. G. (2012). Conservation of pattern and process: Developing an alternative paradigm of rangeland management. *Rangeland Ecology & Management* 65(6), 579–589. https://doi.org/10.2111/REM-D-11-00109.1

Gaskin, J. F., Espeland, E., Johnson, C. D., Larson, D. L., Mangold, J. M., McGee, R. A., Milner, C., Paudel, S., Pearson, D. E., Perkins, L. B., Prosser, C. W., Runyon, J. B., Sing, S. E., Sylvain, Z. A., Symstand, A. J., and Tekiel, D. R. (2021). Managing invasive plants on Great Plains grasslands: A discussion of current challenges. *Rangeland Ecology & Management* 78, 235–249. https://doi.org/10.1016/j.rama.2020.04.003

Hatfield, J. L., Wright-Morton, L., and Hall, B. (2018). Vulnerability of grain crops and croplands in the Midwest to climatic variability and adaptation strategies. *Climatic Change* 146(1), 263–275. https://doi.org/10.1007/s10584-017-1997-x

Helzer, C. (2010). *The Ecology and Management of Prairies*. University of Iowa Press.

Hendricks, N. P., and Er, E. (2018). Changes in cropland area in the United States and the role of CRP. *Food Policy* 75, 15–23. https://doi.org/10.1016/j.foodpol.2018.02.001

Jackson, W. (1980, 1984). *New Roots for Agriculture*. New edition. University of Nebraska Press.

Kaye, F. W. (2011). *Goodlands: A Meditation and History on the Great Plains*. AU Press, Athabasca University.

Kukal, M. S., and Irmak, S. (2018). Climate-driven crop yield and yield variability and climate change impacts on the US Great Plains agricultural production. *Scientific Reports* 8(1), 1–18. https://doi.org/10.1038/s41598-018-21848-2

Lark, T. J. (2020). Protecting our prairies: Research and policy actions for conserving America's grasslands. *Land Use Policy* 97, 104727. https://doi.org/10.1016/j.landusepol.2020.104727

Lavin, S. J., Shelley, F. M., and Archer, J. C. (2011). *Atlas of the Great Plains*. University of Nebraska Press.

Manier, D. J., Carr, N. B., Reese, G. C., and Burris, L. (2019.) Using scenarios to evaluate vulnerability of grassland communities to climate change in the Southern Great Plains of the United States. U.S. Geological Survey, Open-File Report 2019–1046. https://doi.org/10.3133/ofr20191046

Morris, P. S. (2011). Breadbasket of North America. In D. J. Wishart (Ed.), *Encyclopedia of the Great Plains*. University of Nebraska Press. http://plainshumanities.unl.edu/encyclopedia/doc/egp.ii.006

O'Brien, D. *Great Plains Bison*. (2017). University of Nebraska Press.

Roos, C. I., Zedeño, M. N., Hollenback, K. L., and Erlick, M. M. H. (2018). Indigenous impacts on North American Great Plains fire regimes of the past millennium. *Proceedings of the National Academy of Sciences* 115(32), 8143–8148. https://doi.org/10.1073/pnas.1805259115

Seager, R., Feldman, J., Lis, N., Ting, M., Williams, A. P., Nakamura, J., Liu, H., and Henderson, N. (2018b). Whither the 100th meridian? The once and future physical and human geography of America's arid-humid divide. Part II: The meridian moves east. *Earth Interactions* 22(5), 1–24. https://doi.org/10.1175/EI-D-17-0012.1

Streit Krug, A., and Tesdell, O. I. (2021). A social perennial vision: Transdisciplinary inquiry for the future of diverse, perennial grain agriculture. *Plants, People, Planet* 3(4), 355–362. https://doi.org/10.1002/ppp3.10175

Seager, R., Lis, N., Feldman, J., Ting, M., Williams, A. P., Nakamura, J., Liu, H., and Henderson, N. (2018a). Whither the 100th meridian? The once and future physical and human geography of America's arid-humid divide. Part I: The story so far. *Earth Interactions* 22(5), 1–22. https://doi.org/10.1175/EI-D-17-0011.1.

Swineheart, J. B. (2011). Geology. In Wishart, D. J. (ed.), *Encyclopedia of the Great Plains*. University of Nebraska Press. http://plainshumanities.unl.edu/encyclopedia/doc/egp.pe.028

Thóren, H., and Olsson, L. (2018). Is resilience a normative concept? *Resilience* 6(2), 112–128. https://doi.org/10.1080/21693293.2017.1406842

Union of Concerned Scientists and HEAL Food Alliance. (2020). Leveling the fields: Opportunities for Black people, Indigenous people, and other people of color. Union of Concerned Scientists Policy Brief. https://www.ucsusa.org/sites/default/files/2020-06/leveling-the-fields.pdf

Wishart, D. J. (2016). *Great Plains Indians*. University of Nebraska Press.

World Wildlife Fund. (2021). *Plowprint Report*. https://www.worldwildlife.org/projects/plowprint-report

Worster, D. (2004). *Dust Bowl: The Southern Plains in the 1930s*. Oxford University Press.

6

RESILIENT FOOD SYSTEMS IN THE CONTEXT OF INTERSECTIONAL DISCRIMINATION

Successful strategies of women and Indigenous Peoples in Mesoamerica

Tania Eulalia Martinez-Cruz, Magdalena Reynoso Martinez and Rachael Ann Cox

Socio-ecological resilience and intersectionality in food systems

Socio-ecological resilience

While resilience in its original definition in ecology refers to a system's ability to survive a disturbance and return to the same stable state of functioning (Gunderson, 2000; Holling, 1973), when applied in socioeconomic systems, the definition was expanded to include human and institutional aspects of adaptation (Hodbod and Eaken, 2015; Zaccarelli et al. 2008). In ecology, biodiversity is considered to be a key component of resilient systems because the distinct responses and functions of different species contribute towards the return to a stable state and are able to respond to a wider range of conditions after a disturbance (Peterson et al. 1998). In a socio-ecological system, resilience is tied to the biodiversity and the diversity of the human community at the individual, organizational, and institutional levels.

In a resilient community, when a disturbance or shock occurs, such as a natural disaster, diverse individuals with distinct talents and different institutions and organizations with distinct response times and response functions participate in organizing and implementing response initiatives in the immediate-, short-, and long-term to ensure the community and their members survive the natural disaster and return to a stable state after the disaster. While some individuals or organizations may contribute funds for disaster relief, others may contribute time or specific talents to the cause. Some institutions are prepared for immediate disaster response, while others support rebuilding efforts after the disaster has passed. Resilient human communities rely on resilient ecosystems. Figure 6.1 depicts and describes the interrelationship between ecological resilience and a

DOI: 10.4324/9781003266440-6

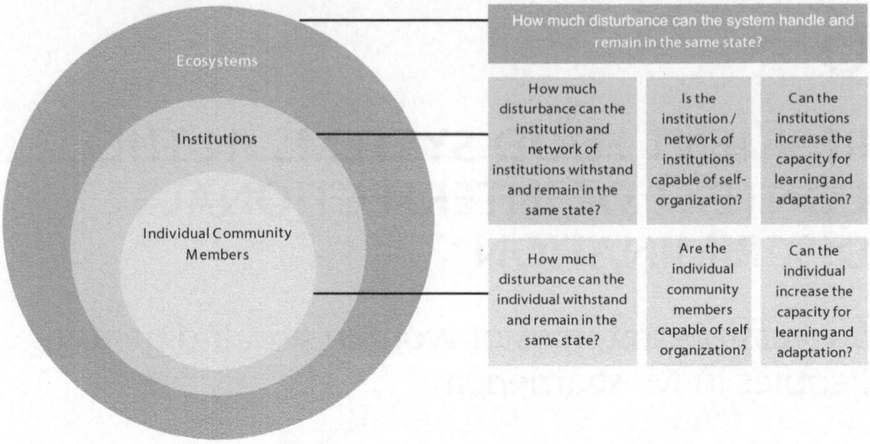

FIGURE 6.1 Key questions for analyzing resilience in socio-ecological systems.
Source: Adapted from socio-ecological models and resilience models (Carpenter et al., 2001; Hodbod and Eaken, 2015).

human community's socio-ecological resilience, based on the core principles of socio-ecological resilience as described by Carpenter et al. 2001.

Introduction to intersectionality

Intersectional identities of individuals and communities influence the participation in, contribution to, and formation of food systems and therefore also affect the individual or community's ability to be resilient. Intersectionality in this context refers to the multiple and overlapping (intersecting) identities that influence an individual or community's reality as they experience discrimination or marginalization (Krenshaw, 1990). By the authors' understanding of intersectionality, we cannot separate the identity, experience, historical, and current oppression of individuals and communities from the diverse factors influencing their existence, including gender, Indigenous identity, rural geography, and migratory status.

Kimberlé Krenshaw coined the term "intersectionality" in her studies of violence against women of color in the United States. She highlighted the numerous forces of oppression that women of color face because of their gender, race, and social class, and that those intersecting oppressions are distinct from the oppressions resulting from each of those factors independently (Krenshaw, 1990).

We provide an example of intersectionality with quantitative data comparing educational outcomes. In Figure 6.2, we compare the number of years of school completed by Guatemalan and Mexican adults, disaggregated by gender. On average, Mexicans had more years of schooling than Guatemalans, and men had more years of schooling than women. Figure 6.2 masks the intersectionality of race, geography, socioeconomic status, and other factors that impact the number of years of schooling beyond gender.

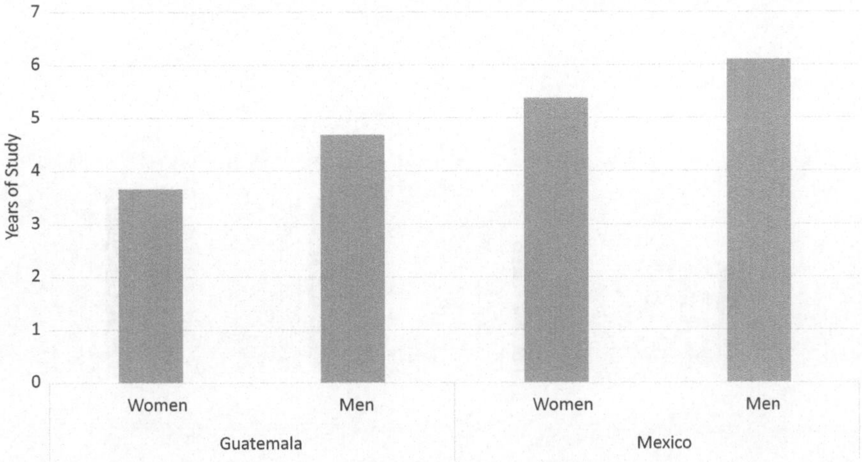

FIGURE 6.2 Average years of study by adult women and men in Mexico and Guatemala.

Source: Data from CEPAL/CELADE Redatam, 2000, Prepared by the authors.

In Figure 6.3, the data is further disaggregated by geography (urban or rural) and the person's identity as Indigenous or not. While the graph does not capture all possible variables of intersectionality, the further disaggregated data highlights much more drastic differences between intersectionalities. As shown in the Figure 6.3, in Guatemala, the intersection of being a woman, living in a rural area, and being an Indigenous person reduced educational attainment by 83% as compared to their male, urban, non-Indigenous counterpart. In contrast, in Figure 6.2, when only comparing men and women in Guatemala, there is a 27% difference in education attainment. This difference, from 83% to 27%, provides a quantitative illustration of the importance of understanding intersectional differences between people and communities when exploring resilience.

Resilience in Mesoamerican food systems – an intersectional framework

To analyze resilience in Mesoamerican food systems, we present a framework that combines the socio-ecological model for resilience as described in Figure 6.1 with an intersectional analysis component to consider the dynamics of oppression impacting the individuals and communities described in the case studies below. The framework presents a series of questions to analyze resilience from an intersectional perspective in a socio-ecological system (Table 6.1).

History and current context

Mesoamerica is the birthplace or origin of maize (*Zea mays*) and other crops and agricultural practices that have now become global commodities and common

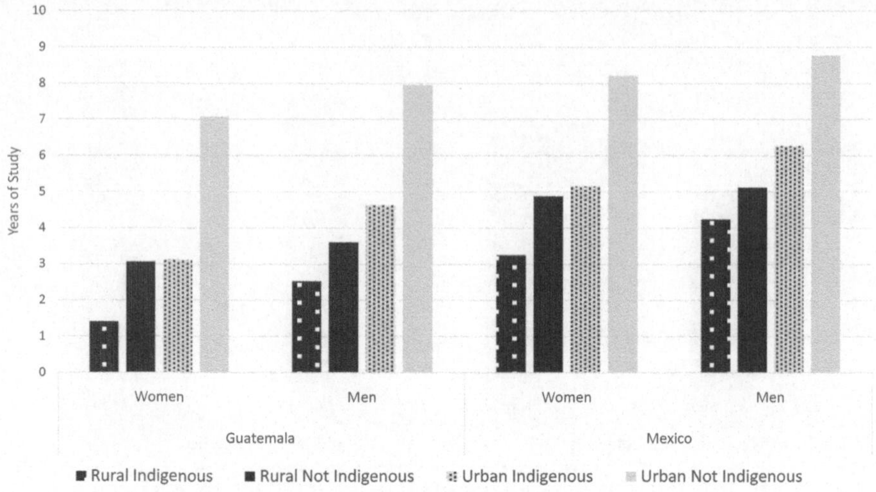

FIGURE 6.3 Average years of study by adults, disaggregated by geography, Indigenous identity, country, and gender.

Source: Data from CEPAL/CELADE Redatam, 2000, prepared by the authors.

practices worldwide.[1] Mesoamerica is the region spanning from Central Mexico to Central America, where maize continues to be the staple or play a key role in food systems. Maize is spread across the landscapes of Mesoamerica, in large monoculture plots in mechanized agricultural production in the flat lowlands to small plots on steep mountainsides, intercropped with beans, diverse squash species, and leafy greens in *milpa*[2] agroecosystems. Two places in Mesoamerica where maize plays a central role in agricultural landscapes, food systems, and cultural traditions are the highlands of Guatemala and in the state of Oaxaca, Mexico (Figure 6.4).

Guatemala is situated in Central America, south of Mexico, nestled between Honduras, El Salvador, and Belize. The country is known for its >60 archaeological sites, 25 actively spoken Indigenous languages, and 30 volcanoes. Guatemala is one of 17 megadiverse countries in the world (Bacon et al. 2019), due to its high species diversity and ecosystem diversity, set between the Pacific and Atlantic oceans, with altitude from sea level to 4,060 m (13,320 ft). Guatemala is a small country (108,889 km^2), approximately the size of the U.S. state of Tennessee, that boasts 14 ecoregions (Tolisano and Lopez, 2010). While 35% of Guatemala's land is covered with forest of which 12% is primary forest (FAO, 2016) and much of the country is delineated in extensive protected areas, threats of illegal wildlife trade, government corruption, and invasion from large-scale commercial agriculture put Guatemala's biodiversity and forests, the foundation of the country's resilience, at constant risk (Forest Carbon Partnership Facility, 2022; Global Forest Watch, 2022).

TABLE 6.1 A framework for analyzing intersectionality and resilience in socio - ecological systems

Components of socio-ecological systems	*A. Socio-ecological resilience analysis*	*B. Intersectional analysis*
1. Ecosystems and agroecosystems	1. A. How much disturbance can the ecosystems and agroecosystems that the food systems rely on and are in proximity to withstand and remain in the same state?	1. B.1. Are the ecosystems and agroecosystems that the food system interconnected with more or less vulnerable due to intersecting historical and present systems of oppression? 1. B.2. What are those intersections?
2. Institutions and economic systems	2. A.1. How much disturbance can the institutions and economic systems that are part of the food system withstand and remain in the same state? 2. A.2. Are the institutions in the food system capable of self-organization? 2. A.3. Can the institutions in the food system increase the capacity for learning and adaptation?	2. B.1. Are the institutions and economic systems in the food system creating more vulnerability for people experiencing intersecting historical and present systems of oppression in the food system? 2. B.2. Are the institutions and economic systems actively combating the intersecting historical and present systems of oppression in the food system?
3. Individual community members	3. A.1. How much disturbance can an individual in the food systems withstand and remain in the same state? 3. A.2. Are the individual community members in the food system capable of self-organization? 3. A.3. Can the individual/s in the food system increase the capacity for learning and adaptation?	3.B.1. Are the individuals in the food systems experiencing intersecting historical and present systems of oppression? 3. B.2. What are those intersections? 3. B.3. How are these intersectional systems of oppression preventing the individuals from withstanding disturbance, self-organizing, or increasing their capacity for adaptation?

Guatemala's northern neighbor, Mexico, has many ecological similarities, sharing protected areas, and is also one of the 17 megadiverse countries in the world (Bray et al. 2008; Bacon et al. 2019). While Spanish is shared as the national language between the two countries, and despite their ecological similarities, they have distinct cultures, foods, and languages. One such location in

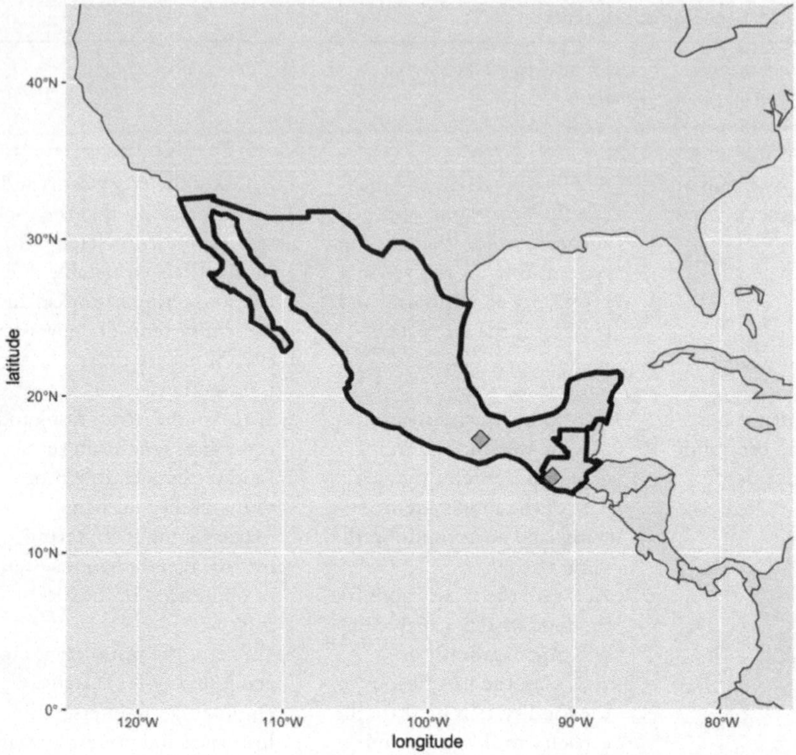

FIGURE 6.4 Locations of the Guatemalan Highlands and Santa María Yavesía, Oaxaca, Mexico, as discussed in the case study.

Source: R version 4.0.3, ggplot2 and sf packages (Pebesma, 2018; R Core Team, 2020; Wickham, 2016).

Mexico with great cultural, ecological, linguistic, and culinary diversity, is the state of Oaxaca, in Southern Mexico. Oaxaca is a large state (93,952 km^2), almost as large as the country Guatemala. In Oaxaca, there are 16 Indigenous languages spoken. Similar to the geography of Guatemala, Oaxaca ranges from sea level to 3,750 m (12,303 ft), which creates a wide range of ecosystems. In Guatemala, there are 13 landraces of maize cultivated and conserved, and in Oaxaca, there are 32 landraces of maize cultivated and conserved in situ (Guzzon et al. 2021, CONABIO, 2021).

Diverging political histories

While Guatemala and Mexico are neighboring countries, they share distinct histories, which continue to influence food systems and their populations' resilience. While the two countries are now separated by a political border that is 871 km (541 mi) long, they were once part of the same Mayan civilization which extended from Southern Mexico through Guatemala, into neighboring Belize, Honduras, and El Salvador (Sharer and Traxler, 2006). While the Mayan empire fluctuated in size and location from its origins in 2000 B.C., their last city was lost to Spanish colonizers in 1697 (Jones, 1998). In 1821, Mexico won independence from Spain after an 11-year war for independence and underwent a political transformation (Rodríguez, 2010). Shortly after, Guatemala was annexed to be part of the First Mexican Empire until Guatemala joined the United Provinces of Central America in 1823 (Kenyon, 1961). Finally, Guatemala became an independent nation in 1840 (UNC Libraries, 2013). While Guatemala gained independence, the country was ruled by a series of presidents who were large coffee-producing landowners or backed by large agricultural companies, like the United Fruit Company (UNC Libraries, 2013). In 1954, the population democratically elected President Jacobo Arbenz Guzmán, who implemented a land redistribution strategy which provoked the Central Intelligence Agency (CIA) of the U.S. government to back a coup d'état implementing a military dictator (Gleijeses, 1989; Streeter, 2000). This led to a civil war from 1960 to 1996 that included a genocide against Indigenous, primarily rural poor farmers (UNC Libraries, 2013).

Current and future stressors

Achieving resilient landscapes and food systems in Mesoamerica is highly connected to land rights, migration policies, international trade laws, historical and cultural relationships with staple crops, and complex intersectional discriminatory forces impacting farmers and rural communities. While the realities of climate change have already become a part of daily life for communities throughout Mesoamerica facing more frequent and more extreme weather events (World Bank, 2011), the way in which these realities impact daily life is highly influenced by an individual or a community's intersectionalities of past and current oppressions.

Ecological and climate stressors

Mexico and Guatemala are two of the world's most vulnerable countries to climate change. In the Global Climate Risk Index by Germanwatch, which analyzed climate vulnerability from 2000 to 2019, Guatemala ranks 16th globally as most vulnerable to climate change and Mexico as 59th for the analysis of economic loss and loss of lives to climate events (Eckstein et al. 2021). Additionally, Guatemala ranks 5th among countries with the highest economic risk to three or more hazards as over 80% of Guatemala's gross domestic product is located in areas that are at-risk to natural hazards (World Bank, 2021a). In both Mexico and Guatemala, there are significant increases in mean annual temperature and maximum annual temperatures (World Bank, 2021b, c).

While both countries are megadiverse, Mexico and Guatemala face large threats of rapid biodiversity loss through deforestation and forest fires due to expanding agricultural frontiers, illegal wildlife trade, pollution,[3] and corruption.[4] Since 2000, Guatemala has experienced a 21% reduction in tree cover, and Mexico has experienced an 8.4% decrease in tree cover (Global Forest Watch, 2022).

Resilience cannot be separated from territories, particularly with Indigenous Peoples since their knowledge and livelihoods depend on their territories (e.g. what to eat through the year, what to use to cure a disease), and their ability gets reduced when they are displaced or their landscapes are affected. In 2020, a third of land defenders who were murdered globally were Indigenous (Global Witnesses, 2020).

Sociopolitical stressors

Numerous social, economic, and political stressors in Guatemala and Mexico add stress or shocks that require resilient systems. Poverty rates in Latin America increased during the public health crisis caused by COVID-19 and now the region faces the highest inflation rate in over 15 years (CEPAL, 2020; Appendino et al. 2022). The current geopolitical situation adds pressure into the access to supplies (including fertilizers and others for agricultural production). Guatemala has a malnutrition problem including that 47% of the children suffer malnutrition (USAID, 2021). In recent years, the "cartels" have been designated as key drivers for deforestation and illegal fishing in rural areas. The "cartels" governance in those areas have generated migration and abandonment of agricultural production units. The stressors previously mentioned (poverty, violence, crime, and corruption) have generated migration, increasing the migration for women, families, and unaccompanied children. The U.S. policies have become stricter, generating pressure on the northern border of Mexico, and endangering migrants in transit.

Resilience strategies led by Indigenous Peoples in Mesoamerica

The chapter highlights case studies of Zapotec farmers in the Northern Sierra of Oaxaca, Mexico, and Indigenous women's farmer groups in Guatemala to illustrate the diverse strategies being employed to create resilient food systems. The chapter reviews the key factors contributing to resilience or challenging resilience in each landscape. We compare the two food systems in the intersectional framework for analyzing socio-ecological resilience. For the sake of these case studies, four questions have been synthesized from the framework described in Table 6.1 and will be explored with each group:

1. In what ways did the individuals and institutions adapt to ecological and sociopolitical shocks?
2. How were those adaptation mechanisms impacted by intersectional systems of oppression?

3 How are these intersectional systems of oppression preventing individuals from withstanding disturbance and self-organizing, or increasing their capacity for adaptation?
4 How did the individuals and communities foster resilient food systems despite these systems of oppression?

Hybridizing culture, knowledge, and technology for resilience: Oaxaca, Mexico

Santa María Yavesía is a Zapotecan community of 448 people, of which only 14 are bilingual, Zapotecan and Spanish speaking, and the rest only speaking Spanish. It is located 1900–2100 MASL (meters above sea level) in the state of Oaxaca. Research for this case study was carried out with Zapotec farmers from 2016 to 2020, when the researchers spent several months in the community and involved 25 semi-structured interviews with key actors and participatory observation to construct a timeline of critical events that affected local farming practices around maize food systems. The research determined the drivers of historical changes in agricultural practices in the community were migration, climatic shocks, and local and international policies (Martinez-Cruz 2020).

One farmer who was interviewed described how the landscape of the community had changed a lot since she was a child, when people lived higher up in the mountains closer to their milpa fields and not so close to each other. As time passed, many men migrated to the United States and the women moved their houses closer together in what is now called the "center" of the town Yavesía. She also said that as time passed, many families reduced the amount of land they were farming, as they could afford to buy food with the money that men earned in the United States, but they still cropped their milpa fields. The fields had to be planted before the men migrated for six months, and the men were always back in time in the fall for the harvest and for the big celebrations of the community. The research with other farmers found that though farming was not economically necessary, it was essential for food sovereignty and thus for resilience. As the remote mountain community had experienced extreme weather events and been cut off from food distribution systems, they had renewed understanding of the importance of cultivating the milpa for a diverse diet throughout the year. Again, during the COVID-19 pandemic when the community maintained a strict quarantine to prevent the spread of the disease, the community members relied heavily on their diverse food production.

In what ways did the individuals and institutions adapt to ecological and sociopolitical shocks?

A central element for the resilience of the Zapotecan community is linked to their rights and sovereignty in territories, their community-based organization and traditional food systems. Although most of the community members were farmers in the 1950s, diverse policies and globalization have re-shaped the community into migrants, where at least one member of each family works as temporary labor in the United States and farming has been reduced. Yet, farmers continue returning to Yavesía as their identity and security links to their community, their territory, and their food systems. For example, in the 1950s, Yavesía experienced different climatic shocks but had enough maize fields and community support to overcome these shocks. In the 2000s, they had higher incomes due to migration and off-farm jobs, but the incomes did not guarantee their food security in the face of climatic shocks. After being cut off from outside food distribution following a hurricane, they started collaborating with other organizations to ensure that in case of any extreme events, they could at least rely on their food systems and be self-sufficient. For example, in the face of COVID-19, despite the strict lockdown, they ensured that everyone had access to food (e.g., elders, children, and women) and used their territory and sovereignty as a tool to prevent the disease from entering their territory.

How were those adaptation mechanisms impacted by intersectional systems of oppression?

One of the criticisms of food policies in Mexico is that they have deliberately excluded traditional food systems, like the milpa system, in order to focus on national food security instead of community food sovereignty. Fox and Haight (2010) suggest that these food policies subsidized inequality because they have been mostly harmful for Indigenous farmers. In the case of Yavesía, many of the policies at national level encouraged migration because they assumed that Indigenous farmers were more useful to the nation's development if turned into laborers while food could be provided by the state through government programs and produced by farmers with their ability to produce maize or food more 'efficiently' (e.g., NAFTA and Programa Braceros). Yavesía Indigenous farmers, as compared to other Indigenous Peoples, are lucky to still preserve rights on their territories and take pride in it because they know that although globalization and national policies can affect them, their ability to be resilient highly relies on their right to self-determination, their rights on their territories, their community-based organization, and their ability to be food and nutrition sovereign.

Gender norms also affect food sovereignty in the community. Testimonies of Yavesía farmers indicate that since most of the men migrated, women had to be relocated to the center of the community as a woman could not live in isolation in the mountains. The belief that women could not live alone in the mountains was a gender norm that impacted food sovereignty as many maize fields historically were located in the mountains where there was more land, so when women were alone, they moved closer to the town center and had less access to land. Additionally, each family had less time to dedicate to food production, as the adult women were solely responsible for food production and also were in charge of taking care of children and elders and taking part in the community activities.

How are these intersectional systems of oppression preventing individuals from withstanding disturbance and self-organizing, or increasing their capacity for adaptation?

In the national context of Mexico, ideally intercultural (i.e. different Indigenous and other cultures within Mexico) food, nutrition, and agricultural policies[5] should be promoted to reinforce the capacity of communities like Yavesía to be resilient. Rights to land tenure, territories, self-determination, and self-organizing should be guaranteed. Yavesía farmers have shown to be resilient but have learned to work around policies at a national level that do not fit their context.

How did the individuals and communities foster resilient food systems?

- Farmers have hybridized their traditional knowledge with modern technologies to achieve resilience in a changing world.
- Ancestral territory is the means for farmers to enact their right to self-determination, which enables hybridization and resilience of maize technology culture.
- Cultivation of milpa and native maize systems is an act of resistance and enactment of their ability to be resilient and ensure farmers' food sovereignty.
- Their social structure and community-based organization is crucial for their resilience because it means they take care of each other but also that their collective actions makes them stronger in the face of extreme shocks.
- Another important dimension is their culture and identity as Zapotecans, despite not conserving the language, their oral traditions, beliefs, and stories on the sacredness of water and interconnectedness with the forests makes them return to their homeland every year after six months as temporary migrants.

Paying for agrobiodiversity conservation by Indigenous women: Totonicapan, Guatemala

In 2017, a small team of Indigenous women from communities surrounding Totonicapan, Guatemala, began to conceptualize how to create economic systems that will pay for, thus recognizing the value of, the conservation of biodiversity by Indigenous women. The team, with support from a nonprofit organization operating in the region, asked participating women farmers to bring them a sample of all the edible plants they had growing around their home, garden, farm, or land. The farmers were paid for their contributions.

The team spent the first days dehydrating the products and the following months creating recipes for tea mixes and other food products that could be sold in high-value markets. The idea was to create products that used the extensive biodiversity cared for by Indigenous women farmers, in order to create economic compensation for the work of agrobiodiversity conservation. The project focused on creating jobs and economic opportunities for women from rural, Indigenous communities in Guatemala while promoting agrobiodiversity conservation.

By 2022, the group of women now operates a formal business, manufacturing and commercializing superfood snacks and tea mixes that use over 50 distinct edible plant species. The company sources their raw materials from a network of farmer cooperatives, Indigenous farmer associations, individual women farmers, and individual Indigenous farmers who use agroecological practices. As a policy, the company only sources from farmers and cooperatives who grow more than one of the species used in the tea production to ensure that the biodiversity that is represented in the tea is also represented on farms and on agricultural landscapes. The company provides training and technical assistance to farmers on best practices to ensure the quality standards required for a commercial food company, but they focus their efforts on empowering the farmers in their network to interchange knowledge and experience on production, as they have been cultivating the plants long before the company was established. While the women and Indigenous farmers who sell their products to the company earn an income from their sales, the crops produced do not compete with their food crops, which is principally the milpa system of maize, beans, and other intermixed food crops.

In 2020, Guatemala faced months of government-imposed lockdown and movement restrictions due to COVID-19. In November and December of 2020, the country, still in the midst of the crisis, was hit with two hurricanes, Eta and Iota, within weeks of one another. The travel restrictions which prevented the use of public transportation and movement between towns to sell crops at markets caused food prices to surge. The natural disasters caused flooding, crop loss, and infrastructure damage across the country.

In what ways did the individuals and institutions adapt to ecological and sociopolitical shocks?

As a response to the movement restrictions, the company created "mini-factories," a smaller extension of the principal factory that are located in the communities of the farmers who sell to the company. These mini-factories are small centers of production that allow producers to sell their crops closer to their farms, continuing both agricultural production and sale. Through this model, the company could continue to operate despite the transportation limitations and without the risk of exposing the farmers to COVID-19 as they moved their products to market.

The economic crisis in Guatemala was particularly difficult for small- and medium-sized enterprises. To prevent the spread of disease, the Government of Guatemala implemented restrictive curfews, mandatory closing of restaurants, and non-essential business closings. While larger enterprises continued to operate and could rapidly pivot to online sales and home delivery, smaller businesses often lacked capital and experience with the necessary technological platforms to make this change. Additionally, not all business models are adaptive to an online and to-go model and Guatemala does not have a culture of online business, so many businesses suffered. To reach larger audiences, the women's tea business invested in learning technological platforms to shift to online sales, sales through social networks, and to prepare for exports to reach larger international markets. Though the markets continue to suffer as of 2022, the company continues to seek creative ways to expand markets while keeping their team and farmer groups safe, conscious of the adverse impacts of poverty and violence on women during the COVID-19 pandemic (Clingain et al. 2021).

Because climate change is causing Totonicapan to become increasingly prone to periods of extreme drought, the company implemented systems that include rainwater harvesting during rainy seasons, which allows women to grow crops during droughts due to the access to water for irrigation. Those systems not only impacted agriculture but also allowed the families access to water for domestic use. The systems were designed considering the physical limitations of older women farmers, economic resource limitations, and for ease of use for farmers that have previously not worked with irrigation systems.

Agrobiodiversity is another resilience mechanism used by the farmers in Totonicapan. The farmers grow more than one species in accordance with agro-ecological practices within each plot and on the farm, contributing to resilience through diverse ecological and economic functions and responses. A typical farm plot may be planted with one or two fruit trees, two or three aromatic herbs, and a crop, such as maize or beans, planted for food security. With multiple similar plots creating the whole farm, a farmer achieves temporal and spatial diversity. The herbs are mature and ready for sale at different moments throughout the year, providing a source of income throughout the year. Fruit trees provide dietary diversity and also an opportunity to sell or trade fruit to neighbors or at local markets. The perennial plants, both fruit trees and aromatic herbs, add root structure diversity to the landscape of annual food crops.

How were those adaptation mechanisms impacted by intersectional systems of oppression?

Indigenous women and girls are amongst the most vulnerable within Indigenous communities, being victims of a triple discrimination based on gender, ethnicity, and socioeconomic status. This discrimination creates fewer opportunities to access higher education, financial services, and economic autonomy, which puts Indigenous women at a disadvantage in business negotiations. The work of the company focuses explicitly on finding economic opportunities and innovative models that address the intersectionality of oppression that the Indigenous women in rural Guatemala face.

The COVID-19 crisis caused women, particularly Indigenous women, to become more vulnerable due to the exacerbation of mechanisms of social exclusion in public health and economic participation (CIM, 2020; IIWF, 2020). While government restrictions impacted all people living in Guatemala, those who had personal vehicles, who knew how to drive, and who had institutional connections to receive letters of permission to circulate between locations were able to adapt more easily to the restrictions and continue operating businesses and economic activities. As rural, Indigenous women have the lowest level of access to education and economic opportunities, they were more adversely affected by these policies.

How are these intersectional systems of oppression preventing individuals from withstanding disturbance and self-organizing, or increasing their capacity for adaptation?

The discrimination that rural, Indigenous women face generates an under-representation at country, community, and family levels, reducing their self-determination. With reduced self-determination, rural Indigenous women face difficulties in self-organizing, such as in the formalization of their businesses or cooperatives.

How did the individuals and communities foster resilient food systems?

- By relying on diverse agroecosystems, farmers can sell diverse products while maintaining most of their land for food security and food sovereignty by continuing the production of milpa systems, thus combining agricultural activities for income with agricultural activities for nutrition and food sovereignty. The diversity maintained on their farms contributes to in situ conservation of agrobiodiversity and contributes to resilient agroecosystems in the face of economic or ecological shocks.
- By working with technologies that are adapted to the specific needs of women's time, bodies, and scale of production, Indigenous women farmers can maintain their traditional practices while adopting climate-smart technologies and practices. With an emphasis on existing, traditional, and ancestral knowledge, women share and learn together, encouraging further empowerment to seek solutions for climate shocks, like drought and other extreme weather events.

Path to the future

Women and Indigenous Peoples are recognized by scientific communities as key contributors to biodiversity conservation and climate change adaptation, yet seldom is the contribution recognized through economic compensation or non-monetary incentives (i.e. public recognition and celebration of contribution, leadership positions and decision making roles) in national or international socioeconomic systems. For example, Indigenous Peoples are acknowledged to be guardians of 80% of the world's remaining biodiversity in less than 25% of the world's surface (Sobrevila, 2008). Despite their claim on land rights of 50% of the world's lands, they are only the legal landholders or have legal rights to use 10% of the world's land (RRI, 2016).

The Indigenous Peoples' Knowledge of using the land and protecting biodiversity is dying in many parts of the world. In one paragraph, we highlight the importance of documenting Indigenous Peoples' Knowledge to be transferred to other regions (space) and next generations (time).

Based on the challenges faced and the successes in developing resilience strategies, we recommend that following resilience strategies that function at the landscape, national, and international levels.

Governance of information and communication about climate change needs to be radically socially inclusive. Intercultural policies on a broad level should be designed and implemented to support Indigenous Peoples and women. These policies should address food sovereignty, land tenure, education, and health. These policies also mean that society and the academic community needs to challenge the hierarchies of knowledge and be able to embrace Indigenous Peoples' Knowledge and Indigenous Peoples' expertise to co-produce policies that support their resilience and do not threaten them.

To better consider Peoples from Indigenous communities, agricultural research and development (AR&D) practitioners should have a more contextual understanding of peoples' needs and aspirations. Before prescribing solutions, we need to learn what other values are tied to their livelihoods and strategies. For example, in the case of Yavesía, income is not the main goal of farmers. They have a cultural linkage to their territory, and therefore, despite all the policies implemented, they have chosen to always return and be connected to their homeland. If we as researchers are serious about changing our practices in AR&D, it also means we should engage Indigenous Peoples and women in active policy making. We should reinforce that they participate in the decisions and planning that affects their livelihoods.

Women and Indigenous Peoples need to be recognized and compensated for protection of and delivery of ecosystem services derived from conservation of biodiversity. Many Indigenous Peoples call themselves living alternatives to climate change because it is their knowledge, livelihoods, and practices that have made them the guardians of most of the biodiversity of the planet (Etchart, 2017). However, Indigenous Peoples and their rights constantly are threatened by forced displacement in the name of development, when projects are implemented in their territories without respecting Free and Prior Informed Consent.[6] If we aim to build a sustainable world, we need to ensure their rights are respected and guaranteed. Indigenous Peoples and women are champions in creating resilience strategies despite the marginalization and discrimination they face, we need to learn from them and support many of their ongoing strategies such as those examples highlighted in the case study from Mexico and Guatemala in this chapter.

Acknowledgments

The authors would like to thank the farmers and women who have participated in the projects and the social impact business that has been highlighted in the case studies in this chapter. Our gratitude goes out to all the farmers in Santa María Yavesía, Mario Fernández, and the technicians (especially Humberto Castro) and officials who participated in the study for their time, knowledge, and openness. Additionally, we would like to highlight our colleagues Adriana Gutierrez, Maria Ixchajchal, Manuela Tzunún, and Claudia Morales who lead the women's tea business discussed in the Guatemalan tea business for their unrelenting work to create economic and empowerment opportunities for Indigenous Women Farmers in Guatemala, who inspired this case study.

Notes

1 In addition to maize, Mexico and Central America are known to be the origin of cotton, beans, squash and pumpkins, vanilla, cocoa beans or chocolate, sweet potatoes, chilis, and peppers (Khoury et al. 2016). Additionally, Mesoamerica is the birthplace of the maize-bean-squash intercropping systems that have co-evolved with humans and continues to do so, a process that began over 5,000 years ago (Landon, 2008).

2 Milpa, a word commonly used in Spanish with Nahuatl origin (REAL ACADEMIA ESPAÑOLA, 2022), used in Mexico and Central America to describe "a traditional agricultural system in which maize is intercropped with other species, such as common beans, faba beans, squashes or potatoes" which was and is in many places is still the backbone of agriculture and food systems since pre-Columbian times in much of Latin America (Lopez-Ridaura et al. 2021).

3 Pollution threatening biodiversity includes contamination to water bodies, dumping of harmful chemicals into land and water, unregulated leaching from mining, and air contamination from forest fires and manufacturing.

4 Corruption that impacts biodiversity loss includes the lack of enforcement of wildlife protection and trafficking, police and government participation in narcotic cartel groups in deforestation, illegal ranching, and illegal wildlife trade, and lack of government funding and enforcement of laws to conserve protected areas and protected species.

5 Intercultural agricultural policies are those do no exclude or discriminate agriculture practices of Mexico's diverse cultures, such as traditional milpa systems and diverse cultural practices throughout Mexico.

6 Free, Prior, and Informed Consent is a component of The Declaration on the Rights of Indigenous Peoples that requires States to consult and cooperate with Indigenous Peoples in good faith regarding legislation and policy development and adoption or projects that affect Indigenous Peoples' land, territory and resource utilization or occupation, which requires they obtain the free, prior, and informed consent of the Indigenous Peoples (Pillay, 2013).

References

Appendino, Maximiliano, Goldfajn, Ilan y Samuel Pienknagura. (2022). América Latina sufre un shock inflacionario tras otro. International Monetary Fund. https://www.imf.org/es/News/Articles/2022/04/15/cf-latin-america-hit-by-one-inflationary-shock-on-top-of-another

Bacon, E., Gannon, P., Stephen, S., Seyoum-Edjigu, E., Schmidt, M., Lang, B.,... & Gidda, S. B. (2019). Aichi Biodiversity Target 11 in the like-minded megadiverse countries. *Journal for Nature Conservation*, 51, 125723.

Bray, D. B., Duran, E., Ramos, V. H., Mas, J. F., Velazquez, A., McNab, R. B.,... & Radachowsky, J. (2008). Tropical deforestation, community forests, and protected areas in the Maya Forest. *Ecology and Society*, 13(2). https://www.jstor.org/stable/26267987

Carpenter, S., Walker, B., Anderies, J. M. & Abel, N. (2001). From metaphor to measurement: Resilience of what to what? *Ecosystems*, 4(8):765–781

CEPAL (Comisión Económica para América Latina y el Caribe). (2020). "América Latina y el Caribe ante la pandemia del COVID-19: efectos económicos y sociales," Informe Especial COVID-19, N° 1, 3 de abril, Santiago.

CEPAL/CELADE Redatam. (2022). *Economic Commission for Latin America and the Caribbean*. United Nations.

CEPAL, COVID-19 Observatory in Latin America and the Caribbean Economic and social impact. Guatemala. https://statistics.cepal.org/forms/covid-countrysheet/index.html?country=GTM&theme=3

Clingain, C., Jayasinghe, D., Hunt, A., & Gray Meral, A. (2021). *Women's economic empowerment in the face of Covid-19 and displacement: restoring resilient futures. HPG commissioned report.* London: ODI (https://odi.org/en/publications/womens-economicempowerment-in-the-face-of-covid-19-and-displacement-restoring-resilient-futures).

CONABIO. (2011). Proyecto Global de Maíces Nativos. Comisión Nacional para el Conocimiento y Uso de la Biodiversidad; Instituto Nacional de Investigaciones Forestales, Agrícolas y Pecuarias; Instituto Nacional de Ecología y cambio Climático. México. https://biodiversidad.gob.mx/diversidad/proyectoMaices

Eckstein, D. E., Kunzel, V. K., Schafer, L. S., & Germanwatch. (2021, January). *GLOBAL CLIMATE RISK INDEX 2021: Who Suffers the Most from Extreme Weather Events? Weather Related Loss Events in 2019 and 2000–2019.* Germanwatch e.V. https://www.germanwatch.org/sites/default/files/Global%20Climate%20Risk%20Index%202021_2.pdf

Etchart, L. (2017). The role of indigenous peoples in combating climate change. *Palgrave Communication,* 3, 17085. https://doi.org/10.1057/palcomms.2017.85

Forest Carbon Partnership Facility. (2022). "Guatemala." Accessed on 21/06/2022. https://www.forestcarbonpartnership.org/country/guatemala.

Fox, J. & Haight, L. (2010). "Mexican agricultural policy: multiple goals and conflicting interests." In J. Fox & L. Haight (Eds). *Subsidizing inequality: Mexican corn policy since NAFTA* Santa Cruz, CA: Woodrow Wilson International Center for Scholars, pp. 9–50.

Gunderson, L. H. (2000). Ecological resilience—in theory and application. *Annual Review of Ecology and Systematics*, 31: 425–439.

Hodbod, J., & Eakin, H. (2015). Adapting a social-ecological resilience framework for food systems. *Journal of Environmental Studies and Sciences*, 5(3), 474–484. https://keep.lib.asu.edu/_flysystem/fedora/c160/Hodbod__Eakin_Accepted_Adapting_a_SER_framework_for_food_systems.pdf

Holling, C. S. (1973). "Resilience and stability of ecological systems." *Annual Review Ecology Systematics,* 4, 1–23.

Gleijeses, P. (1989). "The Agrarian Reform of Jacobo Arbenz." *Journal of Latin American Studies*, 21(3), 453.

Global Forest Watch. "Primary forest loss in Guatemala." Accessed on 21/06/2022 from www.globalforestwatch.org

Guzzon, F., Arandia Rios, L. W., Caviedes Cepeda, G. M., Céspedes Polo, M., Chavez Cabrera, A., Muriel Figueroa, J.,... & Pixley, K. V. (2021). "Conservation and use of Latin American maize diversity: Pillar of nutrition security and cultural heritage of humanity." *Agronomy*, 11(1), 172. https://doi.org/10.3390/agronomy11010172Inter-American Commission of Women [CIM]. (2020). COVID-19 in Women's lives. Reasons to Recognize Differentiated Impacts [PDF file] https://www.oas.org/es/cim/docs/ ArgumentarioCOVID19-ES.pdf

International Indigenous Women's Forum (IIWF). (2020). THE IMPACT OF COVID-19 on the lives of INDIGENOUS WOMEN and their strategies to deal with the pandemic. Available at: https://fimi-iiwf.org/the-impact-of-covid-19-on-the-live-of-indigenous-women/?lang=en

Jones, G. D. (1998). *The conquest of the last Maya Kingdom* (en inglés). Stanford, CA, EE. UU: Stanford University Press. ISBN 978-0-8047-3522-3

Kenyon, G. (1961). Mexican Influence in Central America, 1821–1823. *Hispanic American Historical Review, 41*(2), 175–205.

Khoury, C. K., Achicanoy, H. A., Bjorkman, A. D., Navarro-Racines, C., Guarino, L., Flores-Palacios, X., Engels, J. M. M., Wiersema, J. H., Dempewolf, H., Sotelo, S., Ramirez-Villegas, J., Castaneda-Alvarez, N. P., Fowler, C., Jarvis, A., Rieseberg, L. H., and Struik, P. C. (2016). Origins of food crops connect countries worldwide. *Proceedings of the Royal Society* 283, 20160792. Available online at: https://dx.doi.org/10.1098/rspb.2016.0792

Landon, A. J. (2008). "The "How" of the three sisters: The origins of agriculture in Mesoamerica and the human niche." *Nebraska Anthropologist*, 40. https://digitalcommons.unl.edu/nebanthro/40

Lopez-Ridaura, S., Barba-Escoto, L., Reyna-Ramirez, C. A., Sum, C., Palacios-Rojas, N., & Gerard, B. (2021). "Maize intercropping in the milpa system. Diversity, extent and importance for nutritional security in the Western Highlands of Guatemala." *Scientific Reports*, 11(1), 1–10. https://doi.org/10.1038/s41598-021-82784-2

Martinez-Cruz, T. E. (2020). *Chapter 3 in On continuities and discontinuities: the making of technology-driven interventions and the encounter with the MasAgro Programme in Mexico.* Wageningen University. https://doi.org/10.18174/508387

Pebesma, E., 2018. "Simple features for R: Standardized support for spatial vector data." *The R Journal* 10(1), 439–446, https://doi.org/10.32614/RJ-2018-009

Peterson, G., Allen, C. R., & Holling, C. S. (1998). Ecological resilience, biodiversity, and scale. *Ecosystems, 1*(1), 6–18.

Pillay, Navi. 2013. *Free, prior and informed consent of indigenous peoples.* United Nations Human Rights. https://www.ohchr.org/sites/default/files/Documents/Issues/IPeoples/FreePriorandInformedConsent.pdf

R Core Team. (2021). *R: A language and environment for statistical computing.* R Foundation for Statistical Computing, Vienna, Austria. URL https://www.R-project.org/

REAL ACADEMIA ESPAÑOLA: *Diccionario de la lengua española*, 23.ª ed., [versión 23.5 en línea]. <https://dle.rae.es> 22 May 2022.

Rights and Resources Initiative (RRI). 2016. Land Mark & Woods Hole Research Center. "Towards a Global Baseline of Carbon Storage in Collective Lands." November 2016: https://rightsandresources.org/wp-content/uploads/2016/10/Toward-aGlobal-Baseline-of-Carbon-Storage-in-Collective-Lands-November-2016-RRIWHRC-WRI-report.pdf

Rodríguez, J. E. (2010). "Nosotros somos ahora los verdaderos españoles." El proceso de la independencia de México. *Histórica, 34*(1), 13–37.

Sharer, Robert J., & Traxler, L. P. (2006). *The Ancient Maya* (en inglés) (6.ª edición, completamente revisada). Stanford, CA, EE. UU: Stanford University Press. ISBN 0-8047-4817-9.

Sobrevila, C. 2008. *The role of indigenous peoples in biodiversity conservation: The natural but often forgotten partners (English).* Washington, DC. World Bank Group.

Streeter, S. M. (2000). Interpreting the 1954 US intervention in guatemala: realist, revisionist, and postrevisionist perspectives. *The History Teacher*, 34(1), 61–74.

Tolisano, J., & Lopez, M. M. (2010). *Guatemala biodiversity and tropical forest assessment.* Washington, DC: United States Agency for International Development.

University of North Carolina Libraries. (2013). *Guatemalan independence. Ancient and living Maya in the 19th and 20th centuries: Archaeological discovery, literary voice, and political struggle.* UNC Libraries. Retrieved June 21, 2022, from https://exhibits.lib.unc.edu/exhibits/show/maya/guatemalan-independence

USAID. 2021. *Guatemala: Nutrition profile.* https://www.usaid.gov/sites/default/files/documents/Guatemala-Nutrition-Profile_1.pdf

Wickham, H. (2016). *ggplot2: Elegant graphics for data analysis.* Springer-Verlag New York.

World Bank. (2011). *Climate risk country profile: Guatemala.* World Bank Climate Change Knowledge Portal. Retrieved June 21, 2022, from https://climateknowledgeportal.worldbank.org/sites/default/files/2018-10/wb_gfdrr_climate_change_country_profile_for_GTM.pdf

World Bank. (2021a). *Vulnerability: Guatemala.* World Bank Climate Change Knowledge Portal. Retrieved May 22, 2022, from https://climateknowledgeportal.worldbank.org/country/guatemala/vulnerability

World Bank. (2021b). *Current climate: Trends and significant change against natural variability: Guatemala.* World Bank Climate Change Knowledge Portal. Retrieved May 22, 2022, from https://climateknowledgeportal.worldbank.org/country/guatemala/trends-variability-historical

World Bank. (2021c). *Current climate: Trends and significant change against natural variability: Mexico.* World Bank Climate Change Knowledge Portal. Retrieved May 22, 2022, from https://climateknowledgeportal.worldbank.org/country/mexico/trends-variability-historical

Zaccarelli, N., Petrosillo, I., & Zurlini, G. (2008). Retrospective analysis. In Jørgensen, S.E., Fath, B.D. (Eds.), *Systems ecology*, Vol. 4 of Encyclopedia of Ecology, Vol. 5. Elsevier, Oxford, pp.3020–3029.

7

ECOLOGICAL INTENSIFICATION IN GRASSLANDS FOR RESILIENCE AND ECOSYSTEM SERVICES

The case of beef production systems on the Campos Grasslands of South America

Soledad Orcasberro, Laura Astigarraga, Marta Moura Kohmann, Pablo Modernel and Valentín D. Picasso

Vision: Native grasslands as providers of ecosystem services and resilience

Grassland ecosystems occupy 40% of the global ice-free (Hewins et al., 2018), being the single most extensive form of land use on the planet (Asner et al., 2004). Most of these ecosystems are mainly grazed by ruminants and can be found in the native grasslands of central Asia, Sub-Saharan and southern Africa, North and South America, and Australia/New Zealand (O'Mara, 2012). Grasslands make a significant impact on food security and nutrition, providing feed required by ruminants to produce meat and milk (O'Mara, 2012), and avoiding feed-food competition (i.e., beef cattle grazing on marginal lands, Van Zanten et al., 2018). More than half of the meat and milk produced globally by ruminants comes from grazing systems, and they are more important in terms of food energy than pork and poultry meat (O'Mara, 2012; OECD, 2021).

Grasslands are among the ecosystems with the greatest species richness in the world (Wilson et al., 2012), and they provide a wide range of ecosystem services (Tiscornia et al., 2019) defined as all the benefits given by biodiversity, ecosystem structure, and function to meet the demands of human survival, life, and well-being (Sala and Paruelo, 1997). They include provisioning (e.g., meat, milk, and fuel), supporting (e.g., forage production and nitrogen fixation), cultural (e.g., tourism, recreation, and wildlife habitat), and regulating (e.g., stormwater management, soil conservation, soil carbon storage, aquifer recharge, soil water conservation during drought, and improved soil physical and chemical properties) (Sollenberger et al., 2019). The high biodiversity of grasslands is widely recognized (Habel et al., 2013; Bengtsson et al., 2019), and closely associated with the ecosystem services and the resilience of these systems (Figure 7.1). Species richness increases resistance to disturbance by stabilizing grassland productivity

DOI: 10.4324/9781003266440-7

FIGURE 7.1 Campos grasslands from Rocha and Eastern Sierras of Treinta y Tres, Uruguay. Beef cattle grazing native grasslands coexisting with native fauna. Photo by Miguel Lazaro and Valentín D. Picasso.

(Isbell et al., 2015; Sollenberger et al., 2019). For example, species with particular rooting systems, the presence of certain legumes or highly productive species, can modify the relationship between species richness and soil C accumulation (De Deyn et al., 2009, 2011; Skinner and Dell, 2016; Sollenberger et al., 2019). Most of the biomass in grasslands is belowground with high accumulation rates and slow decomposition of organic material, playing a key role in the global carbon cycle (Gibson, 2009; Andrade et al., 2015).

The vegetation in the Rio de la Plata grasslands, a geographical zone of native grasslands in southeastern South America (Figures 7.1 and 7.2; Soriano et al., 1991), contains around 3,000 plant species including 450 species of grass and 150 species of legumes used as forage for domestic grazing animals (Boldrini, 2002). This abundant biodiversity sustains 385 species of birds and 90 species of terrestrial mammals (Bilenca and Miñarro, 2004; Carvalho et al., 2011).

The resilience of grassland ecosystems relies on the biodiversity of species and their ability to respond against variation (Kuhsel and Bluthgen, 2015). Studies have shown that the diversity of climatic responses across species increases the resilience of species-rich pollinator communities in ecosystems despite land-use intensification (Kuhsel and Bluthgen, 2015). Biodiversity affects ecosystem productivity and buffers the ecosystem against climatic extremes (Hossain et al., 2022).

Research demonstrated that a more diverse biome is often more productive and more resilient to external perturbations (Thompson et al., 2009; Feng et al., 2021) and plant communities with greater species richness need to be maintained to stabilize ecosystem productivity and increase resistance against different climatic extremes (López-Ridaura et al., 2005; López-Ridaura et al., 2005a; Picasso et al., 2019; Hossain et al., 2022).

Management practices influence the provision of ecosystem services and the maintenance of biodiversity (Duru et al., 2019; Sollenberger et al., 2019), affecting the resilience of grassland systems. The degradation of native grasslands has decreased their productivity (Tiscornia et al., 2019), reducing their ability to provide ecosystem goods and services (Wick et al., 2016). The main causes for degradation identified in the Río de la Plata grasslands ecosystems are overgrazing, invasive species, nutrient addition, and the introduction of exotic forage species (Tiscornia et al., 2019).

In the region, research to increase the productivity of natural grasslands has followed two paths. On the one hand, traditional intensification implies the use of external inputs such as the addition of nutrients with synthetic fertilizers and/or the adoption of legumes (Picasso et al., 2014). On the other hand, there is an alternative path called "ecological intensification" (Tittonell, 2021), based on process technology: management of grazing intensity through controlling forage allowance and stocking rate, and knowledge about the plant-animal interactions (Nabinger and Carvalho, 2009). Ecological intensification is a way to increase the efficiency and productivity of grassland-based livestock systems while reducing costs, maintaining the provision of key ecosystem services, reducing the dependence on non-renewable resources, and stimulating adaptability, resilience, and social equity (Albicette et al., 2017; Tittonel, 2021).

Managing grazing intensity and improving the productivity of grasslands have the potential to mitigate almost 1.5 Gt of CO_2 equivalent by 2030 due to carbon storage and carbon sequestration together with potential mitigation associated with the restoration of degraded lands. When management practices that reduce soil carbon stocks in grasslands like overgrazing are reversed the stocks can be reconstructed (Follett et al., 2001).

Additionally, grasslands are also threatened by climate change (Gibson and Newman, 2019). The potential positive effects on primary productivity due to the increase in future CO_2 concentrations can be counterbalanced by greater climate variability and rising temperature associated with more frequent extended drought periods (Bloor et al., 2010). Historical climate records and future climate model projections indicate that variability in annual precipitation is increasing, both at the global and local scales (IPCC, 2013), threatening food security (Godfray et al., 2010; Douxchamps et al., 2016). Consequently, it is crucial to address the resilience of agricultural systems to overcome stressors associated with climate change (Grimm and Wissel, 1997; López-Ridaura et al., 2005; Picasso et al., 2019).

Applying the principles of ecological intensification to livestock grazing systems in grasslands by improving the grazing management (i.e., adjusting forage mass and grazing intensity) could enhance biomass production and promote more efficient use of resources, rehabilitation of degraded grazing lands, and greater profitability, while also mitigating greenhouse gas emissions and increasing resilience to extreme climatic events (Oldeman, 1994; O'Mara, 2012).

History and current context of livestock systems in Campos grasslands

The Río de la Plata grasslands occupy 70 million ha between eastern Argentina, Uruguay, and southern Brazil (Soriano, 1991; Royo Pallares et al., 2005; Carvalho et al., 2011), and represent the largest native grassland biogeographic unit in South America, and one of the largest in the world (Figure 7.2). They are a

FIGURE 7.2 Schematic representation of the location of the Río de la Plata grasslands and sub-regions Pampas and Campos, in southeastern South America. Adapted from Gorosábel et al. (2020).

biodiversity hotspot (Bilenca and Miñarro, 2004; Overbeck et al., 2007; Modernel et al., 2016) that provides provisioning, supporting, regulating, and cultural ecosystems services of local and global importance (Modernel et al., 2016). Average annual temperature in this region varies from 16.6°C in southern Uruguay to 21.1°C in the northern Corrientes, Argentina, while the average annual rainfall varies from 1,000 mm in southern Uruguay to 1,600 mm in the northern Campos in Brazil (Jaurena et al., 2021).

The Río de la Plata grasslands comprise the Campos and the Pampas (Soriano et al. 1991; Carvalho et al., 2008). The Campos vegetation consists mostly of grasses and forbs, but small shrubs and trees can also occur (Berretta, 2001; Royo Pallarés, 2005). These areas are covered by spatially heterogeneous temperate and subtropical grasses in a complex variety of species combinations related to soil type, and grazing intensity as the main factor that keeps the grasslands in an herbaceous pseudo climax phase (Berretta et al., 2000; Royo Pallares et al., 2005; Jaurena et al., 2021). The climatic conditions stimulate the dominance of C3 and C4 grass species (Burkart, 1975), although C3 is more common in the Pampas of Argentina, and C4 in the Campos of Brazil and Uruguay (Berretta et al., 2000).

Río de la Plata grasslands were altered by the introduction of livestock at the beginning of European colonization four centuries ago (Piñeiro et al., 2006), evolving without the participation of large herbivores before that moment. Livestock presence altered the climax condition and transformed it into a disclimax with different growth and life habits (Royo Pallares, 2005). The domestic herbivorous introduction has been the foundation of the most important economic activity in the region (beef cattle and sheep production), utilizing these diverse plant communities as their basic food. The farming systems of these regions have been managed extensively for livestock production to obtain products (meat, wool, and leather) with very low inputs (Viglizzo et al., 2001; Bindritsch, 2014; Picasso et al., 2014). The modification of this zone with crops begins in the 20th century, with the expansion of the fertilization around 1970, which is more recent than in the temperate areas of the Old World or North America (Hall et al., 1991; Paruelo et al., 2007). The area of Campos ecosystems has been declining since then due to conversion to row crops (mostly soybean), cultivated pastures, and forestry plantations (Viglizzo and Frank, 2006; Baeza and Paruelo, 2020). The increase in soybean prices during the 2000–2010 decade was the major driver of soybean expansion in southern South America (Modernel et al., 2016). From 2000 to 2019, the area cultivated with soybean more than doubled and most of the expansion occurred on pastures originally converted from natural vegetation (Song et al., 2021). Currently, native grasslands occupy 36% of their original extension in the state of Rio Grande do Sul in Brazil (Trindade et al., 2018), 64% in Uruguay (Cortelezzi and Mondelli, 2014), and 26% and 72% in the Entre Ríos and Corrientes regions of Argentina, respectively (INDEC, 2018; Jaurena et al., 2021). Of the 27,000 livestock farms in Uruguay, about 60% are family farming and 40% are cow-calf systems (MGAP, 2013; Tommasino et al., 2014; Modernel et al., 2016).

On the other hand, cattle grazing is seen as the most promising conservation tool for the Río de la Plata grasslands, as it maintains floral and faunal diversity (Soussana, 2009; Brandão et al., 2012). However, overgrazing, which occurs when stocking rates exceed the carrying capacity of grasslands, could lead to animal weight loss and losses in soil cover, being the most widespread anthropogenic driver of degradation of these native grasslands (Carvalho, 2009; Overbeck et al., 2009; Soussana, 2009; Ruviaro et al., 2016; Tiscornia et al., 2019). Thus, managing grazing through ecological intensification, controlling the forage allowance, and adjusting the stocking rate, has the potential of increasing socio-economic outcomes without raising the environmental impacts of these systems, which is crucial to support the rural development in the Río de la Plata grasslands (Ruggia et al., 2021).

Current and future stressors and proposed solutions

The Río de la Plata grasslands have been seriously threatened in the past decades by land-use transformation, habitat loss, fragmentation, species loss, and invasive species as a consequence of the use of cultivated pastures. Additionally, overgrazing is degrading natural grasslands due to weed invasion, loss of biodiversity and native species, increased soil erosion, and water pollution (Coffin et al., 1996; Carvalho and Batello, 2009; Belesky et al., 2016; Gibson and Newman, 2019; Cezimbra et al., 2021). Greenhouse gas emissions and anthropogenic climate change leading to warming and changes in precipitation patterns have been recognized as an additional threat to this ecosystem (Carvalho and Batello, 2009; Gibson and Newman, 2019; Carvalho et al., 2011).

Ecological intensification at the paddock level

Ecological intensification through grazing management presents an opportunity to buffer the effects of temperature and precipitation variability in biodiverse grasslands (Cobon et al., 2009; Thurow and Taylor, 1999). Forage allowance defined as the adjustment of the relationship between forage mass and animal live weight per unit area of the specific unit of land being grazed at any one-time (kg of dry matter (DM) kg live weight $(LW)^{-1}$; Sollenberger et al., 2005; Allen et al., 2011) has been identified as the simplest management indicator of forage use efficiency with the greatest impact on grazing systems (Holmes et al., 1984; Animut et al., 2005). Grazing management can control the degree of the modification of the natural structural and floristic heterogeneity that is a response to soil variability in order to maintain ecosystem services and at the same time offers an adequate nutritional environment to the herbivores (Nabinger et al., 2011).

Research in the Campos grasslands has compared "traditional" and "improved" grazing management through manipulation of forage allowance. Here, we summarize the findings from published literature to date (Table 7.1). Traditional management refers to relatively low forage allowance. In cow-calf systems, traditional forage allowance has been reported as 2.5–2.9 kg DM kg LW^{-1} (Table 7.1). In backgrounding systems (i.e., growing heifers and steers), traditional forage allowance has been reported as 4% to 8% (kg DM 100 kg LW^{-1}; which can be converted to 0.9 to 1.5 kg DM kg LW^{-1} as defined by Sollenberger et al., 2005) (Table 7.1). On the other hand, improved forage allowance

TABLE 7.1 Mean and standard deviation (SD) for production and environmental variables of experiments on the Campos grasslands comparing traditional vs improved forage allowance published in the literature

	Traditional		*Improved*		*References*
	Mean	SD	Mean	SD	
Forage allowance cow-calf (kg DM kg LW^{-1})	2.8	0.1	4.7	1.2	1–7, 18
Forage allowance background (kg DM kg LW^{-1})	1.2	0.5	3.4	1.2	8
Forage mass (kg DM ha^{-1})	944	324	1682	465	1–11, 18
Sward height (cm)	5.1	2.2	8.4	3.2	3, 7, 10, 18
Herbage accumulation (kg DM ha^{-1} day^{-1})	12.8	3.8	13.5	5.0	3,9–11, 12
Stocking rate (kg LW ha^{-1})	462	78	368	76	3, 5–8
Body condition score	4.1	0.2	4.4	0.1	1–6, 18
Weaning rate (%)	66.0	24.0	78.5	0.7	2
Milk yield (kg day^{-1})	5.7	0.6	4.6	0.6	18, 19
Pregnancy (%)	62	0.1	88	0.1	6
Calf weight at weaning (kg)	168	9	196	3	1–6, 18
Calf weight per cow (kg)	74	28	96	1	2
Dry matter intake (kg DM cow^{-1} day^{-1})	5.5	1.6	6.8	2.1	6–8
Organic matter digestibility (g kg DM^{-1})	405	11	524	172	7,9
Average daily gain (kg LW $animal^{-1}$ day^{-1})	0.3	0.2	0.4	0.2	1–5, 8, 9, 12, 13
PAR/primary production[a]	0.27	0.09	0.34	0.03	14
PAR/secondary production[b]	0.01	0.005	0.02	0.003	14
Energy efficiency (g calf MJ^{-1})[c]	4.7	1.5	5.8	0.4	2
Meat productivity (kg ha^{-1})	95	10	125	10	20
Methane emissions (g CH_4 $animal^{-1}$ day^{-1})	118.5	0.7	172.6	31.2	7,8
Methane yield (g CH_4 kg ADG^{-1})	1.7	1.0	0.7	0.1	8
Carbon stock (Mg ha^{-1})[d]	96	63	180	0.0	13
CO_2 emissions (kg CO_2 eq. $ha^{-1}year^{-1}$)	2028	167	1653	180	8
Grasslands species diversity index[e]	3.0	0.4	3.4	0.2	15, 16
Resistance herbage accumulation to drought[f]	0.5	...	0.7	...	17
Recovery herbage accumulation to drought[g]	3.7	...	3.1	...	17
Resistance live weight to drought	0.9	...	1.0	...	17
Recovery live weight to drought	18.1	...	20.8	...	17

References: 1. Carriquiry et al. (2012); 2. Do Carmo et al. (2016); 3. Do Carmo et al. (2018); 4. Do Carmo et al. (2021); 5. Soca (2013); 6. Claramunt et al. (2017); 7. Orcasberro et al. (2021); 8. Cezimbra et al. (2021); 9. Moojen and Maraschin (2002); 10. Da Trindade (2011); 11. Soares et al. (2005); 12. Maraschin et al. (1997); 13. Guterres et al. (2006); 14. Nabinger et al. (2000); 15. Carvalho et al. (2003); 16. Nabinger et al. (2011); 17. Modernel et al. (2019); 18. Casal et al. (2016); 19. Gutierrez et al. (2013); 20. Claramunt (2015).

a Efficiency of PAR (photosynthetically active radiation) conversion in dry matter (primary production).
b Efficiency of PAR (photosynthetically active radiation) conversion in live weight gain (secondary production).
c Efficiency = g of calf/MJ of energy consumed by the cow in the cycle.
d Carbon stocks from the upper 40 cm layer of natural grasslands according to grazing intensities.
e The higher the index, the more diverse the species are in the habitat.
f Resistance is a ratio of values during drought and before drought with same units, therefore unitless.
g Recovery is calculated as difference between after drought and during drought response variable over time (season). Units: kg DM ha day^{-1} $season^{-1}$; and kg animal live weight $season^{-1}$ (slope of the variable against time after the perturbation).

Abbreviations: DM: dry matter; LW: live weight; ADG: average daily gain.

above 2.5 kg DM kg LW^{-1}, with seasonal stocking rate variations, resulted in higher productivity (Nabinger et al., 2000; Do Carmo et al., 2018; Do Carmo et al., 2019).

Experiments in the Campos region have shown that controlling forage allowance is the main variable that can be adjusted in grazing management to improve productivity and profitability while preserving biodiversity and enhancing resilience (Carvalho et al., 2003; Soares et al., 2003; Overbeck et al., 2007; Da Trindade et al., 2012; Sollenberger, 2015; Do Carmo et al., 2016; Claramunt et al., 2017; Do Carmo, et al., 2018; Modernel et al., 2019). In Campos grasslands, the production relies on spring-summer growing of plants, with little production during the winter period (Royo Pallares et al., 2005). Thus, experimental evidence confirmed that optimizing herbage management by controlling forage allowance with seasonal variations generates changes in the production and accumulation of herbage that increases biomass during the grass growing season (spring and summer) to be consumed during the period when herbage growth decreases (winter) (Claramunt et al., 2017; Do Carmo et al., 2018; Orcasberro et al., 2021). Improved management applies this strategy, resulting in greater availability of forage above potential intake, allowing grazing animals to select plant species and morphological components of greater nutritive value (Nabinger et al., 2000), increasing their performance (Table 7.1). Furthermore, plants growing in areas under moderate grazing have better growth and development than plants growing in areas with high-intensity grazing, allowing them to flower and reseed, which ensures pasture longevity and stability while improving system-level resilience (Nabinger et al., 2000). When compared to overgrazed grasslands, greater aboveground forage mass resulting from increased forage allowance stimulates deeper and denser rooting systems, which increases grassland resilience to extreme events, including drought (Bartaburu et al., 2009; Van Ruijven and Berendse, 2010; Norton et al., 2016).

According to Modernel et al., 2019, resistance to a crisis (drought in this case), one of the attributes to measure resilience of the system, which illustrates how much the productivity change during the drought compared to the situation before the crisis, is enhanced due to greater herbage accumulation and animal live weight gain when the grazing management is improved (Table 7.1). This management contributes to preserving natural biodiversity as it is highlighted by the grasslands species diversity index and thus, enhances the system's resilience, which has been shown to moderate the effects of extreme perturbations like drought or overgrazing (Table 7.1, Carvalho et al., 2011).

Improved grazing management resulted in greater daily methane emissions per animal, but lower methane yield (g CH_4 kg LWG^{-1}) and CO_2 emissions per unit area (Table 7.1). Additionally, carbon stocks were almost double under improved grazing management compared to traditional grazing management (Table 7.1), suggesting the potential to reach a neutral or negative (net sink) carbon balance with improved grazing management. This way we can contribute to achieving multiple Sustainable Development Goals (SDGs), including increased food security and nutrition (SDG 2), the mitigation of and adaptation to climate change (SDG 13), and life on land (SDG 15; FAO, 2018).

Impacts of ecological intensification at the farm level

At the farm level, we summarized reports from the literature comparing the performance of grazing farms in the Campos region, and have reported production, environmental, and socio-economic information for several years. We classified the farms into three groups (Table 7.2). The first group, "traditional management", represents farms that use relatively low forage allowance and have more than 80% of their area covered by native grasslands, which is the widespread management in the region. The second group, "improved management", represents farms using relatively higher (optimal) forage allowance and less than 20% of their productive area with cultivated pastures. The third group, "improved with cultivated pastures", represents farms that use improved grazing management but have more than 20% of their productive area with cultivated pastures (i.e., seeded pastures with exotic grasses and legumes and higher productivity).

Farms analyzed here were "complete cycle" systems, i.e., beef production from cow-calf to fattening. The optimum forage allowance for complete cycle systems in the Campos has been estimated as ≥4.6 kg DM kg LW^{-1} (Do Carmo et al., 2019). Farmers in the improved management group showed greater weaning rate, cow pregnancy rate, and average daily gain per animal, which results in increased meat productivity compared to the traditional system (Table 7.2).

Similarly, the farm net income on improved management systems was greater than on the traditional farms. Farms in the improved management also show the best economic indicators of the three groups. Furthermore, they exhibited lower CO_2 emissions, energy consumption, soil erosion, pesticide contamination risk, nitrogen balance, phosphorus balance, and carbon footprint, than the

TABLE 7.2 Mean of productive, environmental, resilience, and socio-economic variables of representative farms grouped in traditional management, improved management, and improved through cultivated pastures, in the Campos grasslands region

	Traditional management	*Improved Management*	*Improved cultivated pastures*	*References*
Forage allowance (kg DM kg LW^{-1})	3.4	4.5	5.8	2, 4
Forage mass (kg DM ha^{-1})	3,845	4,196	4,539	2
Area of cultivated pastures (%)	7.1	9.6	37.5	1–4
Stocking rate (kg LW ha^{-1})	325	341	418	1–5
Sheep to cattle ratio	2.3	1.0	2.0	1, 2, 4
Meat productivity (kg ha^{-1})	99.5	128.7	152.5	1–4
Weaning rate %	61.0	82.0	78.5	2
Cow pregnancy %	70.6	86.5	88.5	1,4
Calf weaning weight per cow (kg)	101.5	137.6	151.0	4
Dry matter intake (kg MS cow^{-1} day^{-1})	10.3	9.9	9.8	2,3
Dry matter digestibility (g kg DM^{-1})	453	453	571	5
Average daily gain (kg LW $animal^{-1}$ day^{-1})	0.4	0.6	0.7	3
Interval from calving to fattening (days)	846	643	544	3,5
CH4 emission (kg CO_2 eq kg LW^{-1})	11.6	8.7	5.4	3
N2O emission (kg CO_2 eq kg LW^{-1})	5.1	3.9	3.6	3
CO_2 emission (kg CO_2 eq kg LW ^{1})	0.0	0.4	0.5	3
CO2 emission per ha (kg CO_2 eq ha^{-1} $year^{-1}$)	2,061	1,817	2,241	2
Carbon footprint (CO_2 eq kg LWG^{-1})	29.3	20.7	17.2	2, 3, 5
Energy consumption (MJ kg LW^{-1})[a]	0.0	4.9	11.8	3
Soil erosion (Mg ha^{-1})[b]	1.5	2.7	3.4	3
Pesticides contamination risk (index)[c]	0.0	6.3	12.2	3
Bird richness and abundance[d]	3.2	3.5	3.6	4
Ecosystem integrity index (0–5)[e]	3.6	3.7	3.7	4
Environmental impact index[f]	−1.0	−0.3	0.3	3
Recovery of meat production to drought[g]	14.9	4.5	−1.5	1
Recovery of pregnancy rate to drought	17.7	7.5	−6.5	1

(*Continued*)

	Traditional management	*Improved Management*	*Improved cultivated pastures*	*References*
Resistance of meat production to drought[h]	1.2	0.9	0.7	1
Resistance of pregnancy rate to drought	0.8	0.9	1.0	1
Change in workload (hour $year^{-1}$)[i]	0.0	–339.8	–140.0	4
Gross margin (U$S ha^{-1})	26.7	63.6	42.0	4
Costs (U$S ha^{-1})	86.2	105.4	116.0	4
Farm net income (U$S ha^{-1})	686.3	720.8	984.5	1, 5
Input to output ratio[j]	0.4	1.1	1.9	1, 3, 5
Internal rate of return per ha (%)[k]	3.2	6.3	3.1	5
Profitability index per ha (%)[l]	67.3	146.6	100.3	5

References: 1. Modernel et al. (2019); 2. Becoña et al. (2014); 3. Picasso et al. (2014); Modernel et al. (2013); 4. Ruggia et al. (2021); Scarlato et al. (2015); Aguerre et al. (2018); Albicette et al. (2017); 5. Ruviaro et al. (2015); Ruviaro et al. (2016).

a Fossil fuel energy consumption estimated using the model Agroenergia described by Llanos et al. (2013). The energy use per kg of LW was estimated over one growing cycle and expressed in MJ·kg LW^{-1} (Picasso et al., 2014).

b Soil erosion rates were estimated using EROSION 5.0 (Garcia Prechac et al., 2005; Picasso et al., 2014).

c Pesticide ecotoxicity was calculated as the standard LCA method using USEtox (Picasso et al., 2014). Higher values are associated with more ecotoxicity.

d Shannon index: the higher the index, the more diverse the species are in the habitat.

e Scale: 0 (loss of all ecosystem functions) to 5 (best condition; Aguerre et al., 2018).

f Negative values indicate lower integrated environmental impact including energy consumption, pesticide ecotoxicity, nitrogen balance, phosphorus balance, soil erosion, impact biodiversity, and GHG emissions per kg of meat (carbon footprint) (Picasso et al., 2014).

g Recovery is calculated as the difference between after drought and during drought response variable over time (season).

h Resistance is a ratio of values during drought and before drought (slope of the variable against time after the perturbation).

i Change from traditional to improved. Negative values represent reduction in workload (hours).

j Ratio between costs (input) and income (product).

k Hypothetical discount rate that, when applied to a cash flow, causes costs, and investments values to be taken to the present value so that the net present value would be equal to zero (Ruviaro et al., 2016).

l It consists in separating the net present value per project investment unit and, throughout its lifespan, aims to simultaneously solve projects with different investments and terms (Ruviaro et al., 2016).

Abbreviations: DM: dry matter; GHG: greenhouse gases; LW: live weight; LWG: live weight gain.

improved with cultivated pastures group. Likewise, they showed greater resilience to drought in terms of recovery to meat productivity and pregnancy rate, which means that the velocity of the systems to get over the crisis (drought in this case) and return to the pre-crisis values on both variables was higher, compared to the group of improved management with cultivated pastures. The traditional group exhibited low outputs in terms of productivity leading to reduced farm net income. Additionally, they revealed low meat productivity and low pregnancy rates, before and after the drought, explaining the apparently high recovery and resistance (Modernel et al., 2019).

Path to the future: The sustainability and resilience of livestock systems in Campos grasslands

In the context of a growing global population, associated with income growth and climate change, implementing improved management practices for grazing systems on grasslands play a key role in meeting the rising demand for animal products while preserving the ecosystem services (Hodgson et al., 2000, 2021; Boval and Dixon, 2012; O'Mara, 2012). Ecological intensification should be the core of the management practices in grassland systems which include the balance between the demands of the production system and the provision of resources, adjusting the efficiency between the systems demands and the environmental provision, decreasing the utilization of external inputs that produce environmental impacts, and supporting the natural resources and ecosystem services associated, to improve the resilience of these systems (Tittonell, 2021).

Traditional management in the Campos grasslands livestock systems characterized by year-round grazing at relatively constant stocking rates (Royo Pallares et al., 2005) with limited forage supply for ruminants, especially in winter, leads to low productivity (60–70 kg live weight per hectare per year), giving the impression of a low-profit business (Jaurena et al., 2021). However, past and recent research demonstrated that different levels of ecological intensification can significantly increase livestock production in natural grasslands (Carvalho et al., 2011). The adoption of improved grazing management relative to traditional management strategies, optimizing productivity by controlling both forage allowance (Neves et al., 2009) and pasture structure (Soares et al., 2005; Da Trindade et al., 2016; Azambuja et al., 2020), can triplicate beef production, without rising external inputs or environmental impacts (Carvalho et al. 2011; Kuinchtner et al., 2021).

To assess the broad concept of sustainability in the livestock systems on the Campos grasslands, we standardized some key indicators (from Table 7.2) that illustrate the production, environmental, economic, and social dimensions (Figure 7.3). Each variable was standardized by subtracting the average and dividing by the standard deviation of the three systems (Montgomery, 2008). The values for the standardized variables with a negative impact on the sustainability were multiplied by −1.

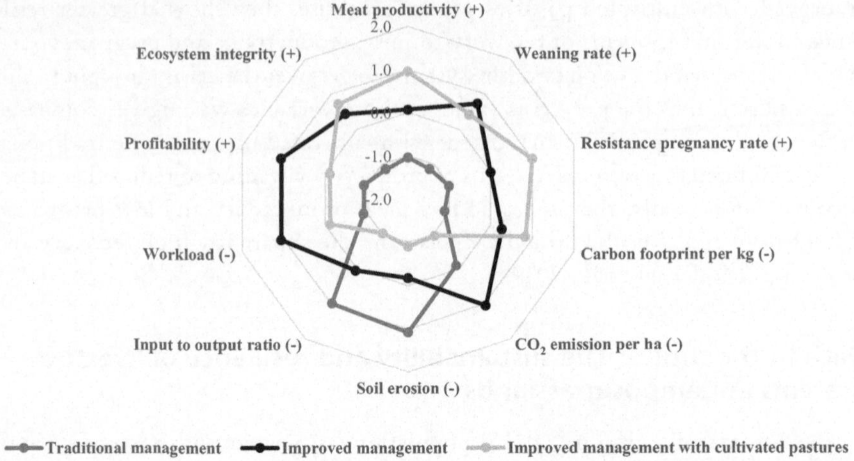

FIGURE 7.3 Standardized sustainability indicators of groups of farms on the Campos grasslands, with traditional management, improved management, and improved management with more than 20% of cultivated pastures.

The improved management system with optimal control of herbage allowance, compared with the traditional one, demonstrates that we can increase the profitability of the system and at the same time obtain lower environmental impacts like lower CO_2 emissions (kg CO_2 eq ha^{-1} $year^{-1}$), carbon footprint (CO_2 eq kg LWG^{-1}), and ecosystem integrity while maintaining high productivity, weaning rate, lower workload, and enhanced resilience to drought (measured as resistance to pregnancy rate). The greater meat productivity in the improved grazing with more than 20% of cultivated pastures was associated with a larger use of inputs, which lead to lower profitability of the farm and more environmental impacts like CO_2 emissions (kg CO_2 eq ha^{-1} $year^{-1}$) and soil erosion, than the others. Although the traditional group seems to indicate less soil erosion, this is probably due to the models used to determine this variable, which could not detect the effects of overgrazing on the degradation of the soils. Additionally, they present higher CO_2 emissions (kg CO_2 eq ha^{-1} $year^{-1}$), lower economic and productive outcomes, and worst social consequences, which result in overall lower sustainability at the system level. Therefore, the group that shows superior sustainability indicators due to better balance in the variables exposed was the improved grazing management.

Optimal management of forage allowance in grasslands also provides several important ecosystem services provisioning food and habitat for pollinators and wildlife, supporting forage production and nutrients cycling, and regulating soil carbon storage, erosion control, quality infiltration water, and cultural services like recreational opportunities, open space, and improve quality of life of the whole society (Peyraud, 2017; Sollenberger et al., 2019; Bellocchi and Picon-Cochard, 2021). Their social and environmental importance is much higher than other crops, including other forage crops, and is progressively distinguished by society (Peyraud, 2017). Several studies also demonstrated that the economic performances of grassland-based systems are similar and sometimes higher than those observed in more intensive systems (Peyraud and Peeters, 2016).

Consequently, grasslands conservation should be directly related to increasing sustainable agricultural practices like improving the management of forage allowance to avoid negative consequences, improving the capacity to adapt to climate change, reducing emissions, and making natural grasslands more competitive relative to other agricultural options such as row crops (Carvalho et al., 2011; Pörtner et al., 2021).

In summary, improving grazing management by optimizing forage allowance in livestock systems in the Campos region can simultaneously increase productivity, socio-economic outcome, ecosystem services, and resilience to drought. The pathway to increasing climate resilience and sustainability in these grasslands landscapes is inexorably linked to the optimization and success of grazing management.

References

Aguerre, V., Albicette, M.M. et al. (2018). Co-innovando para el desarrollo sostenible de sistemas ganaderos familiares de Rocha – Uruguay. Marzo. INIA. Serie técnica 243. http://www.inia.uy/publicaciones/documentos%20compartidos/st-243-2018.pdf

Albicette, M.M., Leoni, C., Ruggia, A., Scarlato, S., Blumetto, O., Albín, A., & Aguerre, V. (2017). Co-innovation in family-farming livestock systems in Rocha, Uruguay: a 3-year learning process. *Outlook on Agriculture* 46(2), 92–98. doi: 10.1177/0030727017707407

Allen, V.G., Batello, C., Berretta, E.J., Hodgson, J., Kothmann, M., Li, X., … Sanderson, M. (2011). An international terminology for grazing lands and grazing animals. *Grass & Forage Science* 66, 2–28. https ://doi.org/10.1111/j.1365–2494.2010.00780.x

Andrade, B.O., Koch, C., Boldrini, I.I., Vélez-Martin, E., Hasenack, H., Hermann, J.M., Kollmann, J., Pillar, V.D., & Overbeck, G.E. (2015). Grassland degradation and restoration: a conceptual framework of stages and thresholds illustrated by southern Brazilian grasslands. *Natureza & Conservação* 13(2), 95–104. doi: 10.1016/j.ncon.2015.08.002

Animut, G., Goetsch, A.L., Aiken, G.E., Puchala, R., Detweiler, G., Krehbiel, C.R., Merkel, R.C., Sahlu, T., Dawson, L.J., Johnson, Z.B., & Gipson, T.A. (2005). Grazing behaviour and energy expenditure by sheep and goats co-grazing grass/forb pastures at three stocking rates. *Small Ruminant Research* 59, 191–201.

Asner, G.P., Elmore, A., Olander, L.P., Martin, R.E., & Harris, A.T. (2004). Grazing systems, ecosystem responses, and global change. *Annual Review of Environment and Resources* 29(1), 261–299. doi: 10.1146/annurev.energy.29.062403.102142

Azambuja, J.C.R., Carvalho, P.C.F., Bonnet, O.J.F., Bastianelli, D., & Jouven, M. (2020). Functional classification of feed items in pampa grassland, based on their near-infrared spectrum. Rangeland Ecology & Management 73, 358–367. doi: 10.1016/j.rama.2020.02.001.

Baeza, S., & Paruelo, J.M. (2020). Land use/land cover change (2000–2014) in the Rio de la Plata grasslands: an analysis based on MODIS NDVI time series. *Remote Sensors* 12, 381. doi: 10.3390/rs12030381

Bartaburu, D., Duarte, E., Montes, E., Morales Grosskopf, H., & Pereira, M. (2009). Las sequías : un evento que afecta la trayectoria de las empresas y su gente. In H. Morales Grosskopf, & F. Dieguez Cameroni (Eds.), *Familias y Campo: Rescatando Estrategias de Adaptación. Instituto Plan Agropecuario* (pp. 155–168). Montevideo: Instituto Plan Agropecuario.

Becoña, G., Astigarraga, L., Picasso, V. (2014). Greenhouse gas emissions of beef cow-calf grazing systems in Uruguay. *Sustainable Agriculture Research* 3(2), 89–105.

Belesky, D., Malinowski, D. (2016). Grassland communities in the USA and expected trends associated with climate change. *Acta Agrobotanica* 69(2), 1673.

Bellocchi, G., & Picon-Cochard, C. (2021). Effects of climate change on grassland biodiversity and productivity. *Agronomy* 11, 1047. doi: 10.3390/agronomy11061047

Bengtsson, J., Bullock, J.M., Egoh, B., Everson, C., Everson, T., O'Connor, T., O'Farrell, P.J., Smith, H.G., & Lindborg, R. (2019). Grasslands—more important for ecosystem services than you might think. *Ecosphere* 10(2), e02582. doi: 10.1002/ecs2.2582

Berretta, E., Risso, D., & Bemhaja, M. (2001). Tecnologias para la mejora de la producción de forraje en suelos de Basalto. In D.F. Risso, & E.J. Berretta (Eds.), *Tecnologias Forrajeras Para Sistemas Ganaderos de Uruguay* (pp. 5–12). Montevideo: Boletín de divulgación 76, Hemisferio Sur.

Berretta, E.J., Risso, D.F., Montossi, F., & Pigurina, G. (2000). Campos in Uruguay. In G. Lemaire, J. Hodgson, A. Moraes, P.C.F. Carvalho, & C. Nabinger (Eds.), *Grassland ecophysiology and grazing ecology* (pp. 377–394). Wallingford-Oxon: CAB Intern. doi: 10.1079/9780851994529.0377

Bilenca, D., & Miñarro, F. (2004). *Identificación de Áreas Valiosas de Pastizal (AVPs) en las Pampas y Campos de Argentina, Uruguay y sur de Brasil.* Buenos Aires: Fundación Vida Silvestre.

Bindritsch, L. (2014). An´alisis de la percepci´on de una poblaci´on rural sobre los servicios ecosist´emicos del pastizal. In: *Tesis de grado.* Facultad de Ciencias. Universidad de la República (Uruguay). https://www.colibri.udelar.edu.uy/jspui/bitstream/20.500.12008/1606/1/uy24-16985.pdf

Bloor, J.M.G., Pichon, P., Falcimagne, R., Leadley, P., & Soussana, J.F. (2010). Effects of warming, summer drought, and CO_2 enrichment on aboveground biomass production, flowering phenology, and community structure in an upland grassland ecosystem. *Ecosystems* 13, 888–900.

Boldrini, I.I. (2002). Campos sulinos: caracterizacao e biodiversidade. In E.A. Araujo, E.V.S.B. Sampaio et al. (Eds.), *Biodiversidade, conservacao e uso sustentavel da flora do Brasil* (pp. 95–97). Recife: UFPe/Soc. Bot. do Brasil.

Boval, M., & Dixon, R.M. (2012). The importance of grasslands for animal production and other functions: a review on management and methodological progress in the tropics. *Animal* 6, 748–762. doi: 10.1017/S1751731112000304.

Brandão, F.S., et al. (2012). O papel do agronegócio brasileiro na reducão de emissão de gases de efeito estufa (GEES). *Revista Agroambiente* 6(1), 84–90.

Burkart, A. (1975). Evolution of grasses and grasslands in South America. *Taxon* 24, 53–66.

Carriquiry, M., Espasandin, A., Astessiano, A., Casal, A., Claramunt, M., Do Carmo, M., & Soca, P. (2012). *La cría vacuna sobre campo nativo: Un enfoque de investigación jerárquico para mejorar su productividad y sostenibilidad,* 48 (pp. 41–48). Montevideo: Veterinaria.

Carvalho, P.C.de F. (2009). Lotac¸ão animal em pastagens naturais: políticas, pesquisas, preservac¸ão e produtividade. In: Campos Sulinos: Conservac¸ão e uso sustentável da biodiversidade. MMA, Brasília, pp. 214–228.

Carvalho, P., Nabinger, C., Lemaire, G., & Genro, T. (2011). Challenges and opportunities for livestock production in natural pastures: the case of Brazilian Pampa Biome. En: Feldman S, Oliva G, Sacido M (Eds.). *Diverse Rangelands for a Sustainable Society*, Rosario, Argentina, April 2 to April 8, 2011, IX International Rangeland Congress. 9–15 p.

Carvalho, P.C.F., & Batello, C. (2009). Access to land, livestock production and ecosystem conservation in the Brazilian Campos biome: The natural grasslands dilemma. *Livestock Science* 120, 158–162.

Carvalho, P.C.F., Paruelo, J., & Ayala, W. (2008). La intensificacion productiva en los pastizales del Rio de la Plata: tendencias y consecuencias ecosistemicas. In Anais da Reuniao Anual do Grupo Tecnico em Forrageiras do Cone Sul – Grupo Campos. Minas, Uruguai: INIA.

Carvalho, P.C.F., Soares, A.B., Garcia, E.N., et al. (2003). Herbage allowance and species diversity in native pastures. In *International Rangeland Congress, 7, Durban. Proceedings....* (pp. 858–859). Durban: Document Transformation Technology Congress.

Casal, A., Astessiano, A. Espasandin, A., Trujillo, A.I., Soca, P. and Carriquiry, M. (2016). Changes in body composition during winter gestation period in mature beef cows grazing different herbage allowance of native pastures. *Animal Production Science*. doi: 10.1071/AN14984

Cezimbra, I.M., de Albuquerque Nunes, P.A., de Souza Filho, W., Ritzel Tischler, M., Moraes Genro, T.C., Bayer, C., Savian, J.V., François Bonnet, O.J., Soussana, J.F., & de Faccio Carvalho, P.C. (2021). Potential of grazing management to improve beef cattle production and mitigate methane emissions in native grasslands of the Pampa biome. *Science of the Total Environment* 780, 146582. doi: 10.1016/j.scitotenv.2021.146582

Claramunt, M., Fernandez-Foren, A., & Soca, P. (2017). Effect of herbage allowance on productive and reproductive responses of primiparous beef cows grazing on Campos grassland. *Animal Production Science*. doi: 10.1071/AN16601

Claramunt, M.R. (2015). *Efecto de la oferta de forraje sobre parámetros productivos, reproductivos y eficiencia de uso del forraje de vacas primíparas en pastoreo de Campos de Basalto* (84 p.). Tesis MS. Montevideo: Facultad de Agronomía.

Cortelezzi, A., & Mondelli, M. (2014). *Censo General Agropecuario 2011: Interpretación de los principales resultados y cambios observados* (pp. 471–490). Montevideo: Anuario OPYPA.

Cobon, D.H., Stone, G.S., Carter, J.O., Scanlan, J.C., Toombs, N.R., Zhang, X., … McKeon, G.M. (2009). The climate change risk management matrix for the grazing industry of northern Australia. *The Rangeland Journal* 31, 31–49. doi: 10.1071/RJ08069

Coffin, D.P., Lauenroth, W.K. (1996). Transient responses of North American grasslands to changes in climate. *Climate Change* 34(2), 69–78.

Da Trindade, J.K. (2011). *Comportamento e consumo de forragem de bovinos de corte em pastagem natural complexa* (208 p.). Porto Alegre, Rio Grande do Sul, Brasil: Universidade Federal do Rio Grande do Sul. Faculdade de Agronomia, Programa de pós-graduação em zootecnia.

Da Trindade, J.K., Neves, F.P., Pinto, C.E., Bremm, C., Mezzalira, J.C., Nadin, L.B., Genro, T.C.M., Gonda, H.L., Carvalho, P.C.F. (2016). Daily forage intake by cattle on natural grassland: response to forage allowance and sward structure. *Rangeland Ecology & Management* 69, 59–67. doi: 10.1016/j.rama.2015.10.002

Da Trindade, J.K., Pinto, C.E., Neves, F.P., Mezzalira, J.C., Bremm, C., Genro, T.C., Tischler, M.R., Nabinger, C., Gonda, H.L., & Carvalho, P.C.F. (2012). Forage allowance as a target of grazing management: implications on grazing time and forage searching. *Rangeland Ecology & Management* 65(4), 382–393. doi:

10.2111/REM-D-11-00204.1De Deyn, G.B., Quirk, H., Yi, Z., Oakley, S., Ostle, N.J., & Bardgett, R.D. (2009). Vegetation composition promotes carbon and nitrogen storage in model grassland communities of contrasting soil fertility. *Journal of Ecology* 97, 864–875. doi: 10.1111/j.1365-2745.2009.01536.x

De Deyn, G.B., Shiel, R.S., Ostle, N.J., McNamara, N.P., Oakley, S., Young, I. et al. (2011). Additional carbon sequestration benefits of grassland diversity restoration. *Journal of Applied Ecology* 48, 600–608. doi: 10.1111/j.1365–2664.2010.01925.xDo Carmo, M., Cardozo, G., Jaurena, M., & Soca, P. (2019). Demonstrating control of forage allowance for beef cattle grazing Campos grassland in Uruguay to improve system productivity. *Tropical Grasslands-Forrajes Tropicales* 7(1), 35–47. Epub April 30, 2020. doi: 10.17138/tgft(7)35-47

Do Carmo, M., Claramunt, M., Carriquiry, M., & Soca, P. (2016). Animal energetics in extensive grazing systems: rationality and results of research models to improve energy efficiency of beef cow-calf grazing Campos systems. *Journal of Animal Science* 94(S6), 85–92. doi: 10.2527/jas2016–0596

Do Carmo, M., Genro, T.C.M., Cibils, A.F., & Soca, P.M. (2021). Herbage mass and allowance and animal genotype affect daily herbage intake, productivity, and efficiency of beef cows grazing native subtropical grassland, *Journal of Animal Science* 99(10), skab279. doi: 10.1093/jas/skab279

Do Carmo, M., Sollenberger, L.E., Carriquiry, M., & Soca, P. (2018). Controlling herbage allowance and selection of cow genotype improve cow-calf productivity in Campos grasslands. *Professional Animal Scientist* 34, 32–41. doi: 10.15232/ pas.2016-01600

Douxchamps, S., Van Wijk, M.T., Silvestri, S., Moussa, A.S., Quiros, C., Ndour, N.Y.B., … Rufino, M.C. (2016). Linking agricultural adaptation strategies, food security and vulnerability: evidence from West Africa. *Regional Environmental Change* 16, 1305–1317. doi: 10.1007/s10113-015-0838-6

Duru, M., Pontes, L.D.A.S., Schellberg, J., Theau, J.P., & Therond, O. (2019). Grassland functional diversity and management for enhancing ecosystem services and reducing environmental impacts: a cross-scale analysis. In: G. Lemaire, et al. (Eds.), *Agroecosystem diversity: Reconciling contemporary agriculture and environmental quality* (pp. 211–230). Cambridge, MA: Academic Press. doi: 10.1016/B978-0-12–811050-8.00013-3

FAO. (2018). *World Livestock: Transforming the livestock sector through the Sustainable Development Goals*. Rome. 222 pp. http://www.fao.org/3/CA1177EN/ca1177en.pdf

Feng, Y.; Su, H.; Tang, Z. et al. (2021). Reduced resilience of terrestrial ecosystems locally is not reflected on a global scale. *Communications Earth & Environment* 2, 88. doi: 10.1038/s43247-021-00163-1

Follett, R.F., Kimble, J.M., Lal, R. (Eds.) (2001). *The potential of US grazing lands to sequester carbon and mitigate the greenhouse effect*. Boca Raton, FL: CRC Press LLC.

Garcia Prechac, F., Clerici, C., Hill, M., & Brignoni, A. (2005). EROSION Version 5.0, Programa de computacion Para el uso de la USLE/RUSLE en la Region Sur de la Cuenca del Plata. Version operativa en Windows DINAMA-UNDP, Proyecto URU/03/G31 y CSICUDELAR. Montevideo: Facultad de Agronomía, UDELAR. Retrieved from www.fagro.edu. uy/manejo/.

Gibson, D., & Newman, J. (2019). Grasslands and climate change: An overview. In D. Gibson & J. Newman (Eds.), *Grasslands and climate change* (Ecological Reviews, pp. 3–18). Cambridge: Cambridge University Press. doi: 10.1017/9781108163941.003

Gibson, D.J. (2009). *Grasses and grassland ecology*. New York: Oxford.

Godfray, H.C., Beddington, J.R., Crute, I.R., Haddad, L., Lawrence, D., Muir, J.F., Pretty, J., Robinson, S., Thomas, S.M., Toulmin, C. (2010). Food security: the challenge of feeding 9 billion people. *Science* 327(5967), 812–818. doi: 10.1126/science.1185383. Epub 2010 Jan 28. PMID: 20110467.

Gorosábel, A., Estigarribia, L., Lopes, L.F., Martinez, A.M, Martínez-Lanfranco, J.A, Adenle, A.A., Rivera-Rebella, C., Oyinlola, M.A. (2020). Insights for policy-based conservation strategies for the Rio de la Plata Grasslands through the IPBES framework. *Biota Neotropica* 20(suppl. 1), e20190902. doi: 10.1590/1676-0611-BN-2019-0902

Grimm, V., & Wissel, C. (1997). Babel, or the ecological stability discussions: an inventory and analysis of terminology and a guide for avoiding confusion. *Oecologia* 109, 323–334. doi: 10.1007/s004420050090

Guterres, D.B., Bayer, C., Castilhos, Z.M.de S., Nabinger, C. (2006). Carbono organico em Chernossolo sob pastagem nativa no RS. In *Reuniao Anual da Sociedade Brasileira de Ciencia do solo,* 16, Sao Cristovao, SE. Anais.... Sao Cristovao: UFSE/SBSC. CD-ROM.

Gutierrez, V., Espasandin, A.C., Astessiano, A.L., Casal, A., López-Mazz, C., Carriquiry, M. (2013). Calf fetal and early life nutrition on grazing conditions: metabolic and endocrine profiles and body composition during the growing phase. *Journal of Animal Physiology and Animal Nutrition* 97, 720–731. doi: 10.1111/j.1439-0396.2012.01314.x

Habel, J.C., Dengler, J., Jani_sov_a, M., T€or€ok, P., Wellstein, C. and Wiezik, M. (2013). European grassland ecosystems: threatened hotspots of biodiversity. *Biodiversity and Conservation* 22, 2131–2138.

Hall, A.J., Rebella, C., Ghersa, C., & Culot, J. (1991). Field crop systems of the Pampas. In: C.J. Pearson (Ed.), *Field crop ecosystems, ecosystems of the world 19* (pp. 413–450). Amsterdam: Elsevier.

Hewins, D.B., Lyseng, M.P., Schoderbek, D.F., Alexander, M., Willms, W.D., Carlyle, C.N. et al. (2018). Grazing and climate effects on soil organic carbon concentration and particle-size association in northern grasslands. *Scientific Reports* 8. doi: 10.1038/s41598-018-19785-1

Hodgson, J., Da Silva, S.C. (2000). Sustainability of grazing systems: goals, concepts and methods. In G. Lemaire, J. Hodgson, A. Moraes, P.C.F. Carvalho, & C. Nabinger (Eds.), *Grassland ecophysiology and grazing ecology* (pp. 1–14). Wallingford, Oxon: CABI Publishing, CAB International.

Holmes, C.W., Wilson, G.F., Mackenzie, D.D.S., Flux, D.S., Brookes, I.M., & Davey, A.W.F. (1984). *Milk Production from Pasture* (p. 319). Wellington: Butterworths Agricultural Books.

Hossain, M.L., Li, J., Hoffmann, S., & Beierkuhnlein, C. (2022). Biodiversity showed positive effects on resistance but mixed effects on resilience to climatic extremes in a long-term grassland experiment. *Science of The Total Environment* 827, 154322 doi: 10.1016/j.scitotenv.2022.154322

INDEC. (2018). *Censo Nacional Agropecuario.* Available online at: https://cna2018.indec.gob.ar/

IPCC. (2013). *Fifth Assessment Report – Climate Change 2013* [WWW Document]. Retrieved from http://www.ipcc.ch/repor t/ar5/wg1/

Isbell, F., Craven, D., Connolly, J., Loreau, M., Schmid, B., Beierkuhnlein, C. et al. (2015). Biodiversity increases the resistance of ecosystem productivity to climate extremes. *Nature* 527, 315–316. doi: 10.1038/nature15374

Jaurena, M., Durante, M., Devincenzi, T., Savian, J., Bendersky, D., Moojen, F.G. et al. (2021). Native grasslands at the core: a new paradigm of intensification for the campos of Southern South America to increase economic and environmental sustainability. *Frontiers in Sustainable Food Systems* 5, 547834. doi: 10.3389/fsufs.2021.547834

Kuinchtner, B.C., Ferreira De Quadros, F.L., Trindade Casanova, P., Marin, L., Seeger, B.B., de Moura Steinhorst, D., & Ongaratto, F. (2021). Thermal sum (degrees-days) rest period as a grazing management tool in natural grasslands: effects on animal performance and forage production. *Environmental and Sustainability Indicators* 10, 100125. doi: 10.1016/j.indic.2021.100125

Kuhsel, S., & Bluthgen, N. (2015). High diversity stabilizes the thermal resilience of pollinator communities in intensively managed grasslands. *Nature Communications* 6, 7989. doi: 10.1038/ncomms8989

Llanos, E., Astigarraga, L., Jacques, R., Picasso, V. (2013). Eficiencia energética en sistemas lecheros del Uruguay. *Agrociencia Uruguay* 17, 99–109. http://refhub.elsevier.com/S1161-0301(17)30164-8/sbref0120

López-Ridaura, S., van Keulen, H., van Ittersum, M.K., & Leffelaar, P.A. (2005). Multiscale sustainability evaluation of natural resource management systems: quantifying indicators for different scales of analysis and their trade-offs using linear programming. *International Journal of Sustainable Development & World Ecology* 12, 81–97. doi: 10.1080/13504 50050 9469621

López-Ridaura, S., Keulen, H.V., Ittersum, M.K.V., & Leffelaar, P.A. (2005a). Multiscale methodological framework to derive criteria and indicators for sustainability evaluation of peasant natural resource management systems. *Environment, Development and Sustainability* 7, 51–69. doi: 10.1007/s10668-003-6976-x

Maraschin, G.E., Moojen, E.L., Escoteguy, C.M.D, Correa, L., Apezteguia, E.S., & Boldrini, II. (1997). Native pasture, forage on offer and animal response. En J.G. BuchananSmith, L.D. Bailey, P. McCaughey (Ed.), *Proceeding of the XVIII International Grassland Congress*, Winnipeg and Saskatoon, Canada. Calgary, Canada. 8–19 June 1997 (pp. 27–29).

MGAP. (2013). *Censo General Agropecuario 2011.* Montevideo, Uruguay. https://www.gub.uy/ministerio-ganaderia-agricultura-pesca/datos-yestadisticas/estadisticas/censo-general-agropecuario-2011

Modernel, P., Astigarraga, L., & Picasso, V. (2013). Global versus local environmental impacts of grazing and confined beef production systems. *Environmental Research Letters* 8, 035052. doi: 10.1088/1748-9326/8/3/035052

Modernel, P., Picasso V., Do Carmo M., Rossing, W.A.H, Corbeels, M., Soca, P., Dogliotti, S., & Tittonell, P. (2019). Grazing management for more resilient mixed livestock farming systems on native grasslands of southern South America. *Grass Forage Science* 74, 636–649. doi: 10.1111/gfs.12445

Modernel, P., Rossing, W.A.H., Corbeels, M., Dogliotti, S., Picasso, V., & Tittonell, P. (2016). Land use change and ecosystem service provision in Pampas and Campos grasslands of southern South America. *Environmental Research Letters* 11, 113002. doi: 10.1088/1748-9326/11/11/113002Moojen, E.L., & Maraschin, G.E. (2002). Potencial produtivo de uma pastagem nativa do rio grande do sul submetida a níveis de oferta de forragem. *Ciência Rural* 32, 127–132.

Montgomery, D.C. (2008). *Design and Analysis of Experiments.* Wiley. https://www.bookfinder.com/isbn/9780470128664/

Nabinger, C., de Faccio Carvalho, P.C., Cassiano Pinto, E., Mezzalira, J.C., Martins Brambilla, D., & Boggiano, P. (2011). Servicios ecosistémicos de las praderas naturales: ¿es posible mejorarlos con más productividad? *Asociación Latinoamericana de Producción Animal* 19(3–4), 27–34.

Nabinger, C., & Carvalho, P.C.F. (2009). *Ecofisiología de Sistemas Pastoriles: Aplicaciones para su Sustentabilidad* (pp. 18–27). Uruguay. *Agrociencia* XIII.

Nabinger, C., de Moraes, A., & Maraschin, G.E. (2000). Campos in southern Brazil. En G. Lemaire, J. Hodgson, A. de Moraes, C. Nabinger, & P.C.F. Carvalho PCF. (Eds.), *Grassland ecophisiology and grazing ecology* (pp. 355–376). Cambridge: University Press.

Neves, F.P., Carvalho, P.C.F., Nabinger, C., Jacques, A.V.A., Carassai, I.J., & Tentardini, F. (2009). Estrat´egias de manejo da oferta de forragem para recria de novilhas em pastagem natural. *Revista Brasileira de Zootecnia* 38(8), 1532–1542. doi: 10.1590/S1516-35982009000800018

Norton, M.R., Malinowski, D.P., & Volaire, F. (2016). Plant drought survival under climate change and strategies to improve perennial grasses. A review. *Agronomy for Sustainable Development* 36, 29. doi: 10.1007/s13593-016-0362-1

OECD. (2021). *Making Better Policies for Food Systems*. Paris: OECD Publishing.

O'Mara, F.P. (2012). Review: part of a highlight on breeding strategies for forage and grass improvement. The role of grasslands in food security and climate change. *Annals of Botany* 110, 1263–1270. doi: 10.1093/aob/mcs209, available online at www.aob.oxfordjournals.org.

Oldeman, L.R. (1994). The global extent of soil degradation. In D.J. Greenland, & I. Szabolcs (Eds.), *Soil resilience and sustainable land use* (pp. 99–118). Wallingford: CAB International.

Orcasberro, M.S., Loza, C., Gere, J., Soca, P., Picasso, V., & Astigarraga, L. (2021). Seasonal effect on feed intake and methane emissions of cow-calf systems on native grassland with variable herbage allowance. *Animals* 11, 882. doi: 10.3390/ani11030882

Overbeck, G.E., et al. (2009). Os Campos Sulinos: um bioma negligenciado. In Campos Sulinos – conservac¸ ão e uso sustentável da biodiversidade (pp. 24–41). Ministério do Meio Ambiente, Brasil, Brasília.

Overbeck, G.E., Mueller, S.C., Fidelis, A., Pfadenhauer, J., Pillar, V.D.D., Blanco, C.C., … Froneck, E.D. (2007). Brazil's neglected biome: the South Brazilian Campos. *Perspectives in Plant Ecology, Evolution and Systematics* 9, 101–116. doi: 10.1016/j.ppees.2007.07.005

Paruelo, J., Jobbágy, E. et al. (2007). *The Grasslands and Steppes of Patagonia and the Río de la Plata Plains. In The Physical Geography of South America.* Oxford University Press. https://oxford.universitypressscholarship.com/view/10.1093/oso/9780195313413.001.0001/isbn-9780195313413-book-part-22

Peyraud, J.L. (2017). *The role of grassland based production system for sustainable protein production.* 54 Annual meeting of the Brazilian society of animal science, Jul 2017, Foz do Iguaçu, Brazil.

Peyraud, J.L., & Peeters, A. (2016). *The role of grassland based production system in the protein security.* 26. General meeting of the European Grassland Federation (EGF), Sep 2016, Trondheim. hal-02743435.

Picasso, V.D., Casler, M.D., & Undersander, D. (2019). Resilience, stability, and productivity of Alfalfa cultivars in rainfed regions of North America. *Crop Science*. doi: 10.2135/crops ci2018.06.0372

Picasso, V.D., Modernel, P., Becoña, G., Salvo, L., Gutiérrez, L., & Astigarraga, L. (2014). Sustainability of meat production beyond carbon footprint: a synthesis of case studies from grazing systems in Uruguay. *Meat Science* 98, 346–354.

Piñeiro, G., Paruelo, J.M., & Oesterheld, M. (2006). Potential long-term impacts of livestock introduction on carbon and nitrogen cycling in grasslands of Southern South America. *Global Change Biology* 12, 1267–1284. doi: 10.1111/j.1365-2486.2006.01173.x

Pörtner, H.O., Scholes, R.J., Agard, J., Archer, E., Arneth, A., Bai, X., Barnes, D., Burrows, M., Chan, L., Cheung, W.L., Diamond, S., Donatti, C., Duarte, C., Eisenhauer, N.,

Foden, W., Gasalla, M.A., Handa, C., Hickler, T., Hoegh-Guldberg, O., Ichii, K., Jacob, U., Insarov, G., Kiessling, W., Leadley, P., Leemans, R., Levin, L., Lim, M., Maharaj, S., Managi, S., Marquet, P.A., McElwee, P., Midgley, G., Oberdorff, T., Obura, D., Osman, E., Pandit, R., Pascual, U., Pires, A.P.F., Popp, A., Reyes-García, V., Sankaran, M., Settele, J., Shin, Y.J., Sintayehu, D.W., Smith, P., Steiner, N., Strassburg, B., Sukumar, R., Trisos, C., Val, A.L., Wu, J., Aldrian, E., Parmesan, C., Pichs-Madruga, R., Roberts, D.C., Rogers, A.D., Díaz, S., Fischer, M., Hashimoto, S., Lavorel, S., Wu, N., & Ngo, H.T. (2021). *Scientific outcome of the IPBES-IPCC co-sponsored workshop on biodiversity and climate change.* Bonn: IPBES secretariat. doi: 10.5281/zenodo.4659158

Royo Pallares, O., Berretta, E.J., & Maraschin, G.E. (2005). The South American Campos ecosystem. In J. Suttie, S.G. Reynolds, & C. Batello (Eds.), *Grasslands of the World* (pp. 171–219). Roma: FAO.

Ruggia, A., Dogliotti, S., Aguerre, V., Albicette, M.M., Albin, A., Blumetto, O., Cardozo, G., Leoni, C., Quintans, G., Scarlato, S., Tittonell, P., & Rossing, W.A.H. (2021). The application of ecologically intensive principles to the systemic redesign of livestock farms on native grasslands: a case of co-innovation in Rocha, Uruguay. *Agricultural Systems* 191, 103148. doi: 10.1016/j.agsy.2021.103148

Ruviaro, C.F., da Costa, J.S., Florindo, T.J., Rodrigues, W., de Medeiros, G.I.B., & Vasconcelos, P.S. (2016). Economic and environmental feasibility of beef production in different feed management systems in the Pampa biome, southern Brazil. *Ecological Indicators* 60, 930–939. doi: 10.1016/J.ECOLIND.2015.08.042

Ruviaro, C.F., de Léis, C.M., Lampert, V.D.N., Barcellos, J.O.J., & Dewes, H. (2015). Carbon footprint in different beef production systems on a southern Brazilian farm: a case study. *Journal of Cleaner Production* 96, 435–443. doi: 10.1016/j.jclepro.2014.01.037.

Sala, O.E., & Paruelo, J.M. (1997). Ecosystem services in grasslands. In G.C. Daily (Ed.), *Nature's services: societal dependence on natural ecosystems* (pp. 237–251). Washington, DC: Island Press.

Scarlato, S., Albicette, M.M., Bortagaray, I., Ruggia, A., Scarlato, M., & Aguerre, V. (2015). *Co-innovation as an effective approach to promote changes in farm management in livestock systems in Uruguay.* 5th International Symposium for Farming Systems Design 7–10 September 2015, Montpellier, France.

Skinner, R.H., & Dell, C.J. (2016). Yield and soil carbon sequestration in grazed pastures sown with two or five forage species. *Crop Science* 56, 2035–2044. doi: 10.2135/cropsci2015.11.0711

Soares, A., Carvalho, P.C.F., Nabinger, C., Semmelmann, C., da Trindade, J.K., Guerra, E., Stella de Freitas, T., Pinto, C.E., Fontoura Júnior, J.A., & Frizzo, A. (2005). Animal and forage production on native pasture under different herbage allowance. *Ciência Rural* 35(5), 1148–1154.

Soares, A.B.P., Carvalho, C.F., Nabinger, C., Frizzo, A., Pinto, C.E, Junior, J.A.F., Semmelmann, C., & da Trindade, J. (2003). Effect of changing herbage allowance on primary and secondary production of natural pasture. En N. Allsopp, A.R. Palmer, S.J. Milton, G.I.H. Kerley, K.P. Kirkman, R. Hurt, & C.J. Brown (Eds.), *Proceedings of the 7th International Rangeland Congress*; Durban, South Africa. 26th July–1st August 2003; 966–968.

Soca, P. (2013). *Aportes del campo natural a las mejoras del resultado y sostenibilidad de la cría vacuna en Uruguay. Producción animal sostenible en pastoreo sobre campo natural* (p. 74). Ministerio de Ganadería Agricultura y Pesca.

Sollenberger, L.E., Kohmann, M.M., Dubeux, Jr., J.C.B., & Silveira, M.L. (2019). Grassland management affects delivery of regulating and supporting ecosystem services. *Crop Science* 59, 441–459. doi: 10.2135/cropsci2018.09.0594

Sollenberger, L.E., Moore, J.E., Allen, V.G., & Pedreira, C.G.S. (2005). Reporting forage allowance in grazing experiments. *Crop Science* 48, 896–900. doi: 10.2135/cropsci2004.0216

Song, X.P., Hansen, M.C., Potapov, P., et al. (2021). Massive soybean expansion in South America since 2000 and implications for conservation. *Nature Sustainability* 4, 784–792. doi: 10.1038/s41893-021-00729-z

Soriano A. (1991). Río de la Plata Grasslands. En R.T. Coupland (Ed.), *Natural grasslands. Introduction and western hemisphere* (pp. 367–407). Amsterdam: Elsevier.

Soussana, J.F. (2009). ¿Os desafios da ciência das pastagens européias são relevantes para os Campos Sulinos? In *Campos Sulinos: Conservaç ão e uso sustentável da biodiversidade* (pp. 331–344). Brasília: MMA.

Thompson, I., Mackey, B., McNulty, S., & Mosseler, A. (2009). Forest resilience, biodiversity, and climate change. In *Secretariat of the convention on biological diversity*, Montreal. Technical Series 43, 1–67.

Thurow, T.L., & Taylor, C.A. (1999). Viewpoint: the role of drought in range management. *Journal of Range Management* 52, 413–419. doi: 10.2307/4003766

Tiscornia, G., Jaurena, M., & Baethgen, W. (2019). Drivers, Process, and Consequences of Native Grassland Degradation: insights from a Literature Review and a Survey in Río de la Plata Grasslands. *Agronomy 9*, 239. doi: 10.3390/agronomy9050239

Tittonell, P. (2021). Beyond CO_2: multiple ecosystem services from ecologically intensive grazing landscapes of South America. *Frontiers in Sustainable Food Systems* 5, 664103. doi: 10.3389/fsufs.2021.664103

Tommasino, H., Cortelezzi, A., Mondelli, M., Bervejillo, J., & Carrazzone, S. (2014). Tipología de productores agropecuarios: caracterizaci´on a partir del Censo Agropecuario 2011. In *Anuario 2014* (pp. 491–508). Uruguay: OPYPA.

Trindade, J.P.P., Rocha, D.S., & Volk, L.B.S. (2018). *Uso da terra no Rio Grande do Sul: ano 2017.* Bagé: Embrapa Pecuária Sul.

Van Ruijven, J., & Berendse, F. (2010). Diversity enhances community recovery, but not resistance, after drought. *Journal of Ecology* 98, 81–86. doi: 10.1111/j.1365-2745.2009.01603.x

Van Zanten, H.H.E., Herrero, M., Hal, O.V., et al. (2018). Defining a land boundary for sustainable livestock consumption. *Global Change Biology* 24, 4185–4194. doi: 10.1111/gcb.14321

Viglizzo, E.F., & Frank, F.C. (2006). Land-use options for Del Plata Basin in South America: tradeoffs analysis based on ecosystem service provision. *Ecological Economics* 57, 140–151. doi: 10.1016/j.ecolecon.2005.03.025

Viglizzo, E.F., L´ertora, F., Pordomingo, A.J., Bernardos, J.N., Roberto, Z.E., & Del Valle, H. (2001). Ecological lessons and applications from one century of low external-input farming in the pampas of Argentina. *Agriculture, Ecosystems & Environment* 83, 65–81. doi: 10.1016/S0167-8809(00)00155-9

Wick, A.F., Geaumont, B.A., Sedivec, K., & Hendrickson, J. (2016). Grassland degradation. In J.F. Shroder, & R. Sivanpillai (Eds.), *Biological and environmental hazards, risks and disasters* (Vol. 8, pp. 257–276). New York: Elsevier.

Wilson, J.B., Peet, R.K., Dengler, J., & Pärtel, M. (2012). Plant species richness: The world records. *Journal of Vegetation Science* 23, 796–802.

8

HOW DOES GARDENING REDUCE VULNERABILITY FOR THE URBAN POOR IN SMALL ISLAND DEVELOPING STATES? A CASE STUDY OF PORT VILA, VANUATU

Andrew MacKenzie

Introduction

The challenges faced by Pacific Small Island Developing States (SIDS) in the planning for and managing of competing uses on peri-urban landscapes are compounded by poor institutional and regulatory regimes. SIDS refers to Pacific, Caribbean, and Indian Ocean nations and territories that share similar features such as small populations, geographic remoteness, and high transport and transaction costs. SIDS are vulnerable to global economic shocks, biodiversity loss, and climate change impacts because communities depend on natural resources and often lack economic alternatives to exploiting fragile marine and terrestrial resources (UN, 2020).

The landscapes on the edge of cities in the Pacific face pressures to accommodate urban expansion like many cities around the world. They also face specific pressures particular to SIDS. In the absence of robust planning and land management regimes, residents that depend on these landscapes for their livelihoods face ongoing uncertainty and lack of investment. These pressures impact on the resilience of these landscapes in the face of increasing frequency and intensity of disasters. In this context, this chapter explores how peri-urban landscapes face ongoing pressures to accommodate changing demands from rapidly growing communities.

On the rural-urban fringe of Pacific cities, residents who benefit from the infrastructure and housing that create urban sprawl also benefit from the productive agricultural landscapes as sources of food and income. These peri-urban landscapes have become sites of contest between the productive uses of the land for growing food and the provision of essential services and infrastructure. These communities live a hybrid existence; adapting to urban life and the cash economy while continuing traditional gardening practices in and around informal

DOI: 10.4324/9781003266440-8

settlements on the urban fringe. During times of hardship such as economic shocks or disaster, this contest for landscape resources is exacerbated as Pacific Island people turn to gardening to supplement income or in some cases, stave of extreme poverty and starvation. This has been the case in Port Vila since the beginning of the COVID-19 pandemic.

This chapter explores how gardening is an adaptive action that supports vulnerable communities during hardships caused by disasters. It identifies some of the regulatory, political, and cultural barriers to effective management of peri-urban landscapes in the Pacific. Gardening has shaped the landscapes of Pacific Islands for centuries and is central to the identity of many Pacific Island people. Urban communities in the Pacific supplement cash incomes by gardening to provide additional food and to reinforce cultural and social ties (Lindstrom, 2012). It is one of the key activities that peri-urban residents practice if they have access to land and sufficient resources and time over and above formal employment (James, 2018; Lindstrom, 2012). Gardening helps to reduce the vulnerability of the urban poor who live on the peri-urban fringe; however contemporary pressures from climate change population growth, and the proliferation of informal settlements present new challenges for communities and governments to preserve and protect these productive landscapes for gardening.

Adaptive landscapes on the rural-urban fringe

Exploring how communities on the rural-urban fringe access and shape the landscape through gardening provides us with a better understanding of the capacity of vulnerable communities to adapt to more intense and more frequent disasters. In this context, it is important to understand how opportunistic gardening on the rural-urban fringe supports these communities during disaster. Opportunistic gardening was an adaptive response to the hardship caused by the loss of alternative forms of income as the tourism sector shut down following the closure of the international borders in March 2020.

Research undertaken in 2021 analyzed the land use changes of selected study areas in Port Vila following the March 2020 closure of international borders in response to the COVID-19 pandemic. COVID-19 was a highly consequential disaster impacting Vanuatu's economy; however, it was not an isolated event. The pandemic was a part of a continuum of disasters that have affected the capital between 2014 and 2021; including cyclones, drought, volcanic ash fall, floods, earthquakes, and tsunami. The research investigated whether the observed changes to landscape patterns were temporary and reflected short-term adaptive responses of gardeners to hardship caused by the pandemic or the changing landscape pattern reflected more permanent adaptive actions of communities in response to compounding disasters.

Port Vila's experience is comparable to the increase in gardening activity detected in other developing nations following a sustained economic shock.

Following the collapse of the Soviet Union in 1989, Havana, the capital of Cuba, experienced economic shock as exports to the former Soviet Union collapsed (Saunders, 1999). The example of Havana is instructive for Vanuatu from a policy perspective. What started as a community-level urban agricultural movement to adapt to hardship became a municipal, and then national-level goal to maximize the amount of urban agricultural production (Koont, 2009). The shift toward urban agriculture in Cuba was so profound that the government encouraged and supported a wide range of gardening and animal husbandry activities and enterprises across the entire city, not just on the urban fringes (Premat, 2009). While lessons can be learned from the response of the municipal government in Havana, it should be noted that the factors that caused the hardships in Cuba were permanent and in response to a global geopolitical reordering that triggered a permanent change to Cuban economic policies. The challenge for policy makers in Vanuatu is that the hardship caused by the pandemic is viewed as temporary. Once the borders open and tourism returns; so too will the jobs, the incomes, and the livelihoods that depend on paid work, however tenuous these livelihoods may be.

From a land use policy perspective, two problems face these communities. The first is that there is no way of knowing at this stage how long will these communities have to depend on gardening, and second, in the absence of robust planning policies, there is no certainty that these landscapes be available and suitable for gardening in the event of another future disaster that will require Ni Vanuatu (the vernacular name for Indigenous citizens of Vanuatu) to turn to gardening to reduce their vulnerability to disaster. Further research is needed to examine how institutions can support a more sustainable, flexible and resilient peri-urban landscapes.

The reasons for gardening receiving little attention from an environmental and land use planning perspective may be because, unlike agriculture or forestry, urban gardening is more diffuse and opportunistic, temporary, and integrated into the expanding urban footprint. There does not appear to be a robust model for evaluating the economic value of urban gardening similar to agriculture or forestry, yet gardening provides meaningful alternatives to formal employment in times of hardship. Gardening generates essential fresh produce and makes it available to poor urban resident's thorough community networks that exchange goods using barter along with cash. It is an essential adaptive action during disaster for vulnerable urban communities. The diffuse, fragmented nature of gardening also reveals many of the difficulties for disaster researchers to evaluate the social and economic value of gardening. Equally difficult for land use planners is the lack of policies to protect landscapes from urban sprawl. The experiences of gardeners in Port Vila during the pandemic reveal two underlying barriers to a more sustainable urban agricultural sector. The first limiting factor is the government's ambivalence toward urban planning policy. This allows for a more dispersed power relationship between those with control of over urban landscapes through tenure and their exercising of property rights in the absence of legal enforcement (McDonnell, 2018). As disasters become more frequent and more intense, the dynamic pressures on the landscape from rural-urban migration are further compounded by the increasing vulnerability of urban edge communities from the effects of climate change and their ability to recovery from disaster. The second limiting factor is the opportunistic nature of gardening as an effective adaptive action during hardship. Custom landowners and lessees are not incentivized to invest in infrastructure or services to support and promote gardening once the pandemic has passed and the cash-based jobs return.

The interaction between urban sprawl and gardening will come into further conflict as land suitable for gardening becomes scarcer as the city expands over productive agricultural landscapes, and sea level rise renders coastal land unsuitable. The increased flooding due to rising sea levels and more intense rainfall events in Port Vila will impact cash crops such as yams and sugar cane that are grown on the flood plains (ADB, 2019). Likewise, the high levels of migration will place further pressure on the land available for gardening within the urban footprint. In some regards, gardening is an excellent adaptive response to landscapes that are not suitable for more permanent settlement. Despite this, in times of hardship induced by disasters, these communities will become more dependent on outside sources of food as land for gardening becomes scarce.

Regulating land use on the urban fringe

An issue raised in this chapter is the reduction of available productive land for gardening for peri-urban communities during disaster. The social and economic implication of this loss of land is cause for concern for regulators in Vanuatu. Land regulation, formal and customary, has been a volatile and politically sensitive topic since the early missionary settlements (McDonnell, 2018). Urban and peri-urban land today is not subject to the same level of contest that is experienced in the rural areas; however, conflicts over land ownership are common in the villages around Port Vila. The Vanuatu constitution states that all land belongs to Ni Vanuatu. No foreigner or non-citizen can own land. Instead, the land, while remaining under ownership of the custom land owner (Custom Land Management Act 2014), can be leased under certain criteria. In effect, the leasing of land affords the holder of the lease exclusive rights for the duration of the lease. Under the Land Leases Act (2006), residential land is leased for up to 70 years and commercial and rural properties are leased for up to 40 years; guaranteeing exclusive use to the lessee. While this is the case in Vanuatu, the custom landowner will often negotiate ongoing access to land, particularly in peri-urban and rural areas to allow villagers to continue to harvest land and marine resources. During the pandemic, access to land for gardening was at the behest of lessees or custom land owners.

The development of regulation since independence has aimed to reduce land speculation that can often exclude Indigenous communities from enjoying economic benefits from their land. However, this is not the case in practice. Prior to the 2014 Land Reform Act, many of these agreements were verbal and with the blessing of the Chief who made the original agreement (McDonnell, 2018). The period following independence to the early 2000s saw an effective land grab of peri-urban land by mainly expatriate investors who took advantage of the weak planning regulation and non-binding verbal agreements (Lindstrom, 2012). While the reforms implemented following the 2006 Land Reform Commission has significantly limited the exploitation of lease agreements; residual conflicts over custom land ownership and lease agreements exist. Since 2014, the passing of legislation to reduce the incidence of opaque and exploitative leases has resulted in more orderly lease agreements in otherwise undeveloped land. The Land Reform Act (2006) sets out rules for a range of matters pertaining to land. These include negotiations and agreements relating to custom land (Part 4), management of land (Part 5), registered leases (Part 7), and rights of entry (Part 8). The legislation, along with the Custom Land Management Act (2014), has created checks and balances that have allowed for more orderly land releases to occur on otherwise uncontested land. Where land ownership disputes remain, very few leases have been issued since 2014; effectively slowing the spread of the urban edge in some peri-urban areas while compounding the expansion of informal settlements on land not subject to the new legislation. The impact of unregulated expansion of informal settlements will have a greater impact on the productive peri-urban landscapes. In subdivisions with comparatively large lots, residents are able to use their own land for gardening rather than depending on the adjacent unleased land. This is counterintuitive as large lot rural residential style development is generally considered to be damaging (Gude, Hansen, Rasker, & Maxwell, 2006). In the case of Port Vila, this type of large lot development may be suitable for preserving land for gardening.

Conclusion

This chapter argues that land use regulation plays an important role in enhancing or inhibiting individuals' access to peri-urban landscapes. Socio-physical factors are major determinants of the resilience of peri-urban landscapes; however, the role of formal and informal modes of regulation should be taken into consideration. This chapter highlights the need to explore how institutions interact with individuals in accessing landscapes in the face of drivers such as urbanization and socio-economic changes driven by disaster. Planners and policy makers should consider how institutional factors affect landscape resilience, particularly for vulnerable communities. In the case of Port Vila, this isn't simply a case of more enforcement of existing rules, but rather a more hybrid approach to how formal and informal regulations should operate within a strategic urban planning framework. This is most evident when we consider how gardeners mobilized landscape resources during the COVID-19 pandemic.

The primary instruments for regulating land use in the past have focused on defining ownership rights rather than embedding lease holder responsibilities set out in formal planning tools such as development control plans and strategic urban planning. Recent planning projects such as the Port Vila Urban Greening Masterplan highlight the potential for Port Vila to structure urban planning around a green infrastructure framework. However significant reforms are needed to give this plan the regulatory authority needed to implement the necessary changes.

In between disasters, the co-dependence of Port Vila's economy and the workers living on the rural-urban fringe disguises the vulnerability of these communities to economic shocks (unemployment, loss of resources) caused by disaster. We can gain a better appreciation of how gardening fits into a broader understanding of the drivers of changes brought about by disaster. Gardening as an adaptive action is well recognized as an important cultural and economic activity in Vanuatu (Lebot & Simeoni, 2015); however, its value is rarely incorporated into land use planning.

On the rural-urban fringe, those that benefit from the infrastructure and housing that create urban sprawl also benefit from the productive agricultural landscapes as sources of food and income (Magee, Verdon-Kidd, Kiem, & Royle, 2016). From an urban policy perspective, limiting urban expansion or displacing productive landscapes in favor of growth can be potentially both beneficial and detrimental to those affected. The diffuse nature of the peri-urban interface demonstrates that there is a degree of co-existence between the gardening communities and the informal settlements, particularly in the urban edge. Gardening has to some extent buffered these communities (and the Port Vila economy) in times of hardship. However, without adequate attention being paid to the combined stressors of population increase (and the associated infrastructure development that goes with it) and the reduction of suitable land for gardening, the resilience of these landscapes will be at risk of in the aftermath of a future disaster.

Acknowledgment

This chapter reports on research from a larger research project supported by the University of the South Pacific. No funding was provided by a third party.

References

Gude, P. H., Hansen, A. J., Rasker, R., & Maxwell, B. (2006). Rates and drivers of rural residential development in the Greater Yellowstone. *Landscape and Urban Planning, 77*(1–2), 131–151.

James, S. (2018). *Food security in Port Vila, Vanuatu.* Port Vila, Vanuatu: World Vision Vanuatu.

Koont, S. (2009). The urban agriculture of Havana. *Monthly Review, 60*(8), 44–63.

Lebot, V., & Simeoni, P. (2015). Community food security: Resilience and vulnerability in Vanuatu. *Human Ecology, 43*(6), 827–842.

Lindstrom, L. (2012). Vanuatu migrant lives in village and town. *Ethnology: An International Journal of Cultural and Social Anthropology, 50*(1), 1–15.

Magee, A. D., Verdon-Kidd, D., Kiem, A. S., & Royle, S. A. (2016). Tropical cyclone perceptions, impacts and adaptation in the southwest pacific: An urban perspective from fiji, vanuatu and tonga. *Natural Hazards and Earth System Sciences, 16*(5), 1091–1105. http://doi.org.virtual.anu.edu.au/10.5194/nhess-16-1091-2016

McDonnell, S. (2018). Selling "sites of desire": Paradise in reality television, tourism, and real estate promotion in Vanuatu. *Contemporary Pacific, 30*(2), 413–436.

Premat, A. (2009). State power, private plots and the greening of Havana's urban agriculture movement. *City & Society, 21*(1), 28–57. http://doi.org/10.1111/j.1548-744X.2009.01014.x

Saunders, J. (1999, Jun 1999). Seizing the reins [Green city projects in Tblisi, Russia, Luton, UK, Havana, Cuba, and Berkeley, US]. *New Internationalist*, 28–29.

UN. (2020). *World social report 2020: Inequality in a rapidly changing world*. New York: UN.

9

THE CASE OF THE KHAYELITSHA WETLANDS PARK, SOUTH AFRICA

Securing biodiversity and social benefits from urban greenspace

Fezile Mathenjwa, Pippin Anderson and Patrick O'Farrell

Vision

This chapter focuses on the Khayelitsha Wetlands Park, an urban park located in one of South Africa's historical apartheid townships situated on the periphery of the City of Cape Town (Turok, 2001; Standing, 2003; Ernstson et al., 2010; Malan et al., 2015). A study of the contribution of this wetland park to the broader social and ecological landscape demonstrates a case where an innovative intervention in a poor urban neighborhood could contribute to a more resilient future. Meerow et al. (2016) provide a useful reflection on the concept of resilience in an urban landscape which, we draw on here, refers to the ability of an urban system, in all its complexity and across temporal and spatial scales, to maintain or rapidly return to the desired function, to adapt, to change or to transform, all relevant factors speak to its adaptive capacity. This speaks to the macro-scale vision for this site, to be one that engages the complexities of its social and biological context and contributes to a resilient future in the face of change.

The case study sits in a landscape fraught with social and ecological stresses and presents current and potential solutions that can be learnt from and built upon. Themes of ecological functioning that speak to both mitigation in response to global change and adaptation to the impacts of global change (Betsill & Bulkley, 2007; Brown et al., 2015), as well as themes of social redress and access to greenspace (Anderson et al., 2020), also underpin this case study. Elmqvist et al. (2018) present the notion of 'urban tinkering', a proposed mode of operation relating to all aspects of urban systems ranging from policy to design and management that allows a degree of flexibility. The idea is that the complexity of urban systems, combined with the high degree of unpredictability around the manifestation of the impacts of global change, means we need systems that foster resilience in the face

DOI: 10.4324/9781003266440-9

of uncertainty. Therefore, we must replace 'predictable, linear and monofunctional design' with approaches and systems that embrace deep uncertainties of our future and are better able to meet social, economic and ecological needs (Elmqvist et al., 2018). We argue the case of the Khayelitsha Wetlands Park is just such a case where the Park presents a designed intervention that draws together biological and social elements and is not 'hard cast', therefore has the opportunity for augmentation or revision should it be desired or required. However, the wetland park is constrained by finances, urban pressures from both formal and informal landscape features, pollution and invasion by alien plant species that threaten biodiversity gains. A future vision for the wetland park, where it might be akin to parks designed specifically for the amelioration of the impacts of global change, will need to ensure these obstacles are overcome (Brown et al., 2015). The constraints and challenges faced by the park are true to urban greenspaces in many developing regions (Basu & Nagendra, 2021), and so too are the pressures felt (Sudhira & Nagendra, 2013). While this case study is contextually bounded, the need to plan urban greenspaces that are socially and ecologically resilient and can serve to both secure important biological landscape features in the face of future stressors (Betsill & Bulkeley, 2007) and simultaneously allow for the social engagement, benefit, and growth that the literature confirms is critical to human wellbeing makes the vision of the Khayelitsha Wetland Park one that is universal (Browns et al., 2015).

History and current context

The City of Cape Town is located in the Cape Floristic Region, the smallest but most diverse Floral Kingdom, and due to intense pressure from urbanization, conversion for agriculture and the spread of invasive alien plants is also a recognized biodiversity hotspot (Cowling et al., 1992; Myers et al., 2000; Holmes et al., 2012; Goodness & Anderson, 2013; Anderson et al., 2014). Like all South African cities, Cape Town was developed as an apartheid city with segregated development, and hallmarks of the apartheid city persist (Turok, 2001). The Khayelitsha Wetlands Park exists in a low-income township (Wolchet al., 2005; Sutton, 2008; McConnachie & Shackleton, 2010; Shackleton & Blair, 2013) characterized by high rates of unemployment, poverty and extremely low-quality infrastructure and services (Stanvliet et al., 2004; Goodness & Anderson, 2013). Underlying this are the former apartheid laws, which emphasized social and racial segregation (1948–1994), and the apartheid legacy is embedded in social and institutional practices that persist and continue to override a number of more recent policy aspirations (Turok, 2001). This places pressure on green space infrastructure development and maintenance as municipalities struggle to balance demands with constrained budgets, especially as they seek redress the past injustices in these historically underdeveloped areas (Sutton, 2008; Kabisch, 2015). Cape Town remains a starkly polarized city where affluent suburbs (occupied mostly by white residents) and flourishing economic centers are bordered by beautiful mountains and coastal areas that offer wide-ranging opportunities and wellbeing benefits.

In contrast, the poor townships (occupied by primarily people of color[1]) remain densely populated at the urban periphery, with seasonally strong prevailing winds, mobile sand dunes, and flood-prone low-lying areas. Khayelitsha demonstrates how the social inequity of apartheid planning practices have, and continue to, erode the capacity to conserve ecosystems (Ernstson et al., 2010). In particular, the coastal dune and wetland ecosystems that once characterized Khayelitsha continue to be lost to development and associated ecosystems are degraded or lost (Brown & Magoba, 2009).

According to Rebelo et al. (2011), the earliest written conservation plan for the City, Cape Town's 'Greening of the City' report of 1982, included the conservation of the Kuils River and associated wetlands through the proclamation of a large park, which was to be named the False Bay Coastal Park. This Park was planned to connect the Cape Peninsula with the Hottentot Holland Mountains through the Kuils River and associated wetlands (Rebelo et al., 2011). However, the apartheid government decided that this area was to be developed as the Khayelitsha township to cope with population growth (Rebelo et al., 2011). The development of Khayelitsha required bulldozing of the 'Kuils River dune-slack wetland system' to flatten the area and create space for low-income housing (Brown & Magoba, 2009). The resulting development area was left vulnerable to seasonal inundation, with the water table rising in response to winter rainfall (Goodness & Anderson, 2013). Despite the significant degradation of the Khayelitsha Wetlands, what remains of the ecosystem has substantial value as a habitat for aquatic life, water purification, and recharging of the Cape Flats aquifer (Brown & Magoba, 2009).

More recent biodiversity conservation plans, highlighted in the Cities Biodiversity Network (or BioNet), categorize Khayelitsha wetlands as Critical Biodiversity Areas which must be afforded conservation focus at least through protection and preferably rehabilitation (Snaddon et al., 2009). The BioNet forms the backbone of Cape Town's Metropolitan Open Space System and focuses on conserving both the pattern and processes associated with the City's biodiversity (City of Cape Town, 2007). The establishment of the Khayelitsha Wetlands Park is one of the main efforts contributing to the improved protection of the broader wetlands system (Malan et al., 2015).

The Khayelitsha Wetlands Park (Figure 9.1), an urban green space and recreational area, was established in 1998 following a study on the Khayelitsha wetlands that identified various land-use zones, including an urban park for the area (Matthews, 2015; City of Cape Town, 2020; Swift, 2020). The 45 ha Park boasts a beautiful wetland and garden, and has various amenities for leisure and relaxation, such as a play park, an outdoor gym, park benches, a skate park, walking and cycling paths, and access for canoeing (City of Cape Town, 2016, Mathenjwa, 2017). The Park is described as an important meeting place for constructive community activism and a source of pride for the Khayelitsha community (Wilson & Pereira, 2012). The rehabilitation of Khayelitsha Wetlands Park has been described as a resounding success (City of Cape Town, 2020) and

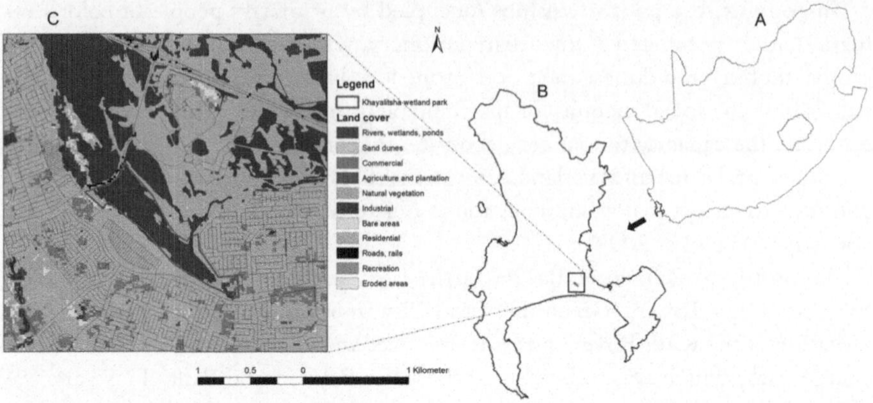

FIGURE 9.1 A land-cover map of the Khayelitsha Wetlands Park showing its position within the City of Cape Town, South Africa. The close proximity of the natural landscape feature of the wetland and dense urban settlement of Khayelitsha is evident.

an outstanding example of sustainable development in the heart of a township (Solomon, 2016). Its management falls to the City Parks Department of the City of Cape Town and various partners, including community members and various organizations (Mathenjwa, 2017; African Centre for a Green Economy, 2018). Several rehabilitation initiatives have been undertaken with the replanting of indigenous fynbos vegetation and fencing of some areas of the wetland (Arendse, 2017). Wetland clean-ups, alien vegetation removal, and environmental awareness programs are also regularly carried out (Pereira, 2012).

The wetland area of the Park is a small section of the greater Khayelitsha Wetlands (City of Cape Town, 2011). The wetlands are of important conservation value (City of Cape Town, 2019) for both their biodiversity contribution but also for a number of ecosystem functions, including flood attenuation (Oonyu, 2001). This ecosystem supports a variety of bird species, including *Macronyx capensis* (Cape Orange throated long-claw) and the *Gallinago nigripennis* (African snipe) (Ewart-Smith, 2015). The wetland also has a number of important plants including a few endemic species, such as *Elegia tectorum* (Cape Thatching Reed) and *Senecio halimifolius* (Tobacco Bush) (Mathenjwa, 2017). It is also believed to have two threatened amphibians, namely the endangered *Amietophrynus pantherinus* (Western Leopard Toad) and the near threatened *Breviceps gibbosus* (Cape Rain Frog) (City of Cape Town, 2019).

Current and future stressors

The Khayelitsha Wetlands Park is an important feature of everyday life of the local community (Govender, 2004). The way people engage with such urban ecosystems is highly complex (Anderson & O'Farrell, 2012) and can present multiple challenges for management. Stressors associated with the Khayelitsha Wetlands Park are both social and ecological and present, in some instances, limitations to adaptive capacity (Meerow et al., 2016).

Social and economic stressors

The Khayelitsha township became the fastest growing township in the City of Cape Town after the termination of restrictive apartheid legislation in 1994 (Kok & Collinson, 2006). Today it is the largest township in this city (Rodina & Harris, 2016). This rapid growth is characterized by a dramatic rise in formal and informal housing (Oonyu, 2001), and today Khayelitsha is sprawling in nature and constantly receiving new arrivals from the Eastern Cape Province of South Africa (Poswa & Levy, 2006). This movement across the country is driven by job prospecting and other forms of livelihood desires and needs perceived as available in Cape Town (Ndegwa et al., 2007). However, many job prospectors struggle to find employment (Poswa & Levy, 2006). They then remain in a state of informality, in shack dwellings on remnant vegetation strips on dunes (Ernstson et al., 2010). As the population grows, issues associated with urban sprawl, such as the continued emergence of shack dwellings, the lack of service infrastructure and associated solid waste disposal and pollution, also increased (Uttara et al., 2012; Haywood et al., 2021). This rapid growth, particularly in informal settlements, manifests in stressors to the ecological functioning of the wetlands (Oonyu, 2001).

In Khayelitsha, where unemployment remains high, there is a significant dependence on the wetlands for livelihood support (Reuther & Dewar, 2006; Govender, 2017). People depend on the wetlands for food resources (e.g., fish), medicinal herbs, firewood, building material and agricultural resources, with these areas being commonly grazed (Oonyu, 2001; Govender, 2004; Mathenjwa, 2017). These factors exert considerable pressure on the wetlands and can hinder their ability to function and provide other services (Lundqvist et al., 2003; Lee & Maheswaran, 2010; Meerow et al., 2016).

Biophysical stressors

Pollution

Pollution is a major problem in the Khayelitsha wetland. Dumping of solid waste, encouraged by inadequate or entirely absent household waste collection services, is one of the main contributors (Pereira, 2012; Greef, 2014). Criminals have been found to dump unwanted goods in the dense wetland vegetation to conceal criminal acts (Govender, 2004; Mathenjwa, 2017). The Khayelitsha wetlands also receive a large influx of urban and agricultural runoff, waste, and direct discharge of raw sewage and treated industrial and urban effluent (Wiseman and Sowman, 1992; Govender, 2004; Oonyu, 2001, Malan et al., 2015). Chemical contaminants include, but are not limited to, ammonium, fluorine, potassium, iron, magnesium, sodium and manganese (Govender, 2004; Mathenjwa, 2017). Microbial contaminants, primarily faecal coliforms (e.g., *Escherichia coli*), also contribute to this poor water quality (Ewart-Smith & Ractliffe, 2002; Govender, 2004). Raw sewage has been observed flowing into the Khayelitsha Wetlands from the poorly structured and underserviced communal toilet facilities of the settlement adjacent to the Park (Mathenjwa, 2017). The various contaminants in the wetland are present in quantities that render it unfit for its current agricultural, recreational and religious (baptism) uses (Pereira, 2012; Brown and Magoba, 2009; Malan et al., 2015). There have been outbreaks of water-related human diseases (Dufour and Bartram, 2012), such as cholera and gastroenteritis (Department of Water Affairs and Forestry, 1996), with similar health concerns for disease among livestock drinking in the area (Brew et al., 2009).

The Kuils River, which flows into the Khayelitsha Wetlands, receives treated effluent from several Wastewater Treatment Works, resulting in the loss of natural seasonal variability in (Griffin & Grobicki, 2000; Govender, 2004; Brown & Magoba, 2009; Thomas et al., 2010; Matthews, 2015; Mathenjwa, 2017), and the system is now permanently inundated (Brown & Magoba, 2009).

Invasive plants

The broader catchments that feed into the wetlands are heavily infested with this both indigenous plants and invasive alien plants (Brown & Magoba, 2009; Potgieter, 2019). The Khayelitsha wetlands are generally invaded by *Typha capensis* (Bulrush), *Phragmites australis* (Common Reed), *Acacia longifolia* (Port Jackson Acacia), *Acacia mearnsii* (Black Wattle), *Fontinalis antipyretica* (Common Water Moss) and *Eichhorinia crassipes* (Common Water Hyacinth) (Govender, 2004; Brown & Magoba, 2009; Malan et al., 2015; Mathenjwa, 2017). These species are aggressively invasive and can survive various anthropogenic changes (McKay et al., 2018). For example, *P. australis* and *T. capensis* thrive in nutrient-enriched and permanently inundated waters (Govender, 2004; Brown & Magoba, 2009), outcompete indigenous species and reduce species diversity (Kercher & Zedler, 2004; McKay et al., 2018). The resultant change in community structure and community composition (Verhoeven et al., 2006; Mavimbela, 2018; Talal & Santelman, 2019) are a concern because they reflect a loss of biodiversity and ecosystem function (Turner et al., 2000, Nolte et al., 2014, McKay et al., 2018). In addition, invasive species that form dense stands in water cause water stagnation and worsen the already poor water quality (Govendor, 2004).

Path to the future

Set in a hard-cast landscape constructed through separatist-racist land settlement policies, situated in a biodiversity hotspot and immediately constrained by the physical nature of the wetland system, the Khayelitsha Wetland Park can be considered a significant success. The Park provides some redress in securing social benefits in access to greenspace for local residents, secures potential biodiversity and contributes to the ecological functioning and associated delivery of ecosystem services such as flood mitigation and certain livelihood benefits. The governance and management of the Park, which is held across a diversity of institutions, provides a further element of resilience (Mathenjwa, 2017). However, social and ecological health issues persist in this urban green space and pose a threat to its sustainability. We propose a vision and path to the future that engages some of these concerns, positioning them in the wider discourse of resilience to change (Meerow et al., 2016), and positioning the Park for this purpose (Brown et al., 2015). In general, the success of these approaches will rely on adequate budgets, compliance enforcement and the fostering of old, and facilitation of new, partnerships including those with funders.

Improving the ecological health for the benefit of the community

Tackling water quality issues

The need to address water quality issues in the Khayelitsha Wetlands has been echoed by various researchers (Wiseman & Sowman, 1992; Govender, 2004; Oonyu, 2001; Day et al., 2020). The improvement of water quality would enhance the ecological state of the wetlands and amplify benefits to the local community, including ecotourism endeavors that could promote job creation.

The City of Cape Town's *State of the Environment Report* (2019) suggests that water quality in the Khayelitsha Wetlands should, at least, be fit for intermediate contact through recreational use (City of Cape Town, 2019). This stated water quality expectation would ensure that ongoing recreational activities such as swimming and canoeing are safe and allow for an expansion of water activities. Improving water quality will require interventions that incorporate upstream factors, particularly a significant reduction in volume of point source discharge and the nutrient loads of both point and non-point source discharges.

While the City of Cape Town has wastewater and industrial effluent by-laws (City of Cape Town: Wastewater and Industrial Effluent By-law, 2013), monitoring and compliance enforcement are currently lacking. More deliberate oversight by relevant authorities is needed (McKay et al., 2018). To address this problem, Brown and Magoba (2009) recommend the involvement of the National Environment Inspectorate and the implementation of the 'polluter pays principle'. The strengthening of regulatory mechanisms as well as monitoring and enforcement for wastewater treatment has recently been stressed by Alabaster et al. (2021) in the Progress on Wastewater Treatment Report on the Global status and acceleration needs for Sustainable Development Goal indicator 6.3.1.

Govender's (2004) research on the Khayelitsha Wetlands recommended striving for improvement in basic sanitation in the broader Khayelitsha area. Poorly structured and underserviced communal toilet facilities are one of the key issues that need to be tackled. The provision of adequate sanitation is not only important for the improvement and protection of the Wetlands Park ecosystem, but it is also crucial for maintaining human dignity and privacy (De Carvalho et al., 2008; Thevenon, 2020). The democratic government of South Africa has among its goals to provide adequate sanitation to all, and in this case, it would serve both social and ecological ends (Khosa, 2000). Dumping waste into the wetlands and upstream river reaches needs to be addressed. While monitoring and compliance enforcement is important, there also needs to be efforts toward shaping the views and perspectives of all Capetonians, such that rivers are viewed and recognized as the foundation of a healthy society (Brown & Magoba, 2009). Environmental education is a vital tool for influencing people towards positive environmental practices and policy compliance (Sarker et al., 2012; Du et al., 2018). The environmental awareness programs that have been implemented along the Kuils River and the Khayelitsha Wetlands have contributed to the reduction of solid waste dumping and have led to more appreciation of the wetlands (Greef, 2014). However, there is still a long way to go, ultimate success is having good water quality in wetland and changing people's perspectives is crucial to achieving that (Brown & Magoba, 2009). The use of mass media platforms can assist in extending environmental awareness programs to the broader community, and ultimately to reduce dumping of waste in the wetland (Sarker et al., 2012).

Tackling invasive plant issues

Management interventions focused on invasive alien plants, such as eradication and control, are often necessary in re-establishing important ecosystem services (Nsikani et al., 2017; Martin et al., 2020). Several authors, including Govender (2004), have called for the control of invasive alien plant species such as *T. capensis* in the Khayelitsha Wetlands. There are a number of methods that can be used to control invasive alien plants (Ailstock et al., 2001; Abonyo & Howard, 2012) including mechanical, chemical and biological control (van Wilgen et al., 2001; Abonyo & Howard, 2012; Hoare, 2021). The issue of invasive alien plant species in this ecosystem is one that can be referred to as a 'wicked problem' as there is no single optimal solution (Potgieter, 2019). Therefore, an integrated approach usually involving a combination of two or more of the above-mentioned methods to optimize the successful eradication and control of invasive alien plants will be necessary given the extent and the number of different species in the Khayelitsha Wetlands as well as in the Kuils River system (van Wilgen et al., 2001; Abonyo & Howard, 2012).

Transferable lessons

As cities in the Global South, and particularly in Africa, grow and develop, they need to ensure that they actively retain both green infrastructure and green open space, as this is critical in facilitating shifts towards more sustainable and resilient urban futures (O'Farrell et al., 2019). The Khayelitsha Wetland Park case study highlights both the benefits of retaining green infrastructure and open spaces but also that these spaces require significant investments of time and money to bring them up to a level where they are able to fulfil their ecosystem service potential. It is vital that these cities start identifying and retaining green space for enhancing future resilience and human wellbeing.

Acknowledgment

We would like to thank the Alliance for Collaboration on Climate and Earth Systems Science, and the National Research Foundation for funding this research. We are grateful to the reviewers for their useful insights which certainly served to improve this chapter.

Note

1 While we find the use of racial terminology distasteful, in the case of South Africa where there has been little redress and spatial reconfiguration of the apartheid city since the end of apartheid these terms remain contextually relevant as they relate to spatial injustice.

References

Abonyo, E., & Howard, G. (2012). *Guide to some invasive plants affecting Lake Tanganyika.* Nairobi: International Union for Conservation of Nature and Natural Resources, 64.

Africa Center For a Green Economy. (2018). *Exploring landscape green innovations to improve aquatic ecosystem services for the benefit of urban and peri-urban communities: A case study of the Khayelitsha Wetlands.* Pretoria: Water Research Commission.

Ailstock, M., Norman, C., & Bushman, P. (2001). Common Reed Phragmites australis: Control and Effects Upon Bio diversity in Freshwater Nontidal Wetlands. *Society for Ecological Restoration*, 9(1), 49–59.

Alabaster, G., Johnston, R., Thevenon, F., & Shantz, A. (2021). *Progress on wastewater treatment: Global status and acceleration needs for SDG indicator 6.3.1.* United Nations Human Settlements Programme (UN-Habitat) and World Health Organization (WHO).

Allsopp, N., Anderson, P. M., Holmes, P. M., Melin, A., & O'Farrell, P. J. (2014). People, the Cape Floristic Region, and sustainability. In N. Allsopp, J. F. Colville, & G. A. Verboom, *Fynbos: Ecology, Evolution, and Conservation of a Megadiverse Region* (p. 337). Oxford University Press.

Anderson, P., Avlontis, G., & Ernstson, H. (2014). Ecological outcomes of civic and expert-led urban greening projects using indigenous plant species in Cape Town, South Africa. *Landscape and Urban Planning, 127*, 104–113.

Arendse, C. (2017). *Makhaze community celebrate International Wetlands Day.* Retrieved from https://www.adaptationnetwork.org.za/2017/03/aimed-enabling-participants-design-facilitate-adaptation-processes-vulnerable-communities-course-will-facilitated-noel-oettle-shannon-parring-please-register-woul/

Brew, M., Carter, J., & Maddox, M. (2009). *The impact of water quality on beef cattle health and performance.* UF IFAS Extension. Gainesville, FL: University of Florida, 1–4.

Brown, C., & Magoba, R. (2008). *Cape Town's rivers and wetlands. Caring for our rich aquatic heritage.* Pretoria: Water Research Commission.

Brown, C., & Magoba, R. (2009). *Rivers and wetlands of Cape Town: caring for our rich aquatic heritage.* Pretoria, South Africa: Water Research Commission.

City of Cape Town. (2007). *The identification and prioritisation of a biodiversity network for the city of Cape Town.* Cape town. Retrieved from https://resource.capetown.gov.za/documentcentre/Documents/City%20research%20reports%20and%20review/BioDNet_Final_Report_02_2007_19122007172753_465.pdf

City of Cape Town. (2011). *Khayelithsa/Mitchells plain district plan. Baseline information and analysis report.* Cape Town: City of Cape Town Government. Retrieved from https://www.capetown.gov.za/Family%20and%20home/See-all-city-facilities/Our-recreational-facilities/District%20parks/Khayelitsha%20Wetlands%20Park

City of Cape Town. (2016). *Khayelitsha Wetlands Park.* Retrieved from https://www.capetown.gov.za/Family%20and%20home/See-all-city-facilities/Our-recreational-facilities/District%20parks/Khayelitsha%20Wetlands%20Park

City of Cape Town. (2019). *Draft Khayelitsha, Mitchells plain, greater blue downs district baseline and analysis report 2019. 1–20.* Cape Town: City of Cape Town Government. Retrieved from https://resource.capetown.gov.za/documentcentre/Documents/City%20research%20reports%20and%20review/Draft_KMPBD_BaAR_2019_Preface.pdf

City of Cape Town. (2020). *A history of Cape Town's district parks.* Retrieved from https://resource.capetown.gov.za/documentcentre/Documents/Graphics%20and%20educational%20material/A%20HISTORY%20OF%20DISTRICT%20PARKS.pdf

Cowling, R., Holmes, P., & Rebelo, A. (1992). Plant diversity and endemism. In R. Cowling, *The Ecology of Fynbos: Nutrients, Fire and Diversity* (pp. 62–112). Oxford University Press.

Day, L., Ollis, D., Ngobela, T., & Rivers-Moore, N. (2020). *Status and historical trends, with a focus on the period. April 2015 To March 2020.* Liz Day consulting.

De Carvalho, S., Carden, K., & Armitage, N. (2008). Application of a sustainability index for integrated urban water management in Southern African cities: Case study comparison – Maputo and Hermanus. *Water Institute of Southern Africa (WISA) Biennial Conference* (pp. 144–151). Retrieved from http://www.wrc.org.za/

Du, Y., Wang, X., Brombal, D., Moriggi, A., Sharpley, A., & Pang, S. (2018). Changes in environmental awareness and its connection to local environmental management in water conservation zones: The Case of Beijing, China. *Sustainability, 10*(2087), 1–24. https://doi.org/10.3390/su10062087

Dufour, A., & Bartram, J. (2012). *Animal waste, water quality and human health.* IWA publishing.

DWAF (Department of Water Affairs and Forestry). (1996). *South African water quality guidelines. Volume 5: Agricultural Use-Livestock watering.* Pretoria, South Africa.: Department of Water Affairs and Forestry. Retrieved from http://www.dwa.gov.za/iwqs/wq_guide/Pol_saWQguideFRESHLivestockwateringvol5.pdf

Ernstson, H., van der Leeuw, S. E., Redman, C. L., Meffert, M. J., Davis, G., Alfsen, C., & Elmqvist, T. (2010). Urban transitions: On urban resilience and human-dominated ecosystems. *The Human Environment., 39*(8), 531–545. https://doi.org/10.1007/s13280-010-0081-9

Ewart-Smith, J. (2015). *Proposed development of the Vuyani precinct, Khayelitsha freshwater ecosystem assessment.* Cape Town: Dity of Cape Town Municipality.

Ewart-Smith, J., & Ractliffe, S. (2002). *Assessment of the potential impacts of the proposed N1/N2 Winelands Toll Highway Project on aquatic ecosystems.* Specialist EIA report to Crowther Campbell & Associates, on behalf of the National Roads Agency.

Goodness, J., & Anderson, P. M. L. (2013). Local assessment of cape town: Navigating the management complexities of urbanization, biodiversity, and ecosystem services in the cape floristic region. In T. Elmqvist, M. Fragkias, J. Goodness, B. Güneralp, P. J. Marcotullio, R. I. McDonald, . . . C. Wilkinson, *Urbanization, biodiversity and ecosystem services: Challenges and opportunities*. Dordrecht: Springer. https://doi.org/10.1007/978-94-007-7088-1_24

Govender, K. (2004). *The effect of development on seasonal wetlands on the Cape Flats, Western Cape, South Africa. MSc Thesis, University of Cape Town*. Cape Town. Retrieved from https://open.uct.ac.za/bitstream/item/15701/thesis_sci_2004_govender_kamal.pdf?sequence=1

Greeff, L. (2014). *Kuils river social audit - The story of an urban river*. Retrieved from https://static1.squarespace.com/static/5a7859a10abd0477ecb31301/t/5c5d8c2ae2c483ba6bc2fa99/1549634663525/Kuils+River+Social+Audit.pdf

Griffin, N. J., & Grobicki, A. M. (2000). Community income generation through cultivation of high value plants in degraded urban wetlands. In *WISA 2000*, Sun City, South Africa, 28 May – 1 June 2000, Abbott Grobicki (Pty) Ltd. (pp. 1–15).

Haywood, L. K., Kapwata, T., Oelofse, S., Breetzke, G., & Wright, C. Y. (2021). Waste disposal practices in low-income settlements of South Africa. *International Journal of Environmental Research and Public Health, 8*, 1–12.

Hoare, D. (2021). *Alien and invasive plant species management plan: For the proposed 99 MW Oya Wind energy facility between sutherland and matjiesfontein in the western and northern cape provinces* (Vol. 83). Cell. Retrieved from https://sahris.sahra.org.za/sites/default/files/additionaldocs/Appendix%20D3b_Alien%20Invasive%20Management%20Plan.pdf

Holmes, P., Rebelo, A., Dorse, C., & Wood, J. (2012). Can Cape Town's unique biodiversity be saved? Balancing conservation imperatives and development needs. *Ecology and Society, 17*(2), 1–13. http://doi.org/10.5751/ES-04552-170228

Kabisch, N. (2015). Ecosystem service implementation and governance challenges in urban green space planning: The case of Berlin, Germany. *Land Use Policy, 42*, 557–567. https://doi.org/10.1016/j.landusepol.2014.09.005

Khosa, M. (2000). *Infrastructure mandates for change*. HSRC Press.

Lee, A., & Maheswaran, R. (2010). The health benefits of urban green spaces: a review of the evidence. *Journal of Public Health., 33*(2), 212–222. https://doi.org/10.1093/pubmed/fdq068

Lundqvist, J., Appasamy, P., & Nelliyat, P. (2003). imensions and approaches for third world city water security. *Philosophical Transactions of the Royal Society: Biological Sciences., 358*, 1985–1996.

Malan, H., Day, J., Ramjukadh, C., & Olivier, N. (2015). *Trajectories of change in wetlands of the Fynbos Biome from the late 1980s to 2014: Status reports for the individual wetlands sampled during the project: Water Research Commission Report no. 2183/1/14*. Water Research Commission. Retrieved from http://www.wrc.org.za/wp-content/uploads/mdocs/2183-1-14.pdf

Martin, P., Shackelford, G., Bullock, J., Gallardo, B., Aldridge, D., & Sutherland, W. (2020). Management of UK priority invasive alien plants: a systematic review protocol. *Environmental Evidence, 9*(1), 1–11. https://doi.org/10.1186/s13750-020-0186-y

Mathenjwa, F. (2017). *Exploring the ecological and social benefits of the Khayelitsha Wetlands Park. MSc Thesis, University of Cape Town*. University of Cape Town., Department of Environmental and Geographical science. Retrieved from https://open.uct.ac.za/handle/11427/27484?show=full

Matthews, S. (2015). Wetlands: fynbos wetlands – then and now. *The Water Wheel*, 1–25.

Mavimbela, L., Sieben, E., & Procheş, Ş. (2018). Invasive alien plant species, fragmentation and scale effects on urban forest community composition in Durban, South Africa. *New Zealand Journal of Forestry Science, 48*(19), 1–14. https://doi.org/10.1186/s40490-018-0124-8

McConnachie, M., & Shackleton, C. M. (2010). Public green space inequality in small towns in South Africa. *Habitat International., 34*, 244–248.

McKay, T., Ndlopfu, W. V., & Ahmed, Q. A. (2018). *The Blesbokspruit wetland, South Africa: A high altitude wetland under threat.* University of South Africa. https://doi.org/10.13140/RG.2.2.12042.98244

Myers, N., Mittermeyer, R. A., Fonseca, G. A., & Kent, J. (2000). Biodiversity hot spots for conservation priorities. *Nature, 403*, 853–858. https://doi.org/10.1038/35002501

Ndegwa, D., Horner, D., & Esau, F. (2007). The links between migration, poverty and health: evidence from Khayelitsha and Mitchell's plain. *Social Indicators Research, 81*(2), 223–234. https://doi.org/10.1007/s11205-006-9008-z

Nolte, S., Esselink, P., Smit, C., & Bakker, J. (2014). Herbivore species and density affect vegetation-structure patchiness. *Agriculture, Ecosystems and Environment, 185*, 41–47.

Nsikani, M., Nova, A., Van Wilgen, B., Keet, J., & Gaertner, M. (2017). Acacia saligna's soil legacy effects persist up to 10 years after clearing: Implications for ecological restoration. *Ecological Society of Australia, 42*, 880–889. https://doi.org/10.1111/aec.12515

Oonyu, J. C. (2001). *Land-use changes and community participation in the conservation and management of Khayelitsha Wetlands, Cape Town.* Cape Town (Unpublished UNESCO Research Report: University of the Western Cape).

Poswa, N., & Levy, R. (2006). *Migration study in Monwabisi Park (Endlovini), Khayelitsha. Cape Town: Strategic Development Information and GIS Department.* Cape Town. Retrieved from https://studylib.net/doc/10389716/migration-study-in-monwabisi-park--endlovini---khayelitsha.

Potgieter, L. (2019). *Assessing the impacts of invasive alien plants on urban ecosystem services. PhD Thesis, Stellenbosch University. Stellenbosch, South Africa.* Retrieved from https://scholar.sun.ac.za/bitstream/handle/10019.1/105742/potgieter_plants_2019.pdf?sequence=1&isAllowed=y

Rebelo, A., Holmes, P. M., Dorse, C., & Wood, J. (2011). Impacts of urbanization in a biodiversity hotspot: Conservation challenges in Metropolitan Cape Town. *South African Journal of Botany.*, 77(1), 20–35. https://doi.org/10.1016/j.sajb.2010.04.006

Reuther, S., & Dewar, N. (2006). Competition for the use of public open space in low-income urban areas: the economic potential of urban gardening in Khayelitsha, Cape Town. *Development Southern Africa, 23*(1), 97–122. https://doi.org/10.1080/03768350600556273

Rodina, L., & Harris, L. M. (2016). Water services, lived citizenship, and notions of the state in marginalised urban spaces: The case of Khayelitsha, South Africa. *Water Alternatives, 9*, 336–355.

Sarker, B. C., Sarker, S., Islam, M., & Sharmin, S. (2012). Public Awareness about Disposal of Solid Waste and its Impact: A Study in Tangail Pourashava, Tangail. *Journal of Environmental Science & Natural Resources, 5*(2), 239–244.

Shackleton, C., & Blair, A. (2013). Perceptions and use of public green space is influenced by its relative abundance in two small towns in South Africa. *Landscape and Urban Planning*, 104–112. https://doi.org/10.1016/j.landurbplan.2013.01.011

Snaddon, K., Day, L., Nel, J., & Maherry, A. (2009). *Prioritisation of city wetlands.* Cape Town: City of Cape Town; Department of Environmental Resource Management.

Solomon, N. (2016, August 23). *Neoliberalization of the commons.* Cape Town, Western Cape, South Africa.

Standing, A. (2003). The social contradictions of organised crime on the Cape Flats. *Institute for Security Studies,* 74, 1–14.

Stanvliet, R., Jackson, J., Davis, G., Swardt, C., Mokhoele, J., Thom, Q., & Lane, B. (2004). The UNESCO biosphere reserve concept as a tool for urban sustainability: The CUBES Cape Town Case Study. *Annals of the New York Academy of Sciences, 1023*(1), 80–104. https://doi.org/10.1196/annals.1319.003

Sutton, C. (2008). *Urban open space: A case study of Msunduzi municipality, South Africa.* MSc Thesis, Queen's University. Retrieved from http://citeseerx.ist.psu.edu/viewdoc/download?doi=10.1.1.526.1700&rep=rep1&type=pdf

Swift, L. (2020). Khayelitsha wetlands park. *Urban spaces.* Retrieved from https://urbanscapes.co.za/2020/06/08/wetland-rehabilitation-and-parks/

Talal, M., & Santelmann, M. (2019). Plant community composition and biodiversity patterns in urban parks of Portland, Oregon. *Frontiers in Ecology and Evolution,* 7(201), 1–16. https://doi.org/10.3389/fevo.2019.00201

Thevenon, F. (2020). *Sustainable sanitation systems: Health, environment and governance challenges. The case of human rights-based policy reform in alternative wastewater management strategies.* Retrieved from https://www.waterlex.org/wp-content/uploads/2021/04/Wastewater-report_WaterLex-UNEP_LR.pdf

Thomas, A., Chingombe, W., Ayuk, J., & Scheepers, T. (2010). *A comprehensive investigation of the Kuils-eerste river catchments water pollution and development of a catchment sustainability plan.* Water Research Commission (WRC).

Turok, I. (2001). Persistent polarisation post-apartheid? Progress towards urban integration in Cape Town. *Urban Studies, 38*(13), 2349–2371. https://doi.org/10.1080/00420980120094551

Uttara, S., Bhuvandas, N., & Aggarwal, V. (2012, February). Impacts of urbanization on environment. *International Journal of Research in Engineering & Applied Sciences, 2*(2), 1637–1645.

van Wilgen, B., Richardson, D., & Higgins, S. (2001). Integrated control of invasive alien plants in terrestrial ecosystems. *Land Use and Water Resources Research, 1*(5), 1–6.

Verhoeven, J., Arheimer, B., Yin, C., & Hefting, M. (2006). Regional and global concerns over wetlands and water quality. *Trends in Ecology and Evolution, 21*(2), 96–103.

Wilson, J., & Pereira, T. (2012). *Water demand management's shadow side: Tackling inequality and scarcity of water provision in Cape Town.* Cape Town: Environmental Monitoring Group.

Wiseman, K., & Sowman, M. R. (1992). *An evaluation of the potential for restoring degraded estuaries in South Africa.* Retrieved from ournals.co.za/doi/pdf/10.10520/AJA03784738_1249

Wolch, J., Wilson, J. P., & Fehrenbach, J. (2005). Parks and park funding in Los Angeles: An equity-mapping analysis. *Urban Geography, 26*(1), 4–35.

10

GREEN INFRASTRUCTURE IN HORNSBY, NSW

A collaborative method toward landscape resilience

Simon Kilbane

Vision

Introduction

Since 2015, more than half of the world's population now reside in urban areas (World Bank, 2014). This exerts pressure on environments including competition for an ever-increasing scarcity of food, space for biodiversity conservation, water and land resources and results in ecological fragmentation, loss of ecological connectivity and species extinction. To this are added an array of social challenges including social cohesion, mental and physical wellness, impacts of over population and urban heat island heat stress. Compounded by climate change, real-world solutions are often scarce, in contrast to an abundance of policy and good intentions that often fail to adequately embrace the views and considerations of citizens who are increasingly becoming disenfranchised from the decision-making processes in their local area. This project aspired to breach this gap in an Australian context and to craft accurate, measurable and visual solutions for one local government area (LGA), formulated through a collaborative methodology with a specific view toward the safeguarding of biodiversity.

The Hornsby biodiversity conservation strategy and its green infrastructure framework

This project highlights several real-world solutions to conceptualizing more resilient landscapes in an era of global climate change to meet the local pressures of urban densification and land-use change, ecological fragmentation and species extinction and in so doing acts as a response to these critical, contemporary C21st city challenges. While policies and dialogue abound, seldom are these made

DOI: 10.4324/9781003266440-10

visual, measurable and accurate through transparent planning processes that also include community. Indeed, in many instances, policies and plans, such as Sydney's *Green Grid* (Office of the Government Architect NSW, 2013), float above the landscapes and peoples they concern and only exist as theoretical or abstract concepts that may never come to fruition as realized plans, on the ground but are instead incapable of application due to the complexity of landscapes below. This is a critical challenge, as the aspirations and promised improvements cannot materialize for the city, its biota and residents. This project takes a different perspective and specifically sought to address this type of planning implementation as series of accurate, measurable and visual plans and illustrations.

Consideration for the protection of biodiversity is the foundation of landscape resilience and increasingly the articulation of policy into reality is not occurring (Kilbane & Kopinski, 2016; Kilbane, Weller, & Hobbs, 2019) and it is this key gap this project sought to address. It asks how can we ensure planning that collaboratively moves from 'theory into practice' (Kilbane, 2013) and demonstrate a shift from 'knowledge to action' (Khirfan & El-Shayeb, 2019). Ultimately, how might an Australian LGA envisage new and creative ways to create the types of sustainable and resilient landscape that are so clearly required by global challenges and to accommodate at all scales resilience thinking as well-articulated responses to intertwined local, regional and international issues. Set against a rapidly changing world where time limits are critical, what types of planning approaches might address the *urgency* of our global condition?

An innovative method and results

This project outlines a long-term, integrated and adaptive approach to manage and help create landscape resilience against a broad range of threats for the Hornsby LGA, a municipality on Sydney metropolitan area's fringe, in New South Wales (NSW), Australia. This project approached the management of biodiversity and its integrity within the LGA through a series of recommendations toward the conservation and protection of remnant vegetation and bushland and protected areas. Informed by the patterns and processes thinking of landscape ecology (Forman & Godron, 1986) as well as specific focus and prioritization of individual species, this entailed a management approach to safeguard and manage both individual species as well as the holistic habitat within which they exist. Of specific interest to this chapter, an innovative methodology underpinned this project. Anchored by three stages – *Mapping and Modelling, Design Charrettes and Community Inputs* and *Design Refinement and Visualization* – this project explored accurate, measurable and visual design solutions and a potential new future for biodiversity in the Hornsby LGA through its specific reliance on the development of a Green Infrastructure Framework (GIF). This methodology is illustrated in Figure 10.1 and these stages and their results will now be briefly introduced.

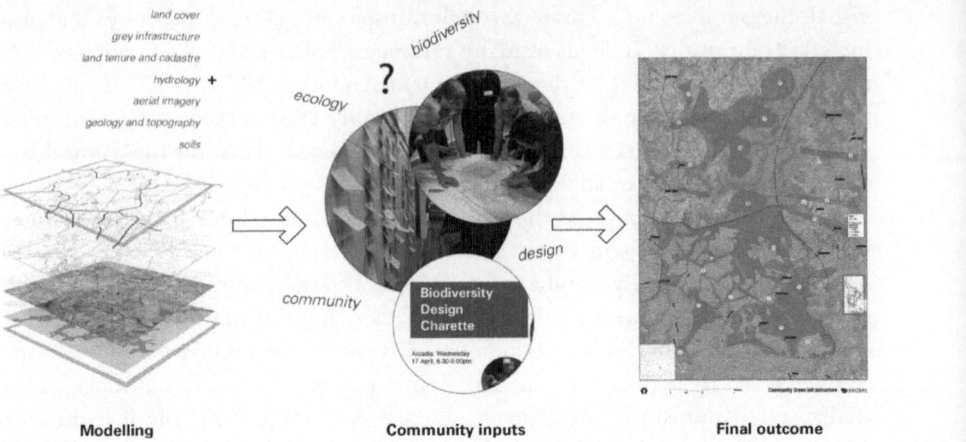

FIGURE 10.1 Illustration of the method underpinning the GI Framework.

Mapping and modeling: Mapping and desktop creation of a model

The first stage 'mapping and modelling' documented the body of policies, knowledge and best practice in urban planning and design for biodiversity, Green Infrastructure (GI) and existing open space. Through an examination of geographical spatial information, a GIF was developed – this was a computer derived model based on numerous layers such as land-use, Threatened Ecological Communities (TECs), land-cover and topography, as illustrated in Figure 10.2. As a framework this quickly provided an overview of a range of potential corridors across the LGA as well as location and prioritization of specific TECs. However, rather being considered as a finished model, this was then considered as a point of departure through iterative adjustment in the project's second stage.

Design charrettes and community inputs: Engagement with community and stakeholder consultation

The project's second stage brought together key stakeholders from Council, local resident citizens and other organizations through a series of discussions and hands-on design 'charrette' workshops. These were held in Hornsby, Arcadia and Pennant Hills in April 2019. The specific focus was upon a series of three test case or 'pilot' planning locations, chosen to sample a range of scales and with varying land-use conditions in order to test and improve the veracity and robustness of the GIF. Through iterative adjustment and redrafting the GIF could now incorporate local expert knowledge and improve accuracy to better suit both the ecological as well as the cultural landscape within which the proposal would operate.

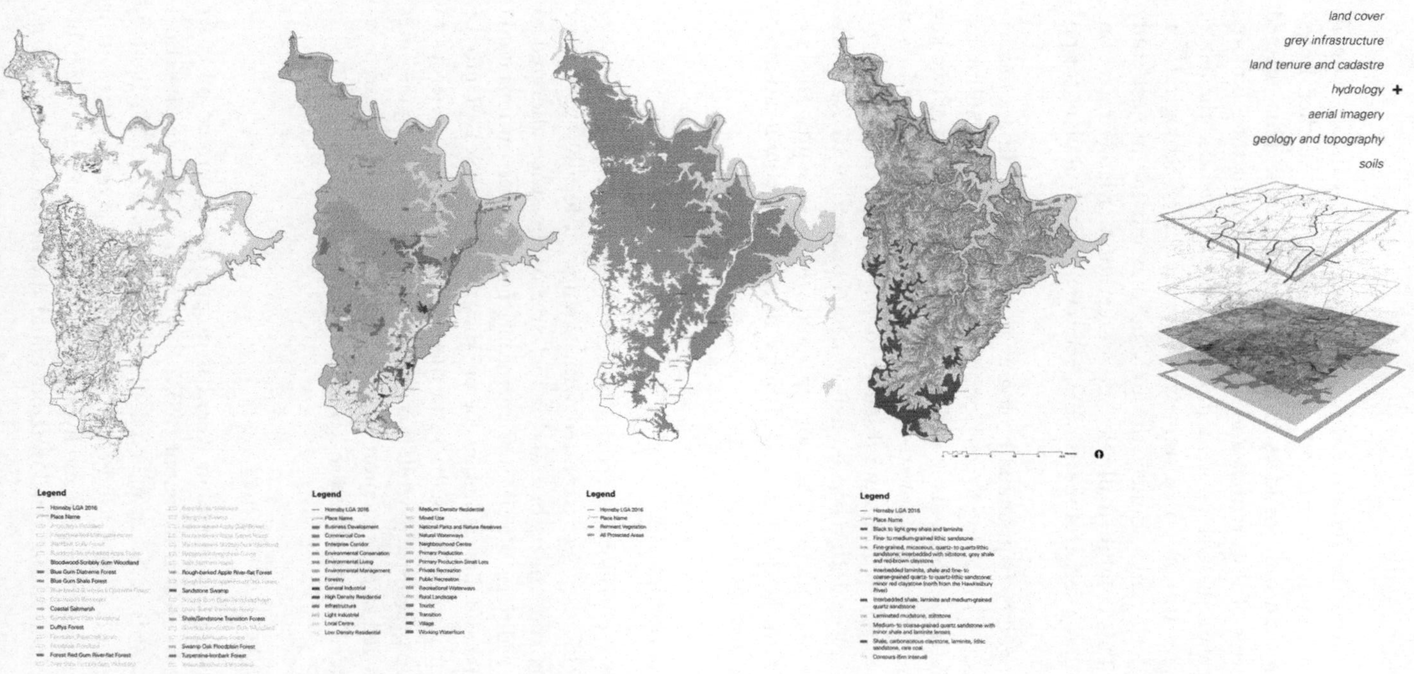

FIGURE 10.2 Datasets compiled to create the model (far right) included Land-use, TEC, Landcover and topography, hydrology and soils (L–R).

Design refinement and visualization: Finalization of accurate, measurable and visual plans

The third stage involved the refinement of the design to best incorporate all stakeholder and community comments in conjunction with further investigations and ground-truthing of ideas *in situ*, where required. This enabled the creation of a series of illustrations and visualizations that further explored and/or helped to explain the GIF's potential. A map set that extended the length of the LGA illustrates the proposed framework as a series of ecological corridors that extend across and connect through and beyond the project study area (Figure 10.3). This combined all relevant conservation or land protection-focused designations as a series of measurable, accurate and visual plans for potential implementation.

Stitching back together fragmented ecologies as well as providing a mechanism to deal with issues including Water Sensitive Urban Design (WSUD), urban heat island, active transport and recreation the GIF was composed of two key elements. First, buffer offsets of 30 metres around TECs (shown in Figure 10.3 in hatch) were proposed as the minimum distance to help protect existing vegetation root structure in line with relevant environmental protection advice (NSW Government, 2014). Second, four corridor types and widths were proposed, dependent upon function, land-use and availability of space for conservation actions. These will be discussed separately.

20 metre 'urban' corridors

The 20 metre 'urban' corridors reflect the average width of Council-managed streets and thoroughfares now retrofitted with a range of locally appropriate street-tree plantings and understory, and is inclusive of the space allocated to road and verge. In addition, WSUD and permeable paving interventions could also harvest rainwater and filter pollutants before they are slowly released into the ground and adjacent waterways. Setbacks in these streets could be generous to facilitate more urban green space, while road width and traffic calming measures might also be narrowed to create a more secluded environment. Overhead powerlines could be placed underground to help build the urban canopy.

30 metre 'hydrological' corridors

The 30 metre 'hydrological' corridors were based upon the minimum offset set for creeklines in rural areas throughout NSW. Both rural and urban creeklines should be granted similar protection and could be correspondingly 'daylighted' (or uncovered as open creekbeds) to better demonstrate the hydrological system and better provide for its ecological and hydrological functions. Additional habitat through revegetation and artificial wetlands could also help to slow and purify these waters.

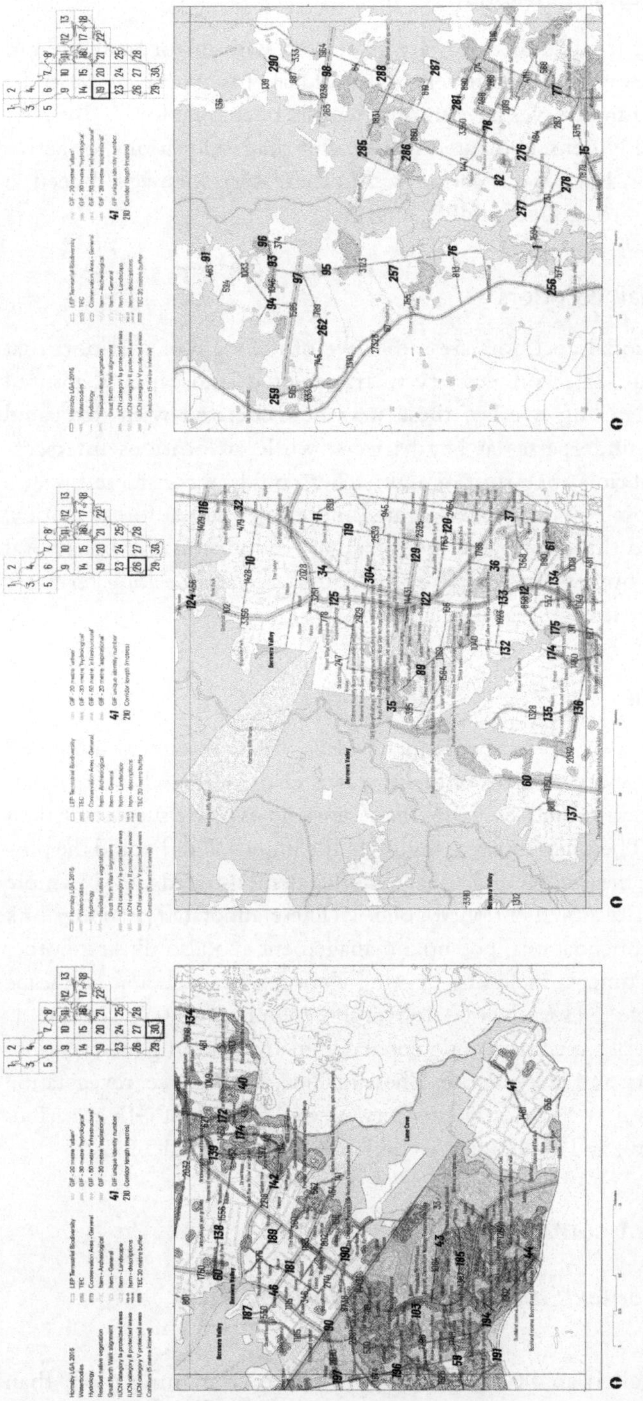

FIGURE 10.3 Sample pages drawn from GIF map set.

50 metre 'infrastructural' corridors

The 50 metre 'infrastructural' corridors are associated with major road and rail routes and correspondingly have different stakeholders. Elsewhere in Australia (for instance in Perth) rail reserves have simultaneously been used for biodiversity protection and as thickly densely planted spaces offer shaded locations for active transport as cycle paths. Barrier mitigation within these corridors would need to be carefully considered to reduce wildlife impacts.

75 metre 'aspirational' corridors

The 75 metre 'aspirational corridors are indicative of the scale of corridors that could provide regional scale connectivity to fragmented landscapes. Captured through a separate planning overlay, these may be associated with dedicated fauna overpasses and underpasses at key barriers; while at locations intersecting with the urban fabric a spectrum of approaches could be considered over a long time period. These could include targeted increase to urban forest canopy, adjusting of fences and the creation of microhabitats. Education and behavioral change regarding pet ownership could also be considered; while long term acquisition of key properties could also be an option (Figure 10.4).

Illustrative case studies

Five further case studies – at Hornsby, Pennant Hills, Cherrybrook, Galston and Berowra Heights – serve as examples of the potential of the GIF and as exemplars sampling the LGA's differing land-use types these case studies hence suggesting their application elsewhere. Their intent was to highlight the implications behind the project's vision as accurate, measurable and visual tools that demonstrated ideas in a more tangible manner than the aforementioned policies. These annotated drawings link back to relevant recommendations, potential management and also illustrate other relevant planning instruments including existing planning heritage and landscape overlays. An example of the Hornsby case study is shown in Figure 10.5. This exemplifies the investigation of a spectrum of potential design and planning interventions drawn from community and best practice. These include, for instance, revegetation of an old quarry site as new 'hub for biodiversity' as well a range of GIF at various widths (20, 50 and 75 m) and the ubiquitous buffer zooming around TECs.

History and current context

Bio-geographical context: Introducing the Hornsby Local Government Area

Hornsby Shire Council is an LGA in NSW, Australia, containing more than 152,419 people (Australian Bureau of Statistics, 2020) over 455 km^2 and is located in Sydney's northern suburbs, approx. 25 km from the city center (Figure 10.6).

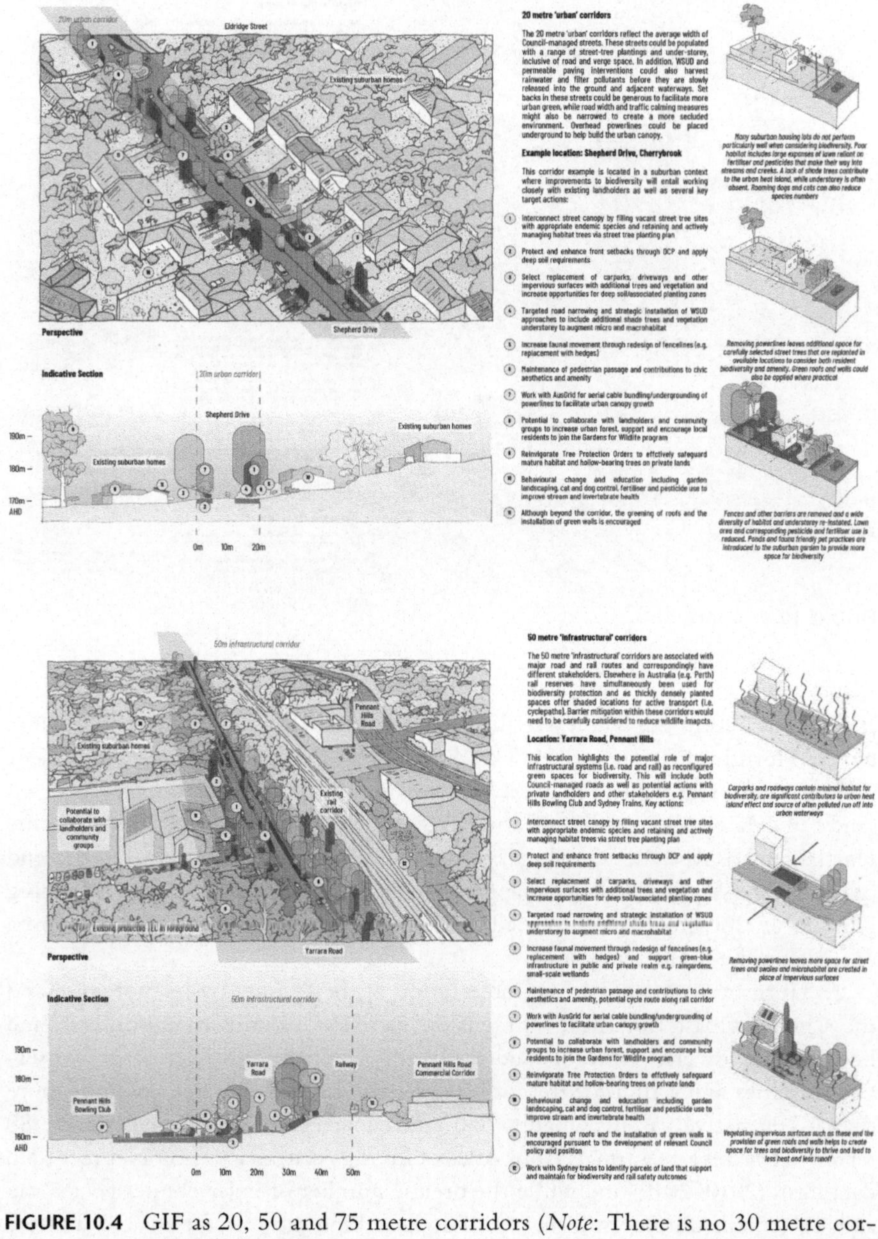

FIGURE 10.4 GIF as 20, 50 and 75 metre corridors (*Note*: There is no 30 metre corridor illustrated).

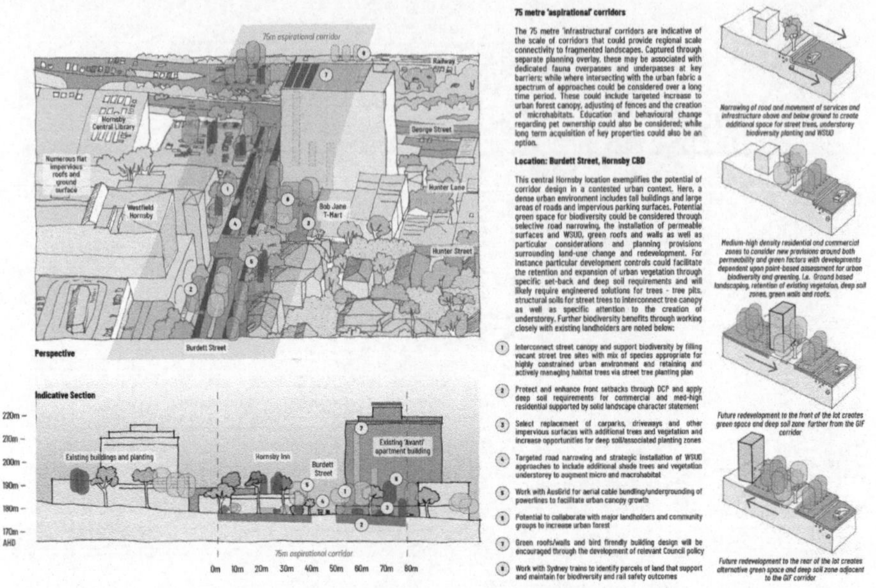

FIGURE 10.4 continued

Located within the Sydney Basin Bioregion, the Hornsby LGA is dominated by the Hornsby Plateau and Hawkesbury River Valley which form the major physiographic regions of the area. Further north and east, the deeply dissected sandstone Hawkesbury valleys feature the drowned river system of the estuarine Hawkesbury River and its tributary creeks. These were formed during the end of the last Ice Age and stabilized approximately 6,000 years ago. The underlying geology of Hornsby LGA is formed predominantly of sandstone, with a capping of shale on the higher ridgelines.

As Hornsby is situated at the intersection of the Sydney Basin Bioregion and the Central Coast Botanical Subdivision, it shares characteristics of each and hosts both a diverse array of landscapes and habitats with significant conservation values and a high degree of biodiversity. Indeed the area contains more than 1,200 native vascular plants, 660 fauna species, 90 fungi species and 388 terrestrial vertebrate animals from Atlas of Living Australia and BioNet recorded data from (2010–2019) and while the precise number of invertebrate species and aquatic remain unknown a survey of aquatic bioindicators found 230 discrete taxa of macroinvertebrates and eight native fish species (Tuft et al., 2000).

History and land-use legacy

The traditional owners of the area where Hornsby LGA now exists are the Darug and Guringai people. Before 1788,[1] traditional ecological knowledge and land management and especially burning practices had taken place for more than

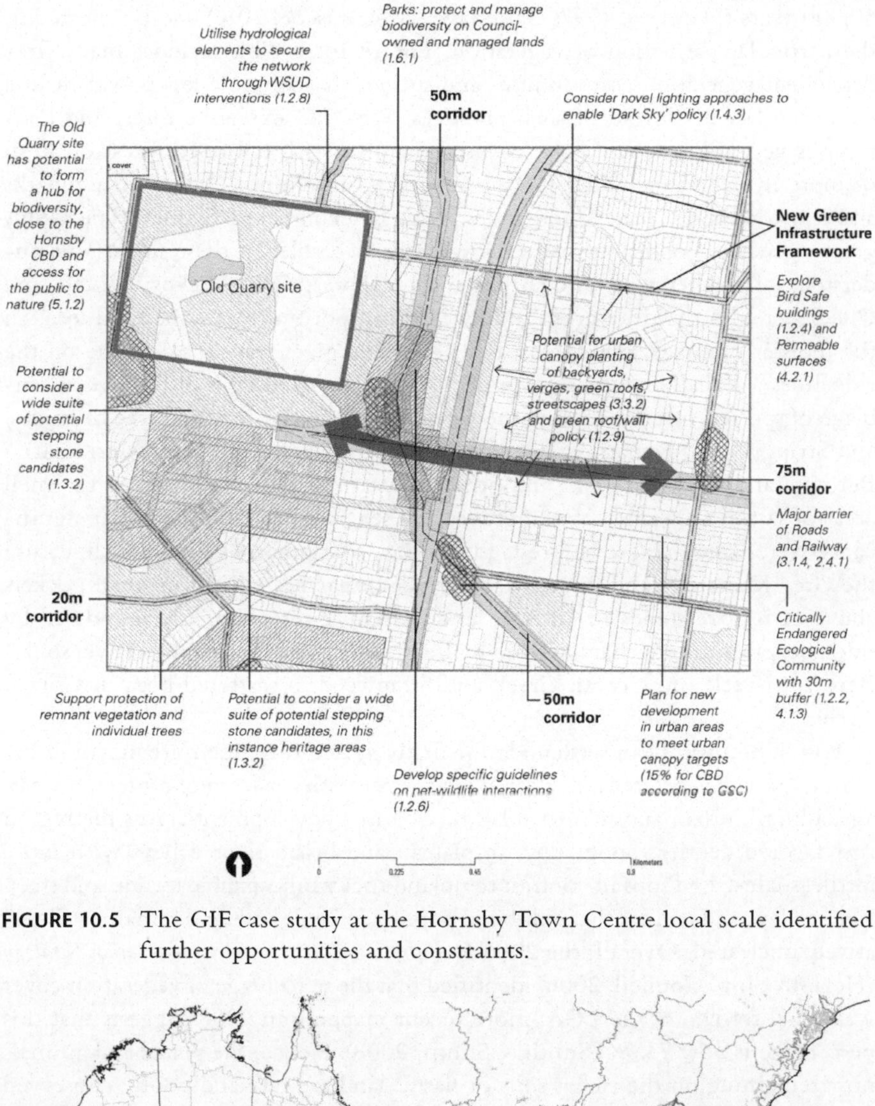

FIGURE 10.5 The GIF case study at the Hornsby Town Centre local scale identified further opportunities and constraints.

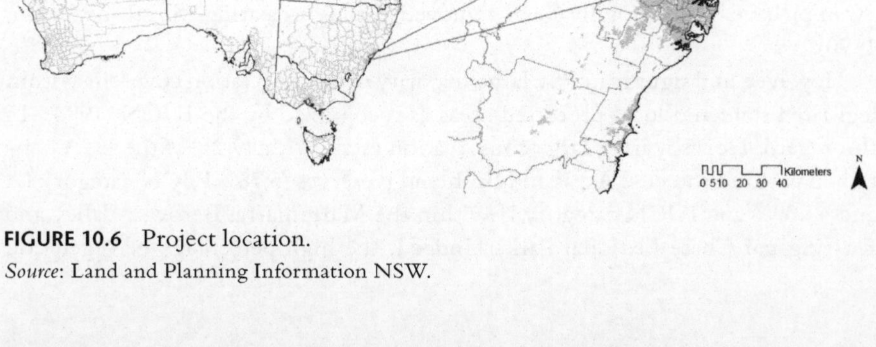

FIGURE 10.6 Project location.
Source: Land and Planning Information NSW.

60,000 years (Flannery, 1997; Gammage, 2011; Pascoe, 2016) and is reflected in the particular patination of vegetation (Figure 10.7). This includes many fire-dependent vegetation communities and species that are now less prevalent in a landscape largely devoid of such practices. Pre-1788 extensive forest and contiguous vegetation would have covered the area with tall open forest of 30 m or more in height including Grey Ironbarks, Turpentines, White Stringybarks and less commonly Red Mahoganies, with Blackbutts and Sydney Blue Gums growing where conditions were particularly favorable. In these forests, an understory of smaller trees and shrubs include Forest Oak, Hickory Wattle and Cheese Tree in the drier areas and Sweet Pittosporum trees, vines and ferns in the moister drainage lines. In areas where shale gives way to sandstone on the ridgelines a transitional area often occurs characterized by a distinctive assemblage of species, often including the Grey Gum (the favored food tree of koalas) and Stringybarks. Estuarine vegetation of the Hawkesbury River, Marramarra, Berowra and Cowan Creeks and other tributaries were characterized by small areas of saltmarsh, stands of mangroves and seagrass beds. Of particular significance were large mangrove forests in Big Bay, Marramarra Creek which feature the Grey Mangrove and River Mangrove. Saltmarshes existed in small pockets above mangrove stands in areas of land that were intermittently inundated by tides. Seagrasses were characterized by Eelgrass (*Zostera* spp.) in the Hawkesbury, Berowra Creek and Cowan Creek and Strapweed in scattered beds in Cowan Creek.

Post-1788, European settlement for firstly agriculture, then urbanization has left a mosaic of vegetated and cleared land cover with a pattern of protected areas, agricultural, urban and suburban lands. Colonial development across the region first favored gentler topography on plateau and broad river valleys with more fertile, shale-based soils in contrast to upland rocky inhospitable terrain and steep river valleys and ravines that exhibited poor or infertile soils and remains still relatively uncleared. Overall, the 2006 *Hornsby Biodiversity and Conservation Strategy* (Hornsby Shire Council, 2006) identified that the percentage of vegetation cover was equal to 69% of the LGA; more recent mapping in 2008 suggests that this percentage is now 73.6% (Smith & Smith, 2008). Hence, the plants and animals now remaining on the richer soils or flatter land are rare and poorly conserved in Hornsby. As a result of these clearing patterns, over 50% of Hornsby's plant communities are not conserved in any parks or reserves and two other vegetation communities (Swamp Sclerophyll Forest on Coastal Floodplains and Freshwater Swamp) have almost totally been removed through clearing (Smith & Smith, 1990).

However, and significantly, a large majority of this vegetation cover sits within legislated state and local protected areas as recognized by the IUCN (1994). In this regard, Hornsby has a large conservation estate with 49.2% of the LGA contained within Protected Areas and Bushland Reserves (5.78% IUCN category IA and 43.39% and IUCN category II within the Marramarra, Berowra Valley and Ku-ring-gai Chase National Parks). Indeed, the high percentage of vegetation

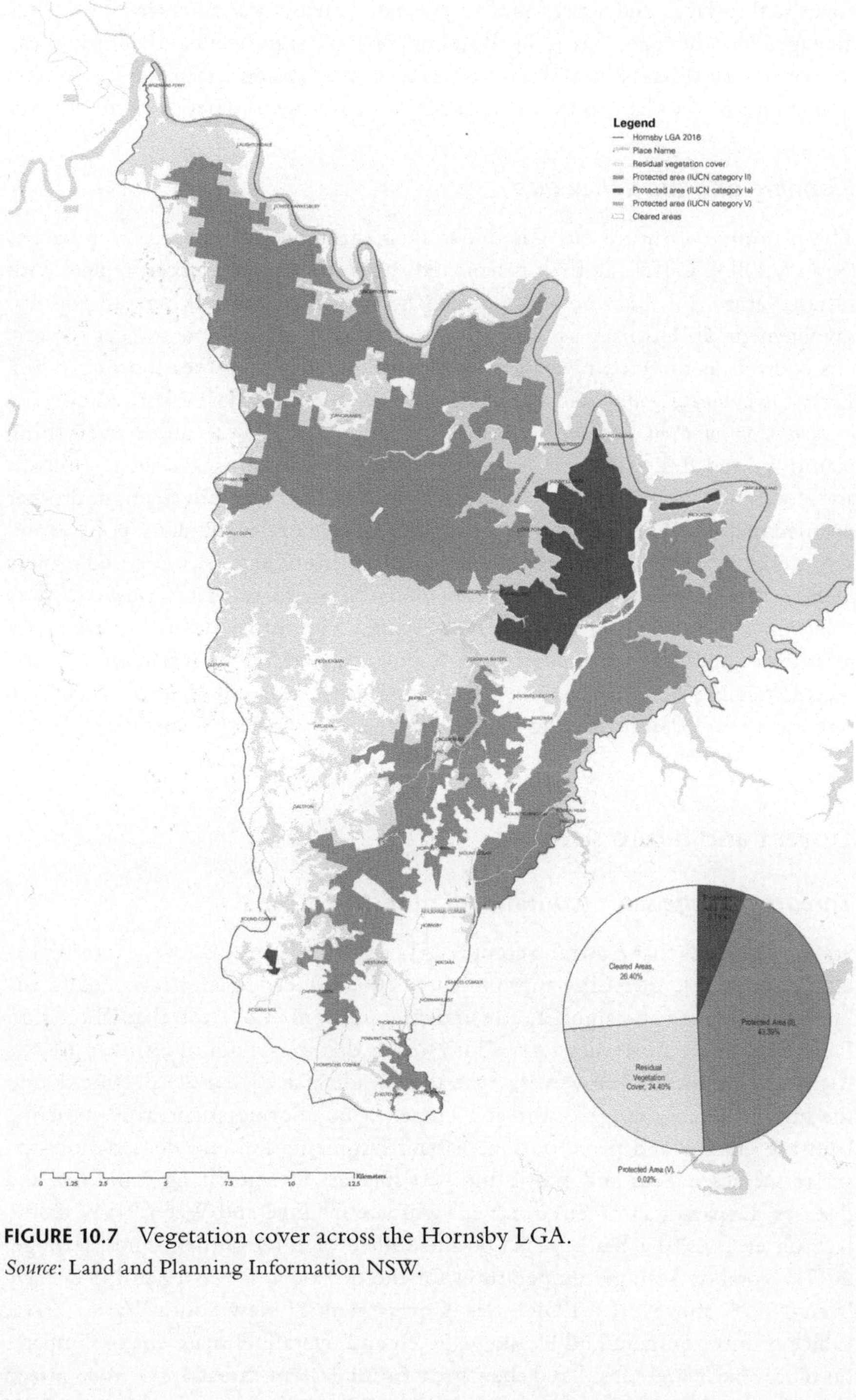

FIGURE 10.7 Vegetation cover across the Hornsby LGA.
Source: Land and Planning Information NSW.

cover within large and representative protected areas, council reserves on lands managed by other agencies and private property is a significant and defining feature of the 'Bushland Shire'. It also offers an enduring connection and is of continued significant place to the area's Indigenous People to practice their culture.

Planning and the policy gap

The planning of future cities is undertaken through a diversity of approaches (Sanoff, 2000). Cities and their urban landscapes are complex spaces, replete with infrastructure, inhabitation and multiple layers of decision-making and political involvement. In Hornsby as in other parts of NSW, planning is subject to various controls as directed by state government including the creation of guiding *Local Environment Plans* (Government of New South Wales, 2013) and *Development Control Plans* (Hornsby Shire Council, 2013). These guide everything planning- and design-related, including compatible land-use, heritage controls and protections as well as height, density and set-back of built form, and other detailed matters. These sit against a backdrop of other relevant plans, policies and acts, noted in Table 10.1. Those most critical to this biodiversity-focused project include the *Biodiversity Act* and recent policies put forward by *Greater Sydney Commission North District Plan* (Greater Sydney Commission, 2018) that aspire to interconnect disparate ecological fragments; and the Government Architects Office NSW's *Bushland and Waterway Guide* (2018) and *Greener Places 2020* framework and the supporting *Greener Places Design Guide* (Government Architect NSW, 2017).

Current and future stressors

Threats at large scale (continental and state)

Steffen estimates that Australia comprises between 7% and 10% of the world's total species (2009), including approximately 200,000 terrestrial species, many unknown to science and significantly endemic (Chapman & Australian Biological Resources Study Australia, 2009). This means that in ecological terms, Australia is globally important. However, Australia's biodiversity is on the decline due to the impacts from a range of threats. These include development, land-clearing, land-use change and urbanization, habitat fragmentation and degradation; invasive species, weeds and exotic and pest species; climate change; and fire and diseases (Department of Environment Climate Change and Water NSW, 2009; Jackson et al., 2017; State of NSW and Office of Environment and Heritage, 2017). Notably, average temperatures for the decade 2000–2010 in NSW have been 0.99°C above 1910–1939 levels (Government of New South Wales, 2019a; Office of Environment and Heritage, 2019) and overall, climate change impacts on biodiversity are considered their greatest long-term threat and is recognized as a key threatening process under the *NSW Biodiversity Conservation Act 2016*

TABLE 10.1 Planning and policy objectives that underpinned the project

Biodiversity Conservation Act 2016 No. 63 (New South Wales Government, 2016)	
Planning Priority N16 Protecting and enhancing bushland and biodiversity	Promotes the value of bushland and biodiversity for its contribution to safeguarding threatened species, investing in 'connected bushland corridors and protecting large pockets of remnant vegetation' and 'a. supporting landscape-scale biodiversity conservation and the restoration of bushland corridors; b. managing urban bushland and remnant vegetation as green infrastructure; c. managing urban development and urban bushland to reduce edge-effect impacts'
Planning Priority N19 Increasing urban tree canopy cover and delivering Green Grid connections	'Outlines a series of key values for increasing the urban forest with tangible benefits for mitigating the UHI effect, providing amenity and air quality as well as connections to the SGG for both recreational, active transport and biodiversity benefits while acknowledging the pressures on the urban forest by increasing densification and the delivery of grey infrastructure through: 71. Expand urban tree canopy in the public realm; 72. Progressively refine the detailed design and delivery of: a. Greater Sydney Green Grid priority corridors; b. opportunities for connections that form the long-term vision of the network; c. walking and cycling links for transport as well as leisure and recreational trips'
State Environmental Planning Policy No 19—Bushland in Urban Areas	'To protect and preserve bushland within urban areas through: (a) to protect the remnants of plant communities which were once characteristic of land now within an urban area, (b) to retain bushland in parcels of a size and configuration which will enable the existing plant and animal communities to survive in the long term, (c) to protect rare and endangered flora and fauna species, (d) to protect habitats for native flora and fauna, (e) to protect wildlife corridors and vegetation links with other nearby bushland'
State Environmental Planning Policy (Koalas No. 44) (Government of New South Wales, 2018a)	
State Environmental Planning Policy No 44—Koala Habitat Protection	Outlines measures of protection for urban and regional remnant wildlife populations and habitat
State Environmental Planning Policy (Vegetation in Non-rural Areas) (Government of New South Wales, 2018b)	
Scope	'Protect the biodiversity values of trees and other vegetation in non-rural areas of the State', and 'to preserve the amenity of non-rural areas of the State through the preservation of trees and other vegetation'

(Continued)

TABLE 10.1 continued

Our Greater Sydney 2056 North District Plan - connecting communities (Greater Sydney Commission, 2018)	
Planning Priority N16 Protecting and Enhancing bushland and biodiversity	'a. supporting landscape-scale biodiversity conservation and the restoration of bushland corridors; b. managing urban bushland and remnant vegetation as green infrastructure; c. managing urban development and urban bushland to reduce edge-effect impacts'
Planning Priority N18 Better managing rural areas	'Creation of protected biodiversity corridors, buffers to support investment in rural industries and protection of scenic landscapes'
Planning Priority N19 Increasing urban tree canopy cover and delivering Green Grid connections	'71. Expand urban tree canopy in the public realm; 72. Progressively refine the detailed design and delivery of: a. Greater Sydney Green Grid priority corridors; b. opportunities for connections that form the long-term vision of the network; c. walking and cycling links for transport as well as leisure and recreational trips'
Planning Priority N20 Delivering high quality open space	'Connect communities to the natural landscape' while recognizing the simultaneous need to manage to 'minimize impacts on biodiversity'
Planning Priority N22 Adapting to the impacts of urban and natural hazards and climate change to	'80. Support initiatives that respond to the impacts of climate change', suggesting to '81. Avoid locating new urban development in areas exposed to natural and urban hazards and consider options to limit the intensification of development in existing urban areas most exposed to hazards; in order to '82. Mitigate the urban heat island effect and reduce vulnerability to extreme heat'
Greener Places Design Guide (Government Architect NSW, 2017)	
	'Conservation of urban habitat and biodiversity in a holistic way that not only directs strategic planning but also acts at the management level... in the form of maps identifying core, transition, and habitat connection areas, with development controls and land management provisions suited to the local area'

(New South Wales Government, 2016) and the *Environment Protection and Biodiversity Conservation Act 1999* (Commonwealth of Australia, 1999). Furthermore, the impacts of climate change and change to fire regimes impact protected and unprotected lands.

In a regional context, Greater Sydney continues to grow in size and its population is projected to increase from 4.5 million in 2013 to 7.9 million by 2053 (Australian Bureau of Statistics, 2015). This growth that will be provided for through ambitious residential infill targets of 70% within existing urban extents, the highest of any Australian capital city (NSW Department of Planning, 2010).

This means that any biodiversity planning in Hornsby will face increasing pressure for additional housing and development. Since 2015, the number of threatened species listings in NSW has increased by 3% (Government of New South Wales, 2019a).[2] According to the NSW State of the Environment Report 2019: '64% of native mammals for which there are sufficient data have experienced long-term decline in range' (Government of New South Wales, 2019a).

Threats specific to the Hornsby LGA

The Hornsby LGA contains a mix of urban areas with freestanding homes, low rise townhouses and high-rise buildings set against a backdrop of rural acreages and a predominance of extensive bushland. This provides a sense of place to Hornsby and commonly referred to as the 'Bushland Shire'. While at first glance it might appear that this is a generous expanse dedicated to biodiversity protection in reality where land is unprotected overwhelmingly, native vegetation that remains is generally as small remnants around the edges of cleared agricultural land or as small backyard patches in urban areas (Figure 10.8).

Across the Hornsby LGA further threatening processes identified (NSW Office of Environment and Heritage, 2019) and listed by the *NSW Biodiversity Conservation Act 2016* (New South Wales Government, 2016) and include changing fire, water cycle and hydrologic regimes and the increase of impervious surfaces increase flow intensity, reduces groundwater recharge as well as increased pressures upon remnant vegetation by recreation users as well as associated impacts of urbanization including water quality, weed encroachment and feral animals.

Currently 26.4% of the Hornsby LGA (or 12,012 ha) has been cleared as inferred by available aerial, meaning a number of plant and animal communities are inadequately conserved. These include two floodplain communities and remnant bushland on the more fertile Wianamatta Shale, volcanic diatremes and the Hawkesbury River floodplain remain unprotected. Historically, such areas had been extensively cleared due to flat topography and arability and are hence now quite rare. Within the Hornsby LGA, 12 ecological communities mapped by Smith and Smith in 2018 are listed as TECs under legislation of which three are critically endangered: *Blue Gum High Forest in the Sydney Basin Bioregion, Sydney Turpentine-Ironbark Forest* and *Shale/Sandstone Transition Forest in the Sydney Basin Bioregion* (New South Wales Government, 2016). Numerous individual species dependent upon these TECs exist including the Heart-leaved Stringybark (*Eucalyptus camfieldii*), Bauer's Midge Orchid (*Genoplesium baueri*) and the Narrow-leaf Finger Fern (*Grammitis stenophylla*) as well as 50 threatened fauna species including the endangered Southern Brown Bandicoot (*Isoodon obesulus obesulus*), Dural Land Snail (*Pommerhelix duralensis*) and the migratory Swift Parrot (*Lathamus discolor*). Of these, 15 species found in the LGA have a Commonwealth Status.

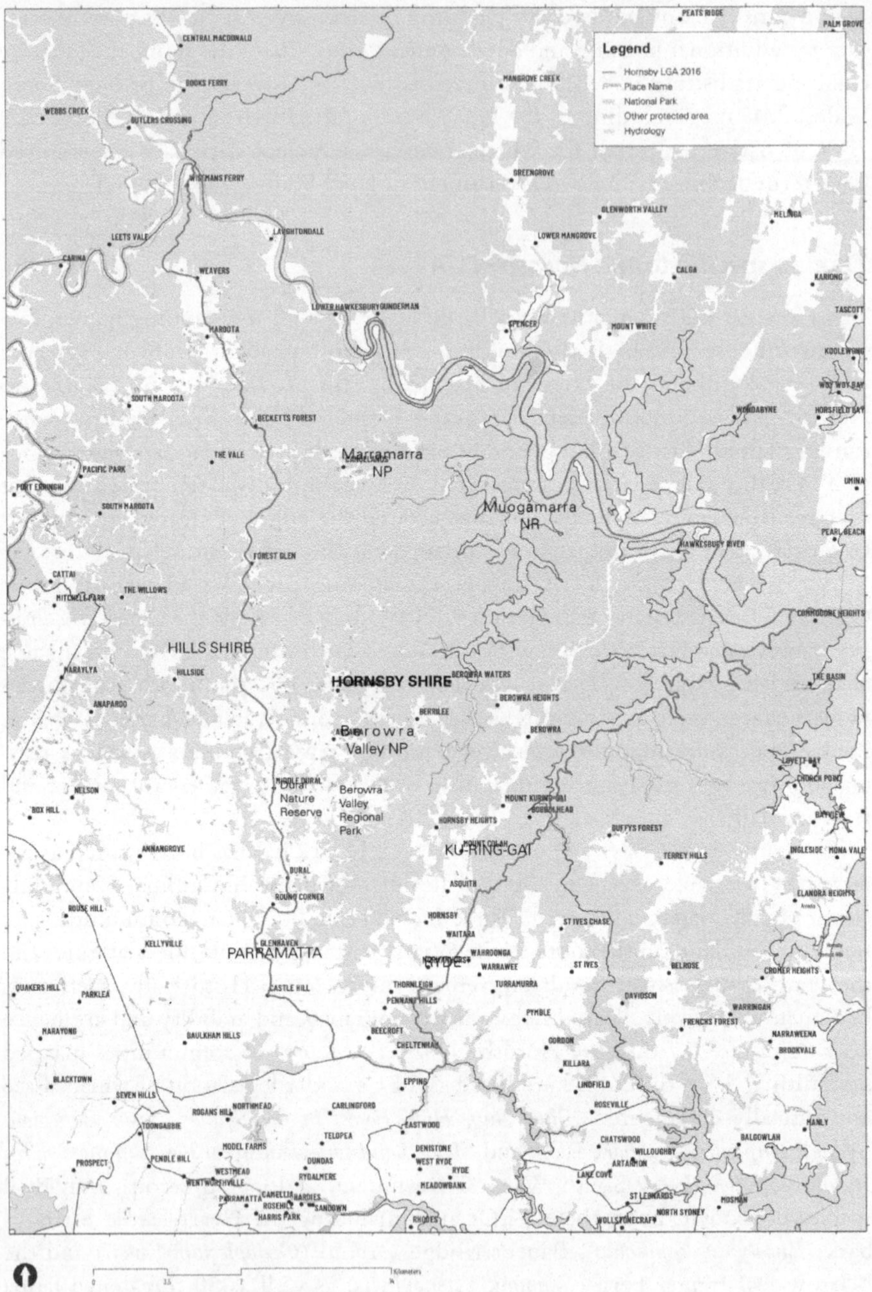

FIGURE 10.8 The Hornsby LGA presents a matrix of protected areas, agricultural, suburban and urban landscapes.

Source: Land and Planning Information NSW.

Maintaining a diversity of habitat is also important and requirements vary considerably between species. For instance, the listed Gang Gang Parrots (*Callocephalon fimbriatum*) preference is for tree hollows in eucalyptus (10 cm in diameter or larger and a minimum of 9 m above the ground), while others can require dense vegetation and debris for habitat such as the Red-crowned Toadlet (*Pseudophryne australis*). Furthermore, a number of vagrant or nomadic species occur in Hornsby such as the endangered species including the Swift Parrot (*Lathamus discolor*) and the vulnerable species Eastern Osprey (*Pandion cristatus*). According to Smith and Smith (1990 and 2006) a number of plant and animal communities are inadequately conserved, even if they may present at all in the major reserves. Plant species which only occur at 2% of sites have been classified as regionally and locally significant in that they may become locally extinct in 20 years if not recognized and afforded conservation status and protection (Hornsby Shire Council, 2006).

The 2017 Hornsby Vegetation Map identified 17,003 ha of vegetation (EcoLogical Australia, 2017) including 666 ha of TEC and a potential further 151 ha of potential relic TEC (subject to field validation). This accounted for 3.9% of the total LGA's vegetation. For example, only 54 ha of (confirmed) *Blue Gum Shale Forest*, or 0.32% of all TEC; and 20 ha of (confirmed) *Blue Gum Diatreme Forest*, or 0.11% of all TEC exist within the LGA, often existing as remnant patches and individual trees in urban landscapes across Council reserves, larger backyards, creek lines and schoolyards. In addition, there are only 323 ha confirmed (and a further143 ha) of *Sydney Turpentine-Ironbark*, or 1.89% of all TEC. Areas such as these are significant as they 'contain genetic material indigenous to the area and provide habitat for native fauna including threatened species and endangered populations [and] form parts of corridors and urban habitat links and contribute to the landscape character of the suburb' (Hornsby Shire Council, 2006).

Critically, a high percentage of remnant vegetation in the Hornsby LGA occurs on private properties across both rural and urban lands such as urban wetlands, waterways, remnant vegetation, solitary native and exotic trees, backyard and verge gardens, and the conservation and management of biodiversity must be undertaken across both of these as critical elements in maintaining biodiversity in a region. These include residents i.e. Long-nosed Bandicoots (*Perameles nasuta*) and Common Ringtail Possums (*Pseudocheirus peregrinus*), migratory species i.e. Swift Parrot (*Lathamus discolor*) and transient species i.e. Grey Headed Flying Fox (*Pteropus poliocephalus*). Such areas may contain species that occur in few other places, although at first glance they may exist as narrow roadsides or apparently weed infested spaces, it must be noted that these offer native smaller trees, shrubs, ground cover plants and grasses refuge in a hostile landscape matrix and may not be found elsewhere. Each portion of this vegetation remnant is important as there is so little of this type of vegetation left. Indeed, across the whole Sydney region, less than 2% of the original *Sydney Turpentine-Ironbark Forest* and *Blue Gum High Forest* remains (Government of New South Wales, 2019b). As such we need to consider these as important, and not just canopy trees often identified (Howell, 2000).

Many of the threats and challenges facing the region's biodiversity were also highlighted in the community and stakeholder workshops conducted as a part of the project. From a total of 321 responses, these included:

- Weeds: i.e. dispersal via public lands and waterways, escape from home gardens and domestic dumping
- Pollution: i.e. stormwater/runoff pollution and impacts to water quality
- Feral and domestic animals: i.e. direct impact of feral (e.g. foxes) and domestic animals (e.g. cats) on wildlife
- Climate Change: i.e. migration and adaption of species due to shifting normal range
- Connectivity: i.e. bushland fragmentation and lack of ecological corridors
- Recreation: i.e. Illegal mountain bike/other access tracks through bushland and protected habitat
- Fire: i.e. mismanaged fire regimes, lack of ecological knowledge used in planning burns
- Resource extraction: i.e. sand mining

Path to the future

An engaging and interactive methodology that included stages of mapping and modelling, design charrettes and community inputs, and, design refinement and visualization created a set of accurate, measurable and visual plans for the Hornsby LGA. This incorporated and executed the aspirations of local and state policy, captured community vision and provided a flexible and iterative approach to GI delivery across multiple scales. The GIF sets out a vision for the longevity of the LGA's biodiversity through the creation of a robust and ground-truthed set of plans, approaches and tactics. This is a framework for landscape resilience in the face of biodiversity loss, rapid climate and land-use change that is inclusive of the people most invested: those who live there. But one that is also science-backed, economically feasible and ecologically sustainable. A discussion is now made of the project and its method, focusing on three discussion points: *The potential of a Green Infrastructure approach; A diverse and inclusive method;* and *Meeting planning aspirations.*

The potential of a Green Infrastructure approach: Useful and incorporates synergistic benefits

At the core of this project is GI. GI is a design approach to create more resilient urban environments and could be considered as the 'ecological framework for environmental, social, and economic health – in short, our natural life-support system' (Benedict & McMahon, 2006, p. 1). GI purports to create a solution-driven approach to the question of creating landscape resilience in a challenging time of climate change, ecological fragmentation and species extinction and urban densification and land-use change. While already familiar with *grey* infrastructure, the idea of GI is less well-known and not something the public may typically appreciate. However, increasingly the adoption of an infrastructural approach to ecological planning and design brings the language and delivery of traditionally recognized grey infrastructures (e.g. sewers, electricity, roads and rail) to a novel green application as a useful spatial framework through which to consider landscapes across diverse land-uses and scales. Offering a design-focused and outcome-focused approach that draws upon both science as well as with clear connections to policy, and the potential to engage with citizens and stakeholders to create accurate, measurable and visual design outcomes, GI also offers a new approach to address environmental challenges and threats. Its utility could be summarized as:

- is policy-ready and offers a way to translate ecological design and planning into spatial reality,
- articulates well-established (best practice) landscape ecological design principles and enables ecological connectivity and
- offers a diverse array of benefits in an ecological, cultural and economic sense.

GI is present at multiple scales, connected and ubiquitous in different scales and contexts and is an approach that offers planning inclusive of the ecological, social, economic and even political perspectives of landscape (Odum, 1986; Opdam, Steingröver, & Van Rooij, 2006) in order to maximize potential for implementation and success. Hill and Johnston (2002) describe this type of approach as considering of 'ecosystem health, biotic integrity and cultural well-being'. While Ahern (2007) describes a need to consider landscapes through an 'abiotic, biotic and cultural' lens. Similarly, Wickham et al. promote GI as an appropriate response to 'integrate natural systems with community well-being' (2010, p. 186). The European Commission named GI as 'one of the main tools to tackle threats on biodiversity resulting from habitat fragmentation, land use change and loss of habitats' (2010). The Government Architect's Office, whose influential policies increasingly drive its adoption in NSW, defines GI as:

> Green infrastructure is the network of green spaces, natural systems and semi-natural systems including parks, rivers, bushland and private gardens that are planned, designed and managed to support a good quality of life in an urban environment.
>
> *(Government Architect NSW, 2017)*

Worldwide enthusiasm for this type of planning has grown and a vast range of initiatives, plans and projects[3] that exist at a wide range of scales and increased reference to the value and potential of GI in urban contexts has become more common at the level of the street, suburb and city. In Australia, examples include the *Perth Biodiversity Project* (West Australian Local Government Authority, 2009), Melbourne's *Urban Forest Strategy* (City of Melbourne, 2012), the *Sydney Green Grid* (Office of the Government Architect NSW, 2013), *Adelaide's GI Guidelines* (Adelaide City Council, 2014) and the *South East Queensland Regional Plan* (Department of Infrastructure Local Government and Planning, 2017). These are bolstered by broader initiatives such as the nationally focused 2020 Vision (Miller & Peacock, 2015) (now called *Greener Spaces, Better Places*). GI's tractability and appeal hails from its origins that did not begin with biodiversity protection, but rather take inspiration from urban planning and specifically the greenway movement (Fabos & Ahern, 1995; Little, 1995). This is a critical point as biodiversity preservation and enhancement has a symbiotic relationship with GI, with both benefiting each other, even if the motivations are mixed.

Like its origins, GI's benefits are diverse and synergistic between conservation-focused and other objectives. Benefits that – as Pickett suggests – entwine the social and the ecological (2001). This is both well-documented in the literature and supported by the findings of this project also. GI initiatives may include a spectrum of benefits from small-scale insertions of habitat and micro-climatic benefit via green roofs and walls at a local street level (Tóth, Halajová, & Halaj, 2015); to active transport and recreation at the neighborhood level; whilst simultaneously benefitting ecological and hydrological systems at regional scales (Matthews, Lo, & Byrne, 2015). This may result in increased biodiversity protection through the safeguarding and re-establishment of ecological connectivity, social and environmental resilience (Meerow & Newell, 2017), mitigating air pollution, urban heat island amelioration (Kubota et al., 2017; Nastran, Kobal, & Eler, 2018), providing recreational opportunities, increasing the potential for carbon sequestration, reducing stormwater run-off erosion and flooding, improving water quality in riparian areas, improving air quality (Government Architect NSW, 2017) and augmenting active transport networks and elevating visual amenity. When scaled up, GI can simultaneously mitigate urban heat island effects of a metropolitan region and help to regulate climate. Indeed, GI's multifunctionality is one of its most recognizable qualities (Ahern, 2007; Kilbane, 2013; Tzoulas et al., 2007) and the concept could be considered as 'a melting pot for innovative planning approaches in the field of nature conservation and green space planning' (Hansen & Pauleit, 2014).

Such benefits are measurable. For instance, Chen et al. found that admissions to hospital and heat-related mortality rate could be halved through doubling vegetation coverage (2014). Taylor et al. found clear relationship between street tree density with use of anti-depressant medication (2015). Even higher property prices have been associated with an increase of green cover, for instance Swinbourne and Rosenwax found that a 10% increase in the size of the canopy in Blacktown (Sydney, NSW) showed 'an increase in the value of property of 7.7 percent, or $AUD 55,000 for the average house' (2017) with similar findings in Perth based on street tree abundance (Pandit, Polyakov, Tapsuwan, & Moran, 2013).

Critically in relation to this study, the ecological corridors such as those that form Hornsby's GIF are considered to be an effective mechanism to maintain the long-term viability of wildlife and remnant vegetation and can play important ecological functions, including: the maintenance of genetic exchange, enabling re-colonization and providing a route for the dispersal of flora and fauna across inhospitable habitats - including migratory and nomadic species (Russell, Herbert, Kohen, & Cooper, 2013). Increasing ecological connectivity is an important consideration when considering fragmented landscapes and may help to mitigate species extinction – especially critical in a time of climate change and shifting distributions.

A diverse and inclusive method: Incorporates stakeholders and community

This project relied on a threefold method with various strengths.

First, the mapping offers accuracy and grounding to ideas and the model ensures that this is specific to the geophysical, hydrological and ecological foundations of place. However, in a departure from common practice, this model was only ever considered as a point of departure for further project stages where this was iteratively adjusted.

Second, design charrette workshops – 'time-limited, multiparty design event organized to generate a collaboratively produced plan for a sustainable community' (2008) – not only identified barriers to implementation but also highlighted synergies and harnessed local skills and knowledge, adding an expert human dimension (Remm, Külvik, Mander, & Sepp, 2004) to otherwise top-down planning (and its limitations). Charrettes are a well-established method in urban design and planning professions (Roggema, 2013; Walker & Seymour, 2008) and have been central to the development of many city design processes now for some time (2015; Karashima, Asano, & Ohgai, 2021; Lennertz & Lutzenhiser, 2006; Low & Plater-Zyberk, 2017; Neuman, Perrone, & Mossa, 2021; Trudeau, 2020). They may offer specific planning responses to climate change (Barrett, Kelly, & Hyde, 2017; Khirfan & El-Shayeb, 2019) and ecological reconstruction (Barrett et al., 2017; Kilbane et al., 2019). Charrettes rely on an *iterative* process to achieve feasible design outcomes (Congress for New Urbanism, 2015; Guerra & Shealy, 2018; Kilbane, 2013). For this reason, this project aimed to visually articulate policy and intent otherwise inaccessible as suite of mapping, illustrations and technology: as a framework for potential implementation. Through the workshop environment, charrettes 'can help in reviewing and confirming or revising the sustainability objectives previously proposed, now informed by local issues and stakeholders' (Julien, Hamilton, & Croxford, 2017, p. 141) and various proponents point to the ability of charrettes to collaboratively move from 'theory into practice' (Kilbane, 2013) or shift from 'knowledge to action' (Khirfan & El-Shayeb, 2019).

The use of design charrettes in this project helped to answer the question of vision for areas such as the Hornsby LGA, i.e. 'what type of city do we want', but also how to navigate a pathway to get to the green imperative that is so clearly required. Without a sound method to deploy this life-giving infrastructure, existing strategies also fall into this trap. The aforementioned *Sydney Green Grid* (Office of the Government Architect NSW, 2013), for instance, is a high-level planning document with broad support across NSW government that has limited adoption as it fails to deal with the realities and complexity of landscapes replete with existing use. This is the strength of the method as it offers a framework of diverse linkages that visualize potential connections between disparate habitat patches. These vary in size according to the process of refinement by community, staff and external advice. These are not theoretical planning overlays, nor static plans, but rather are ground-truthed and through citizen and stakeholder vetting offer a blueprint for a sustainable future, the creation of adjustable frameworks through which to consider the city. By integrating the expert human dimension of the people who actually live there while clearly articulating in a spatial sense the implications, constraints and opportunities of proposed planning outcomes are made visual and more tangible.

However, it is acknowledged as a framework will still require additional ground-truthing of ideas with relevant stakeholders and Doerr suggests that even if the perfect methodology could be agreed upon, it is the actual implementation of GI that is the real challenge (2013). This entails additional buy-in from stakeholders, adjustment to planning policies and mechanisms, sourcing potential avenues for funding and further detailed design to meet landscape specifics. For this reason, community participation was sought as a starting point and used to iteratively adjust and reconfigure the GIF. Here at the largest scale, the Framework knits together stakeholders design efforts and energies, combining biodiversity planning with landscape reality and reconnecting fragmented landscapes and community back together again.

Meeting planning aspirations: accurate, measurable and visual plans breach theory-to-practice gap as real-life solutions

Compatible with the intentions and objectives of the Greater Sydney Commission's North District Plan 2018 (Greater Sydney Commission, 2018) to interconnect disparate ecological fragments, and the Government Architects Office NSW's Bushland and Waterway Guide (Government Architect New South Wales, 2018), this project sets a strategic approach to the management and conservation of the biodiversity of the Hornsby LGA. It responds to the challenge set by the Government Architect of New South Wales in its *Greener Places 2020* framework and the supporting *Greener Places Design Guide* (Government Architect NSW, 2017). This guide suggests that NSW LGAs approach the

> conservation of urban habitat and biodiversity in a holistic way that not only directs strategic planning but also acts at the management level... in the form of maps identifying core, transition, and habitat connection areas, with development controls and land management provisions suited to the local area.
> *(Government Architect NSW, 2017)*

Support for GI already exists in NSW State Planning Policy and the NSW government has incorporated the concept of 'green infrastructure as essential infrastructure' in the latest Sydney Metropolitan Plan and five District Plans. According to the North District Plan 'green infrastructure is the network of green spaces, natural systems, and semi-natural systems that support sustainable communities'; and 'as our cities grow, Australian governments should focus on maintaining and enhancing green infrastructure and the public realm to ensure they remain liveable'. (Government Architect NSW, 2017; Greater Sydney Commission, 2018; Infrastructure Australia, 2018). Similarly, the *Green Grid, Better Placed* and *Greener Places* policies and accompanying manuals offer guidance on urban tree canopy, open space for recreation and connecting bushland and waterways and offer insights into government and community may collaborate. *Greener Places* specifically notes several desired outcomes relating for biodiversity protection and/or management:

- '*Protection and enhancement of natural resources and biodiversity by improving the quality of watercourses, creating green habitat corridors and protecting endangered ecological communities*
- *Promotion of social, cultural, recreational and educational opportunities within natural, cultural and heritage landscapes.*
- *Restoration and enhancement of wetland habitats and increased accessibility to them*
- *Creation of new ecosystems that support biodiversity such as constructed wetlands and green roofs*' (Government Architect NSW, 2017).

Greener Places specifically aims to use the GI to:

- '*create a network of healthy and attractive new and upgraded city environments, sustainable routes and spaces*'
- '*perform essential ecosystem services*'
- '*protect, conserve and enhance NSW's network of green and open natural and cultural spaces*' (Government Architect NSW, 2017).

However, such proposals take time to develop; and, the statutory embrace of GI and its principles is limited to date. For example, further planning of the Sydney Green Grid North District (Tyrrell Studio, 2017) falls well short and offers a bare minimum of ideas with no further detail for the region in question and minimal suggestions for methodology are provided to assist local governments to enact the next level of planning. In contrast, this work offers an example of the further detailed step towards the types of sustainable and resilient environments that we strive for.

Conclusion

Locally and globally, the need to safeguard biodiversity in light of rapid land-use and climatic change pressures has never been more acute than today – and the next 20 years will be critical. This presents a challenge for cities to carefully balance projected population growth targets and associated development against the protection of biodiversity values in the lands and waters under their management. Planning frequently struggles to address and be reconciled with complex landscapes already replete with existing human land-uses and activities, despite many well-intentioned plans and ideas. GI offers a reframing of infrastructure via a design focused and collaborative, visual approach and this Hornsby project evidences a design-based methodology and series of design scenarios, visualizations and accurate, measurable and visual plans through which to deliver this GI. This links to and responds to the international, national and local policy environment and offers an example of a sustainable urban living future and proposes a *Green Infrastructure Framework* that is inclusive of community and stakeholders who live there and have the greatest stake in decision making.

Acknowledgments

Special thanks to Dr. Peter Coad and Mark Hood from Hornsby Shire Council and for the Hornsby Shire Council for funding of the Project 'Hornby Biodiversity Conservation Strategy'.

Notes

1 Note: in Australia, 1788 marks the commencement of European colonization and forms a common ecological benchmark of floral structure.
2 At the broader perspective of NSW, there are currently 497 entities listed under the Biodiversity Conservation Act (NSW Office of Environment and Heritage, 2019). Of these 215 species are vulnerable, 132 endangered, 33 critically endangered and zero extinct; and, 35 endangered populations exist, 50 endangered ecological communities and four vulnerable ecological communities.
3 Such projects may be referred to as GI or as 'ecological', 'landscape' or 'green' networks.

References

Adelaide City Council. (2014). *Adelaide design manual: Green infrastructure guidelines*. Adelaide: Adelaide City Council.

Ahern, J. (2007). Green infrastructure for cities: The spatial dimension. In Vladimir Novotny and Paul Brown (Eds.), *Cities of the future towards integrated sustainable water and landscape management* (pp. 267–283). London: IWA Publishing.

Australian Bureau of Statistics. (2015). Population projections, Australia, 2012 (base) to 2101. Available at: http://www.abs.gov.au/ausstats/abs@.nsf/Lookup/3222.0main+features52012%20(base)%20to%202101

Australian Bureau of Statistics. (2020). *Data by region – Hornsby (A) (LGA)*. Available at: https://dbr.abs.gov.au/

Barrett, J., Kelly, M. R., & Hyde, B. (2017). Living shoreline design Charette: A new twist on the Charette technique. *Journal of Extension, 55*(5).

Benedict, M. A., & McMahon, E. (2006). *Green infrastructure: Linking landscapes and communities*. Washington, DC: Island Press.

NSW Government (2016). Biodiversity Conservation Act 2016 No 63. Available at: https://legislation.nsw.gov.au/view/html/inforce/current/act-2016-063

Chapman, A. D., & Australian Biological Resources Study Australia. (2009). *Numbers of living species in Australia and the world*. (9780642568601 (printed)) 9780642568618 (online)). Canberra: Department of the Environment, Water, Heritage and the Arts. Available at: http://www.environment.gov.au/biodiversity/abrs/publications/other/species-numbers/2009/pubs/nlsaw-2nd-complete.pdf

Chen, D., Wang, X., Thatcher, M., Barnett, G., Kachenko, A., & Prince, R. (2014). Urban vegetation for reducing heat related mortality. *Environmental Pollution, 192*, 275–284. doi:10.1016/j.envpol.2014.05.002

City of Melbourne. (2012). *Urban forest strategy: Making a great city greener 2012–2032*. Available at: melbourne.vic.gov.au/urbanforest

Condon, P. M. (2008). *Design charrettes for sustainable communities*. Washington, DC: Island Press.

Congress for New Urbanism. (2015). *Congress for new urbanism|tools*. Available at: https://www.cnu.org/resources/tools

Department of Environment Climate Change and Water NSW. (2009). *Climate change how will it affect the natural environment in NSW*. Sydney: Department of Environment Climate Change and Water NSW.

Department of Infrastructure Local Government and Planning. (2017). *South East Queensland Regional Plan 2017 – ShapingSEQ background paper 5: Live theme*. Brisbane. Available at: https://planning.dilgp.qld.gov.au/

Doerr, VAJ, Williams, KJ, Drielsma, M, Doerr, ED, Davies, MJ, Love, J, Langston, A, Low Choy, S, Manion, G, Cawsey, EM, McGinness, HM, Jovanovic, T, Crawford, D, Austin, M & Ferrier, S 2013, Designing landscapes for biodiversity under climate change: Final report, National Climate Change Adaptation Research Facility, Gold Coast, 260 pp.

EcoLogical Australia. (2017). *Hornsby vegetation map update 2017 prepared for Hornsby Shire Council*.

European Commission & Directorate-General of the Environment. (2010). *Green infrastructure*. Available at: http://ec.europa.eu/environment/nature/info/pubs/docs/greeninfrastructure.pdf

Fabos, J. G., & Ahern, J. (1995). *Greenways: The beginning of an international movement*. Amsterdam; New York: Elsevier.

Flannery, T. F. (1997). *The future eaters: An ecological history of the Australasian lands and people*. Sydney: New Holland Publishers.

Forman, R. T. T., & Godron, M. (1986). *Landscape ecology*. New York: Wiley.

Gammage, B. (2011). *The biggest estate on Earth: How Aborigines made Australia*. Crows Nest: Allen & Unwin.

Government Architect New South Wales. (2018). *Greener places: Bushland and waterways guide draft for discussion*. Sydney. Available at: https://www.activelivingnsw.com.au/assets/Uploads/Draft-Bushland-and-Waterways-2018-07-31-2.pdf

Government Architect NSW. (2017). *Greener places: Establishing an urban Green Infrastructure policy for New South Wales (Draft for discussion)*, Sydney: NSW Government.

Government of New South Wales. (2019a). *NSW state of the environment*. Sydney. Available at: https://www.soe.epa.nsw.gov.au

Government of New South Wales. (2019b). *NSW threatened species scientific committee – Final determination, Sydney Turpentine-Ironbark Forest in the Sydney Basin Bioregion – Critically endangered ecological community listing*.

Greater Sydney Commission. (2018). *Our greater Sydney 2056 north district plan – connecting communities*. Sydney: Greater Sydney Commission.

Guerra, M., & Shealy, T. (2018). Teaching user-centered design for more sustainable infrastructure through role-play and experiential learning. *Journal of Professional Issues in Engineering Education and Practice, 144*(4), 05018016. doi:10.1061/(asce)ei.1943-5541.0000385

Hansen, R., & Pauleit, S. (2014). From multifunctionality to multiple ecosystem services? A conceptual framework for multifunctionality in green infrastructure planning for urban Areas. *Ambio, 43*(4), 516–529. doi:10.1007/s13280-014-0510-2

Hill, K., & Johnston, B. R. (2002). *Ecology and design: Frameworks for learning* Washington, DC: Island Press.

Hornsby Local Environmental Plan 2013- NSW Legislation, (2013).

Hornsby Shire Council. (2006). *Biodiversity conservation strategy*. Hornsby: Hornsby Shire Council.

Hornsby Shire Council. (2013). *Hornsby development control plan*. Hornsby: Hornsby Shire Council.

Howell, J. (2000). *Bushland in the Hornsby Shire and its significance*. Available at: Royal Botanical Gardens. A brief summary prepared for Hornsby Shire's Rural Lands Incentive Program.

Infrastructure Australia. (2018). *Future cities: Planning for our growing population*. Canberra: Australian Government.

IUCN. (1994). *Guidelines for protected area management categories*. Cambridge: IUCN.

Jackson WJ, Argent RM, Bax NJ, Clark GF, Coleman S, Cresswell ID, Emmerson KM, Evans K, Hibberd MF, Johnston EL, Keywood MD, Klekociuk A, Mackay R, Metcalfe D, Murphy H, Rankin A, Smith DC & Wienecke B (2017). Australia state of the environment 2016: overview, independent report to the Australian Government Minister for the Environment and Energy, Australian Government Department of the Environment and Energy, Canberra.

Julien, A., Hamilton, I., & Croxford, B. (2017). Environmental masterplanning: Defining an integrated approach. In: Carmona, M. (Ed.), *Explorations in urban design: An urban design research primer* (pp. 135–143). Taylor and Francis.

Karashima, K., Asano, J., & Ohgai, A. (2021). Research on the effectiveness of charrette workshop in local cities. *AIJ Journal of Technology and Design, 27*(66), 955–960. doi:10.3130/aijt.27.955

Khirfan, L., & El-Shayeb, H. (2019). Urban climate resilience through socio-ecological planning: A case study in Charlottetown, Prince Edward Island. *Journal of Urbanism: International Research on Placemaking and Urban Sustainability, 13*(2), 187–212. doi:10.1080/17549175.2019.1650801

Kilbane, S. (2013). Green infrastructure: Planning a national green network for Australia. *Journal of Landscape Architecture, 8*(1), 64–73. doi:10.1080/18626033.2013.798930

Kilbane, S., & Kopinski, J. (2016). *Evolution and evaluation of contemporary greenways and green infrastructure in Sydney, Australia*. Paper presented at the Greenways and Landscapes in Change, Budapest.

Kilbane, S., Weller, R., & Hobbs, R. (2019). Beyond ecological modelling: Ground-truthing connectivity conservation networks through a design charrette in Western Australia. *Landscape and Urban Planning*. doi: 10.1016/j.landurbplan.2017.05.001

Kubota, T., Lee, H. S., Trihamdani, A. R., Phuong, T. T. T., Tanaka, T., & Matsuo, K. (2017). Impacts of land use changes from the Hanoi Master Plan 2030 on urban heat islands: Part 1. Cooling effects of proposed green strategies. *Sustainable Cities and Society, 32*, 295–317. doi:10.1016/j.scs.2017.04.001

Lennertz, W. R., & Lutzenhiser, A. (2006). *The charrette handbook: The essential guide for accelerated, collaborative community planning.* Chicago, IL; Washington, DC: American Planning Association.

Little, C. E. (1995). *Greenways for America.* Baltimore, MD: Johns Hopkins University Press.

Low, T. E., & Plater-Zyberk, E. (2017). A new plan for the central neighborhood Cleveland, Ohio. In: Preiser, W.F.E., Russell, F. P., and Varady, D.P. (Eds.), *Future visions of urban public housing (Routledge Revivals): An international forum, November 17–20, 1994* (pp. 411–420). Taylor and Francis Inc.

Matthews, T., Lo, A. Y., & Byrne, J. A. (2015). Reconceptualizing green infrastructure for climate change adaptation: Barriers to adoption and drivers for uptake by spatial planners. *Landscape and Urban Planning, 138,* 155–163. doi:10.1016/j.landurbplan.2015.02.010

Meerow, S., & Newell, J. P. (2017). Spatial planning for multifunctional green infrastructure: Growing resilience in Detroit. *Landscape and Urban Planning, 159,* 62–75. doi:10.1016/j.landurbplan.2016.10.005

Miller, J., & Peacock, B. (2015). *The 202020 vision plan.* Available at: https://202020vision.com.au/media/41955/202020visionplan.pdf

Nastran, M., Kobal, M., & Eler, K. (2018). Urban heat islands in relation to green land use in European cities. *Urban Forestry & Urban Greening.* doi:10.1016/j.ufug.2018.01.008

Neuman, M., Perrone, C., & Mossa, A. (2021). Applied research by design: An experimental collaborative and interdisciplinary design charrette. *European Planning Studies,* 1–21. doi:10.1080/09654313.2021.1911956

NSW Department of Planning. (2010). *Metropolitan plan for Sydney 2036.* Available at: metroplansydney.nsw.gov.au

NSW Government. (2014). Environment protection and biodiversity conservation act 1999 (EPBC Act) approved conservation advice (including listing advice) for Shale Sandstone Transition Forest of the Sydney Basin Bioregion (EC25R).

NSW Office of Environment and Heritage. (2019). Threatened species found in Pittwater IBRA sub-region. Available at: https://www.environment.nsw.gov.au/threatenedspeciesapp/cmaSearchResults.aspx?SubCmaId=380

Odum, E. P. (1986). Introductory review: Perspective of ecosystem theory and application. In N. Polunin (Ed.), *Ecosystem theory and application* (pp. 1–11). New York: Wiley.

State of NSW and Office of Environment and Heritage. (2017). Saving our species key threatening proccesses strategy. *2017 State of NSW and Office of Environment and Heritage.* Sydney: NSW Government. Available at: https://www.environment.nsw.gov.au/-/media/OEH/Corporate-Site/Documents/Animals-and-plants/Threatened-species/saving-our-species-key-threat-process-strategy-draft-june-2017-170093.pdf

Office of Environment and Heritage. (2019). *NSW climate change snapshot* (1837–5650). Sydney: NSW Government.

Office of the Government Architect NSW. (2013). *Sydney green grid.* Available at: http://202020vision.com.au/media/7200/barbara-schaffer-gao-sydneys-green-grid.pdf

Opdam, P., Steingröver, E., & Van Rooij, S. (2006). Ecological networks: A spatial concept for multi-actor plannng of sustainable landscapes. *Landscape and Urban Planning, 143,* 1–8.

Pandit, R., Polyakov, M., Tapsuwan, S., & Moran, T. (2013). The effect of street trees on property value in Perth, Western Australia. *Landscape and Urban Planning, 110,* 134–142. doi:10.1016/j.landurbplan.2012.11.001

Pascoe, B. (2016). *Dark Emu.*

Pickett, S. T. A., Cadenasso, M. L., Grove, J. M., Nilon, C. H., Pouyat, R. V., Zipperer, W. C., & Costanza, R. (2001). Urban ecological systems: Linking terrestrial ecological, physical, and socioeconomic components of Metropolitan Areas. *Annual Review of Ecology and Systematics, 32*(1), 127.

Remm, K., Külvik, M., Mander, Ü., & Sepp, K. (2004). Design of the Pan-European Ecological Network: A national level attempt. In R. H. G. Jongman & G. Pungetti (Eds.), *Ecological networks and greenways*. Cambridge: Cambridge University Press.

Roggema, R. (2013). *The design charrette: Ways to envision sustainable futures*. Available at: http://hdl.handle.net/1959.3/376673

Russell, T. C., Herbert, C. A., Kohen, J. L., & Cooper, D. (2013). The incidence of road-killed possums in the Ku-ring-gai area of Sydney. *Australian Journal of Zoology, 61*(1), 87–94. doi:10.1071/ZO12118

Sanoff, H. (2000). *Community participation methods in design and planning*. John Wiley & Sons.

Smith, P., & Smith, J. (1990). *Hornsby Shire Bushland survey*. Report prepared for Hornsby Shire Council. P&J Smith Ecological Consultants, Blaxland.

Smith, P., & Smith, J. (2008). *Native vegetation communities of Hornsby Shire 2008 update*. Blaxland: P & J Smith Ecological Consultants.

Government of New South Wales (2018a). State Environmental Planning Policy (Koalas no. 44). Sydney: NSW Government.

Government of New South Wales, (2018b). State Environmental Planning Policy (Vegetation in Non Rural Areas). Sydney: NSW Government.

Steffen, W. (2009). *Australia's biodiversity and climate change*. Collingwood: CSIRO Publishing.

Swinbourne, R., & Rosenwax, J. (2017). A brilliant cities report: Green Infrastructure. In AECOM (Ed.).

Taylor, M. S., Wheeler, B. W., White, M. P., Economou, T., & Osborne, N. J. (2015). Research note: Urban street tree density and antidepressant prescription rates—A cross-sectional study in London, UK. *Landscape and Urban Planning, 136*, 174–179. doi:10.1016/j.landurbplan.2014.12.005

Tóth, A., Halajová, D., & Halaj, P. (2015). Green infrastructure: A strategic tool for climate change mitigation in urban environments. *Journal of International Scientific Publications, 9*, 132–138.

Trudeau, D. (2020). Disparate projects, coherent practices: Constructing new urbanism through the charter awards. *Urban Planning, 5*(4), 429–440. doi:10.17645/up.v5i4.3366

Tuft, R., Rooney, W., Bacon, P., Gittins, D., Rawlins, J., & Pyke, G. (2000). *Aquatic bioindicator survey for Hornsby Shire Council*. Sydney: Robyn Tuft & Associates, WS Rooney & Associates, Woodlots & Wetlands, Canonical Solutions, UBMC and The Australian Museum.

Tyrrell Studio. (2017). *North district. Sydney green grid spatial framework and project opportunities*. Sydney Available at: https://www.governmentarchitect.nsw.gov.au/resources/ga/media/files/ga/plans/sydney-green-grid-plan-6-north-district-2017.pdf

Tzoulas, K., Korpela, K., Venn, S., Yli-Pelkonen, V., Kazmierczak, A., Niemela, J., & James, P. (2007). Promoting ecosystem and human health in urban areas using green infrastructure: A literature review. *Landscape and Urban Planning, 81*(3), 167–178. doi:10.1016/j.landurbplan.2007.02.001

Walker, J. B., & Seymour, M. W. (2008). Utilizing the design charrette for teaching sustainability. *International Journal of Sustainability in Higher Education, 9*(2), 157–169.

West Australian Local Government Authority (Cartographer). (2009). *Perth biodiversity project*. Available at: https://walga.asn.au/getattachment/Policy-Advice-and-Advocacy/Environment/Local-Biodiversity-Program/PBP-veg-complex-metadata_updatedJuly-2011.pdf.aspx?lang=en-AU

Wickham, J. D., Riitters, K. H., Wade, T. G., & Vogt, P. (2010). A national assessment of green infrastructure and change for the conterminous United States using morphological image processing. *Landscape and Urban Planning, 94*(3–4), 186–195. Available at: http://www.sciencedirect.com/science/article/B6V91-4XMK1TV-1/2/c5ebb2abfa35b2f33f5f233f063dbc5a

World Bank. (2014). Urban population (% of total). Available at: http://data.worldbank.org/indicator/SP.URB.TOTL.IN.ZS

11

SATOYAMA LANDSCAPES

Creating resilient socio-ecological production landscapes in Japan

Katsue Fukamachi

Vision

Satoyama landscapes comprise an integral social and ecological network of a village and its surroundings, such as agricultural lands, and forests (Fukamachi et al., 2001). They are socio-ecological production landscapes with regular land-use patterns that have been systematically managed by local communities, and similar landscapes can be found in Asia (Ichiawa, 2012).

At COP10 in 2010, Japan drew up the Satoyama Initiative and proposed it to the world (Watanabe et al. 2012). The initiative aims to gather knowledge on living in harmony with nature from areas around the world, to re-create and develop these, and to find practical uses for them. Moreover, it proposes a global framework to promote sustainable land use and use of natural resources. The three approaches proposed for achieving the goal are: (1) consolidating wisdom on securing diverse ecosystem services and values, (2) integrating traditional ecological knowledge and modern science, and (3) exploring new forms of co-management systems (a "New Commons"). It is critical to create synergy between traditional knowledge and modern science in order to promote necessary innovations, and foster new challenges to encourage multi-stakeholder participation. More recently, collaboration in sustainable and multifunctional management of natural resources can be seen.

In the satoyama landscape, local people make diverse use of familiar rivers and springs in their daily lives, and paying attention to these water systems is important for understanding the landscape characteristics of the region (Fukamachi et al., 2016). Prior to the development of public facilities, protection against floods was done on a household basis through traditional flood prevention architecture and site selection. Analyses of historical water systems have shown that both water sources and drainage destinations are diverse, and that water

DOI: 10.4324/9781003266440-11

is returned to the water cycle within a short distance after its use (Ryu et al., 2011). By deciphering these traditional mechanisms of natural resource use, the significance of these systems as ecosystem services and local culture can be clarified, and suggestions for sustainable use of natural resources can be obtained (Fukamachi, 2020).

It has also been pointed out that daily life which makes good use of the blessings of nature can cultivate a respect for nature, fosters disaster prevention wisdom, and passes down a "disaster culture" (Iizuka et al., 2018). The response to floods and intentions for future flood countermeasures are directly related to community activities and the degree of acceptance of flooding risks. It has also been shown that focusing on more "normal" times, rather than on times of disaster, and conducting community exchange programs can lead to proactive support activities in times of emergency. It is also important to make community-based project assessment using the indicators of resilience to improve satoyama landscapes in the future (Dublin and Natori, 2020).

Today, however, the nature of communities is changing, and how to respond to natural disasters such as floods and landslides has become a major issue. There are few cases where the spatial structure and the nature of communities have been discussed, taking a comprehensive view of daily water use and disaster prevention and mitigation. Therefore, it is necessary to accumulate empirical studies, share the significance of water use and management by communities, and develop a system for water system management.

Ecosystem-based Disaster Risk Prevention and Reduction (Eco-DRR) implies maintaining ecosystems and ecosystem services that serve as buffer zone or buffer material against natural disasters, and that can function as a lifeline for the supply of food and water for residents and communities in the case of natural disaster (Nature Conservation Bureau, Ministry of the Environment, 2016). Eco-DRR should be based on the participation of various stakeholders of local community building, and they should be given the opportunity to learn about the ecosystem that existed in their area originally, the local disaster history, and the traditional techniques that were used in the past to respond to natural disasters. For example, by informing about places in the community that are at an increased risk of being impacted by a natural disaster, and about disaster prevention measures that are already in place, residents of the community will be able to actively participate in the discussion of measures needed to create a disaster resilient landscape.

The aim of this chapter is to clarify the local traditional knowledge on natural disaster risk reduction. During the investigation, local historical maps and literature were used and a resident survey using 20 interviews was conducted to examine related documents. This is a case study of community resilience efforts to achieve multiple benefits, including ecological, social, and cultural "commons" frameworks, while respecting traditional communal land tenure and co-management. Based on the local traditional knowledge of natural disaster risk reduction of the water network, spatial structure of satoyama landscape and community activities were examined for creating resilient socio-ecological production landscapes.

Local traditional knowledge of natural disasters depicted in old maps

Figure 11.1 shows the topography and land-use map of the study area, a satoyama landscape in the Hira Mountain range of Otsu City in Shiga Prefecture, Japan in 1893, about 20 km to the northeast of Kyoto. From an altitude of about 1,000 m, steep mountain slopes reach down to the alluvial fan and finally to Lake Biwa (Figure 11.2). Many of the mountain streams and brooks that flow down to the lake have developed riverbeds that are higher than the surrounding ground due to sediment accumulation. Small inland lakes connected to Lake Biwa have

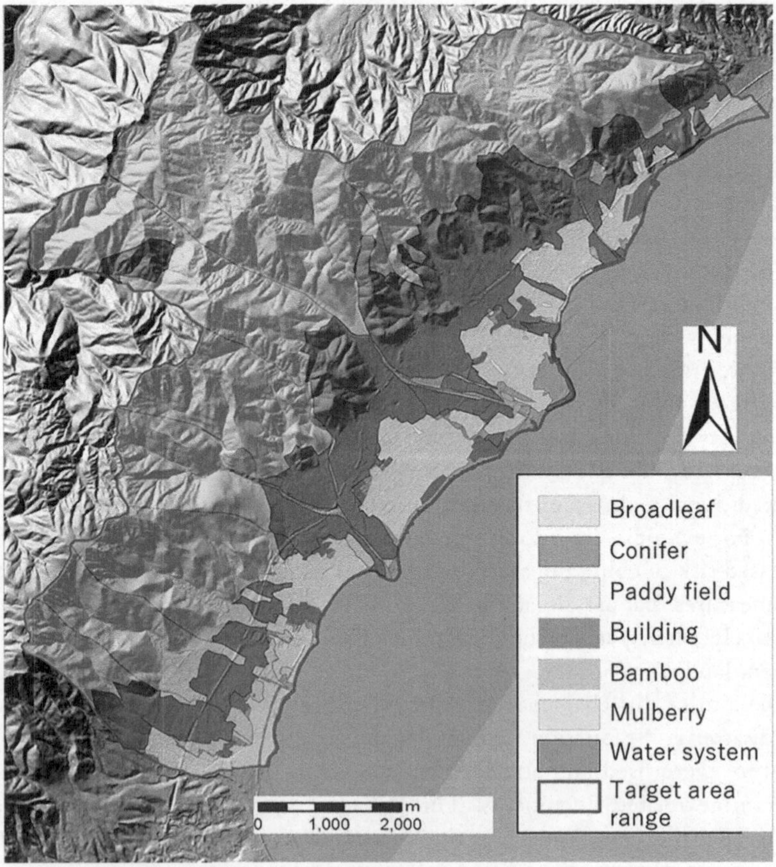

FIGURE 11.1 Topography and land cover of the Hira Mountain range in 1893 (Miyoshi, 2020; partially modified).

FIGURE 11.2 Hira Mountain range and Lake Biwa in Shiga Prefecture.

also been formed. The main soil material in this area consists of granite and sedimentary layers. Settlements were scattered along the lakeshore, and agricultural land, mainly paddy fields, spread out in the vicinity in 1893. Coniferous forests, mainly red pine (*Pinus densiflora*), spread in the low elevation area, and broadleaf forests consisting of a variety of tree species such as oak (*Quercus serrata*) and beech (*Fagus* spp.) were distributed in the higher elevations.

After 1950, road and railway construction, river-improvements, and other land development projects led to urbanization, with many of the former farmlands in the vicinity of the lake having been replaced by new residential areas (Fukamachi et al., 2011). There was also a rapid increase in conifer plantations of Japanese cedar (*Cryptomeria japonica*) and Japanese cypress (*Chamaecyparis obtusa*), and a decrease in broadleaf forests.

There are nine villages in the study area of the Hira Mountain range. Old village maps, land registers, and other historical maps concerning the communal control of mountain forests and streams were found and used in the study. In the villages situated along the lakeshore at the base of the Hira Mountain range, agriculture, fishery, forestry, and stone manufacturing had been the main traditional income sources. The borderlines of water bodies kept changing based on water flow and sediment movement. On the one hand, these areas could be a threat during natural disasters such as mudflows. On the other hand, they were highly appreciated natural resources, both as a source of income and a resource that villagers could use in daily life. The reasons for the distribution of forests along the rivers may be that the areas near the rivers flooded, preventing encroaching land development, and were deliberately maintained as flood prevention forests (Miyoshi, 2020).

Next, using two settlements as case studies, historical maps from the 1600s to 1800s are used to show the characteristics of land use and disaster response. Figure 11.3 illustrates a historical map of Moriyama from the year 1664, the so-called Mt. Fukutani Dispute Approval Illustrated Map. Shown on the map

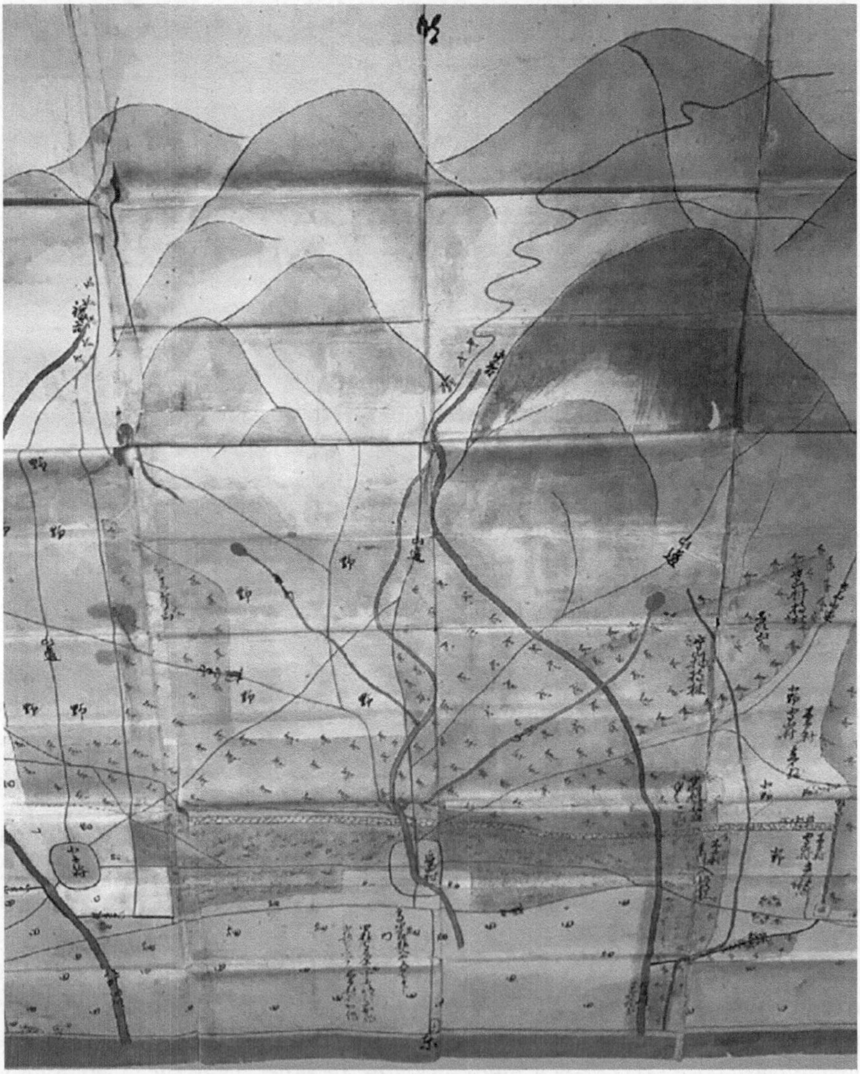

FIGURE 11.3 Historical map of the Moriyama "Mt. Fukutani Dispute Approval" (1664, Moriyama Zaisanku Collection).

are all land uses from the lakeshore to the mountain summit. At the time, there were 57 homes and a total population of 260 people (267 houses and 826 people in 2004). Red lines indicate roads, while blue lines show streams. The land ownership system indicates that forests between the water drawing point for paddy irrigation and the mountain summit, and also the lakeshore, were all communal property.

There was a fence made from locally quarried chert close to the residential area. Located on the border between the residential area and the mountain forest, this stone fence was built to keep game such as wild boars away from the village. The structure was also thicker and built in a sturdier way in certain locations in order to mitigate impacts from mudflows. Usually, it was possible to pass through the opening on a road, yet in times of disaster, the opening could be closed with a sheathing board. The management of natural resources such as forests and water systems involved a variety of local organizations centered around livelihoods and beliefs. A system of joint management functioned to share the responsibility according to the location, size, and importance of the natural resources. Waterway management had two main purposes: Allowing efficient use of natural resources, and reducing as much as possible the impact of natural disasters. Shrines were built at strategic locations, for example, at the confluence of rivers, at water drawing points, or at the borderline between different land uses. Residents were familiar with the locations thanks to regular shrine festivals or events, and land-use change in the surrounding area was strictly controlled.

Figure 11.4 illustrates a map from 1827 in Minamikomatsu. There were 125 homes and a total population of 730 people in 1789 (791 homes and 1,819 people in 2018). The yellow zones stand for residential areas, and the grey zones for wastelands. Wasteland, a natural wilderness environment, was concentrated along the river where pine forests spread out, despite being disturbed by flood damage (Ando et al., 2020). A long embankment along the Hiragawa River is drawn in black. From the wastelands, natural resources used daily such as woods, grass, and stones were collected. The area around the inland lake played a role in flood control, as well as the supply of reed and abundant fish. In three locations next to the river, an inscription mentions the length of the stone embankment.

The traditional knowledge that translated into techniques and practices helped reduce damage in the case of floods and sediment disasters in the Hira mountain range included:

1 The construction of structures using natural resources such as stone and wood, for example the stone embankments and the stone fence.
2 A network of branched-out, man-made canals. Many of these also served as water supply routes for daily life and for agricultural needs.
3 Special care was taken to reduce disaster damage by regulating the ownership and land-use system in the village. Areas prone to disaster damage

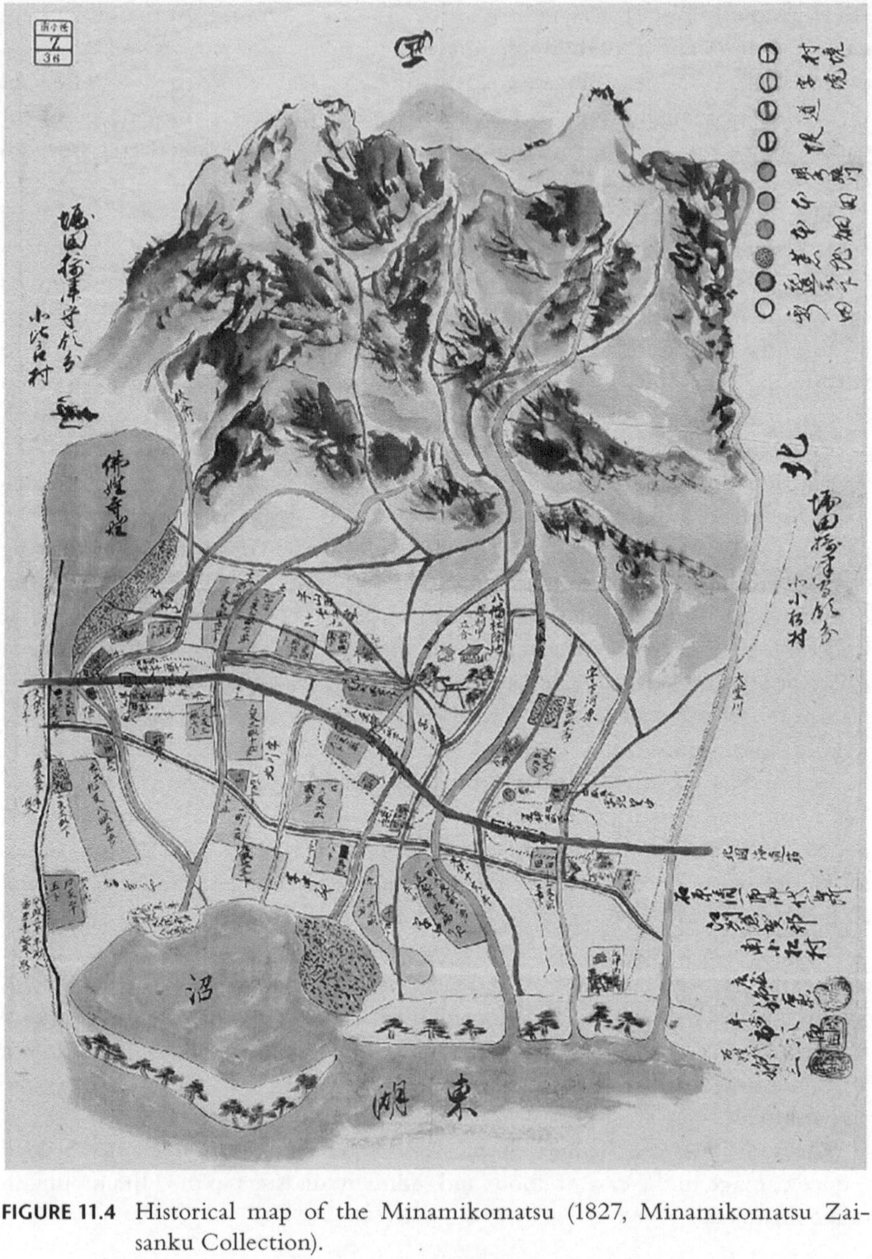

FIGURE 11.4 Historical map of the Minamikomatsu (1827, Minamikomatsu Zaisanku Collection).

were designated as communal land, or shrines were built there, so that they changed into spaces with spiritual meaning.

4 High-risk areas were designated as wasteland zones. In this way, the village as a whole conducted land planning that was closely linked to disaster response. As a result, resident awareness of disaster risk was heightened, and it became difficult to recklessly or thoughtlessly make land-use changes.

Current and future stressors

At the foot of Mt. Hira, there is a high risk of mudslides due to the topographical characteristics of the area. In addition, the riverbed rises higher than the surrounding land, causing the dikes to break and the surrounding area to be flooded or damaged by the inflow of sediment. Figure 11.5 shows the distribution of landslide hazard warning areas and residential areas in 2016. A landslide hazard warning area is defined as a designated area where there is a risk of landslide disasters, such as mudslides, landslides, and collapses, according to the Landslide Disaster Prevention Act. As shown in Figure 11.1, fan-shaped areas that were prone to mudslides and unsuitable for farmland due to the presence of many stones and rocks in the soil were not used as residential areas or farmland, but instead became wastelands or forests in 1893. As a result, most of the residential

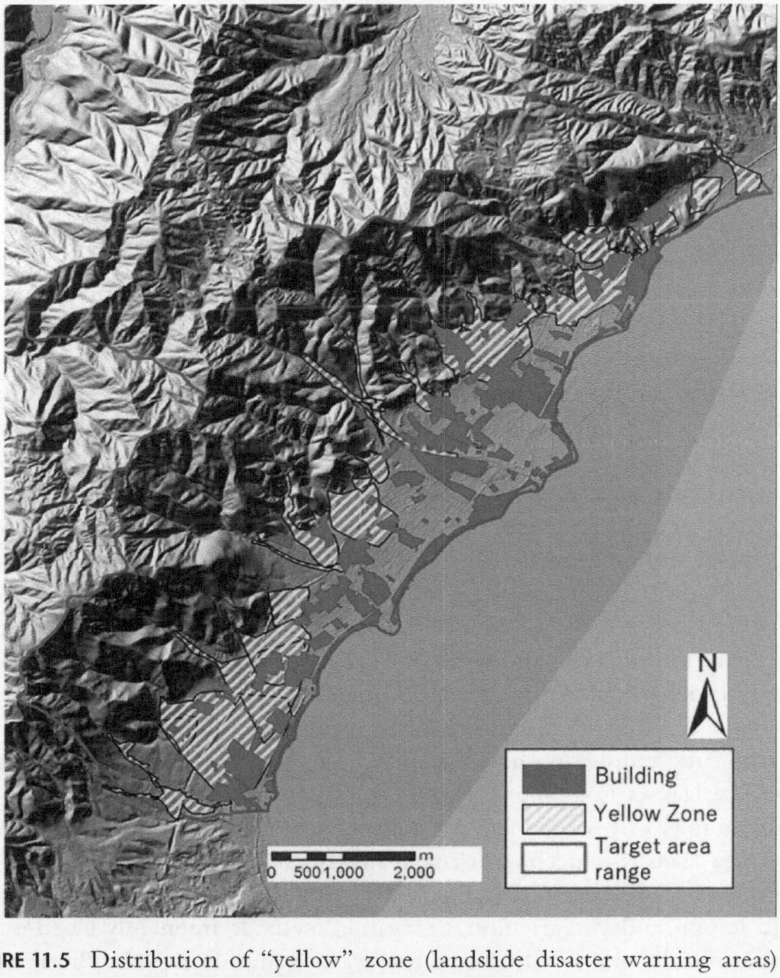

FIGURE 11.5 Distribution of "yellow" zone (landslide disaster warning areas) and residential area in 2016 (Miyoshi, 2020).

areas were outside the landslide disaster hazard areas, but with the progress of urbanization, erosion control projects, river improvement, and other public works since the 1960s, residential areas expanded into farmland and forests including landslide disaster hazard areas in 2016 (Figure 11.5). The total area of residential areas increased 2.42 times during the period from 1893 to 1975, and further 2.64 times during the period from 1975 to 2016. The elevation of residential areas was distributed between 84 m and 154 m on the shore of Lake Biwa in 1893, but this area changed to between 84 m and 331 m in 1975, and residential areas (including villas) became widespread in the high-elevation area. For example, in Moriyama, common land that was used as a natural resource for firewood and organic fertilizer by dividing the surface rights among residents until the 1960s was sold and transformed into a new residential area. There has also been a significant change in forest vegetation, with many of the broadleaf forests that were used as firewood forests being converted into cedar and cypress plantations.

In addition, the water system has been fragmented by the development of fields, transportation networks, residential areas, and tourism development since the 1960s, which has affected the cohesiveness of the landscape. For example, in Minamikomatsu, tourism development projects have progressed around the shore of Lake Biwa and the inner lake, and places that were farmland and wasteland in 1893 have been transformed into large-scale accommodation and leisure facilities and parking lots. Many guest houses were also built on the beach, which is the common land of Minamikomatsu, under lease contracts. As a result of the policy to promote renewable energy during the 2010s, large-scale solar panels have appeared on the shores of lakes and within forests in an unregulated manner.

In the current satoyama landscape at the foot of Mt. Hira, although the population is increasing as a suburb of the city, the population involved in agriculture, forestry, and fisheries is decreasing as well as aging. As people's lives and livelihoods change, the use and management of forests and farmland is becoming less common, and there is a serious shortage of people to manage these common lands. Traditional land-use patterns are changing, and the effects of climate change are increasing the risk of large-scale rainfall and landslides that exceed the planned scale of disaster prevention facilities. Recently, there has been a need to cope with natural disasters that are difficult to predict with conventional experience, such as a large amount of rainfall in a short period of time, causing Lake Biwa, inland lakes, and waterways in residential areas to overflow all at once. Traditional land management techniques, which are important for disaster prevention and mitigation, are no longer being handed down as the younger generation and immigrants have not joined the circle of common land management. The lack of functioning of the lake-sato (settlement)-forest linkage and the community's own rules and organizations for controlling key management points also leads to a decline in risk management capacity to prepare for and adapt to natural disasters such as mudslides and floods.

In addition, today's agricultural irrigation methods are mainly based on the use of artificial water systems using backwater pumps from Lake Biwa, which increases maintenance and management costs and places a heavy burden on the local community. In the future, it will be important to secure and regenerate the continuity of the water system in order to realize basin flood control, as well as a comprehensive water system management system that utilizes rivers and springs. As members of the local community, various actors such as businesses and new residents are required to share roles and collaborate in order to enjoy the benefits of the water system and lead to disaster prevention and mitigation.

The abovementioned issues in the satoyama at the foot of Mt. Hira are common throughout Japan, and have become major problems in terms of the decline of ecosystem services and biodiversity, and the maintenance of disaster prevention functions. In addition, while public works and governmental management systems for natural resources have been established in the modern era, opportunities for joint management of natural resources have decreased, and the active involvement of local communities in utilizing traditional and local knowledge is being lost. While tracing the nature, culture, and history of each region, it is necessary to reaffirm the significance and potential of traditional local organizations and organizations related to livelihood and faith, and to find ways to utilize them in the future.

Resilient landscape regeneration and creation of socio-ecological production landscapes

In the satoyama at the foot of Mt. Hira, citizen group activities such as making use of natural resources in daily life and fostering environmental education have been occurring as far back as the 1950s. For example, landscape architects, farmers, and other residents who are invested in their livelihoods have contributed their wisdom and skills to install wooden fishways in small rivers and revive masonry waterways. In one case, a new place for community networking was created through the sustainable use of forest resources for wood stoves and organic vegetable cultivation from an abandoned mountain forest. In collaboration with researchers, educational programs for parents and children, such as biological surveys in inland lakes and rivers and food culture experiences, are also being implemented. As these activities become more commonplace, young farmers aiming to make a living from organic farming and café owners using local ingredients have emerged. In addition to this, houses built using locally produced stone and wood are also becoming more popular.

Through these civic activities and new livelihoods, and by repeating trial and error, new ingenuity in using natural resources around us is accumulated. In addition, with the participation and cooperation of various actors, long-term craftsmanship in this field has been cultivated. The project can be evaluated as playing an important role in producing important components of resilient landscapes and connecting associated elements. It shows that recognizing the value of existing ecosystems and cultural landscapes from the perspective of livelihoods and making a living in the region, including pioneering actions based on these concepts, will be a major driving force to connect sustainable resource use with culture and diversity in the future.

On the other hand, the major forests and water systems of the local landscape have been considered the common forests of traditional communities or places jointly managed by community organizations, so the opportunities for the activities of citizen organizations (consisting mostly of relocating residents and/or immigrants) have been limited. However, as interest in sudden climate change-fueled natural disasters has increased, community associations and traditional organizations have become the foundation for new activities related to satoyama.

In Moriyama, a voluntary disaster prevention organization was established in the local community association during the 1970s, and joint forest care and river management work aimed at Eco-DRR was carried out in cooperation with experts. In addition, elderly residents and other experts share their knowledge and skills about the nature, culture, and history of the region, which has led to a deeper understanding and attachment to it. By sharing information on disaster response with the governments of Shiga Prefecture and Otsu City, and through discussions and cooperation among residents, a system for local weather monitoring and evacuation according to different disaster risks depending on the location of houses is being established. The practice of Eco-DRR is expected to expand the interest and activities in satoyama from dots to lines, and eventually to entire surfaces, and the enjoyment of the blessings of satoyama as a commons will be the key to the regeneration and creation of resilient landscapes.

In addition, the large-scale development of satoyama by corporations for tourism and other purposes has led to a reevaluation of how to cope with disasters and a reaffirmation of the value of satoyama as a place for the conservation of landscape and biodiversity. In particular, rivers and inland lakes have been maintained and managed mainly by Otsu City and Shiga Prefecture, but interest in their future use and management has increased among local and foreign organizations.

In the Minamikomatsu area, a company started a glamping development project around the inner lake of Omi-Maiko beginning in the late 2010s, which raised issues from the perspective of disaster response, water system management, and sustainable development from the local community. Other issues raised included the conservation of the reed community, rare species and environmental education from citizen organizations and local experts. As for the government, Shiga Prefecture was involved in the use of the law for environmental conservation as a quasi-national park, while Otsu City was involved in the management of the water system of the inner lake. As a result of discussions and planning among various parties with their own positions and opinions, a cooperative system has been established to achieve the major goals of nature restoration and appropriate use while positioning the inner lake and surrounding environment as a "New Commons". The members of the research project on Eco-DRR are playing an important role in this movement by providing academic information and acting as coordinators.

This new trend surrounding the satoyama at the foot of Mt. Hira can be seen as an attempt to revive and create a socio-ecological landscape centered on the water system. It is an attempt to reconceptualize the traditional knowledge that translated into techniques and practices, which have dealt with disasters while enjoying the blessings of nature that have been cultivated over a long history, and to apply them to today. It is important to become aware of the characteristics and values of the various natural resources around us (awareness), to make them function as a network, and to position the key landscape elements as "New Commons". In particular, waterfront areas and areas having undergone land conversion are important for natural resource use, disaster response, and biocultural diversity, and have historically been considered as sacred places and managed as common land. It is necessary to give meaning to key points in the landscape from a modern perspective, and to plan and design in such a way as to create a more developed and continuous relationship with them

Today's natural and social environment surrounding satoyama landscapes is changing direction toward one in which diverse groups such as new residents, experts, schools, administration, and businesses, in addition to traditional residents and traditional organizations, have interests and roles to play. In order to realize resilient landscapes, programs for social implementation are required in which these groups recognize and solve the values and issues that are apparent and inherent in the region. Finally, it is expected that further progress will be made in "local participatory research" for "raising awareness of adjacent natural resources" and "restoration and creation of networks of people and nature" to develop such programs.

Satoyama landscapes can show us that not only do agricultural landscapes and human-dominated landscapes rely on each other but that this system is important for sustainability in developing and developed countries. These landscapes can be used as an example of human nature symbiosis with an emphasis on resiliency. Although Eco-DRR is a specific way of looking at ecosystem functioning in Japanese landscapes, core elements from this concept can also be applied in a variety of places. The International Partnership for the Satoyama Initiative (IPSI) was established in 2010 in order to undertake and facilitate a broad range of activities to implement the concepts of the Satoyama Initiative by diverse stakeholders (Watanabe et al. 2012). In the future, it is important to re-evaluate the elements of satoyama-like landscapes in each country and their connections from a historical-ecological perspective, and apply these traditional and local knowledge to today's sustainable use of resources and disaster prevention while sharing information within an international framework.

Acknowledgment

This research was supported by the Research Institute for Humanity and Nature Eco-DRR Project (14200103) and by JSPS KAKENHI Grant Number 18H02227.

References

Ando, K., Fukamachi, K., Takahashi, D., Azuma, S. (2020). The land-use and disaster management of Minamikomatsu from the Edo to early Meiji Era based on Ezu maps from Minamikomatsu, Otsu City, Shiga Prefecture, Japan. *Journal of the Japanese Institute of Landscape Architecture*, 61(5), 485–490. (In Japanese)

Dublin, D. R., Natori, Y. (2020). Community-based project assessment using the indicators of resilience in SEPLS: Lessons from the GEF-Satoyama Project. *Current Research in Environmental Sustainability*, 2, 543–549.

Fukamachi, K. (2016). Sustainability of terraced paddy fields in traditional satoyama landscapes of Japan. *Journal of Environmental Management*. doi:10.1016/j.jenvman.2016.11.061

Fukamachi, K. (2020). Building resilient socio-ecological systems in Japan: Satoyama examples from Shiga Prefecture, *Ecosystem Services*, Elsevier, 46(C). doi:10.1016/j.ecoser.2020.101187

Fukamachi, K., Miki, Y., Oku, H., Miyoshi, I. (2011). The biocultural link: isolated trees and hedges in Satoyama landscapes indicate a strong connection between biodiversity and local cultural features. *Landscape Ecology and Ecological Engineering*, 7(2), 195–206.

Fukamachi, K., Oku, H., Nakashizuka, T. (2001). The change of satoyama landscape and its causality in Kamiseya, Kyoto Prefecture, Japan between 1970 and 1995. *Landscape Ecology*, 16, 703–717.

Ichikawa, K. (Eds.) (2012). *Socio-ecological Production Landscapes in Asia*. United Nations University Institute for the Advanced Study of Sustainability. 106 p. https://satoyama-initiative.org/wp-content/uploads/2018/01/SEPL_in_Asia_report_2nd_Printing.web_pdf

Iizuka, T., Kuroyamagi, A., Sugahara, R. (2018). Study on belief and disaster tradition as a disaster culture in flood-stricken area. *Journal of the City Planning Institute of Japan*, 53(2), 108–115.

Miyoshi, I. (2020). The landscape and disaster response at the foot of Mt. Hira. In: *Eco-DRR as Learned from Local History: Traditional and Local knowledge of Eco-DRR at the foot of Hira Mountains* (pp. 82–87) Kyoto: Research Institute for Humanity and Nature.

Nature Conservation Bureau, Ministry of the Environment (Eds.) (2016). *Ecosystem-based Disaster Risk Reduction in Japan. A handbook for practitioners*. Nature Conservation Bureau, Ministry of the Environment. https://www.env.go.jp/nature/biodic/eco-drr/pamph04.pdf

Ryu, M., Oguma, K., Kubota. A. (2011). A method to driver typologies of a sustainable water system from historical domestic environments. -Physical form, use and maintenance, and water quality, and their transition-. *Journal of the Housing Research Foundation "Jusoken"*, 38, 257–268. (In Japanese)

Watanabe, T., Okuyama, M., & Fukamachi, K. (2012). A review of Japan's environmental policies for Satoyama and Satoumi landscape restoration. *Global Environmental Research*, 16(2), 123–135.

12

SHIFTING CONCEPTS OF URBAN LANDSCAPE IN HELSINKI

From primary forests to high tech nature-based solutions

Kati Vierikko, Elisa Lähde, Elina Nyberg, Silviya Korpilo and Christopher Raymond

Current and future stressors for resilient landscape in Helsinki

We live in the time of the Anthropocene. The entire world is facing rapid urbanization causing exponential consumptive demand for natural resources, energy, and built infrastructure (McPhearson et al., 2021). For many European countries, including Finland, urbanization depopulates the countryside and many small towns are shrinking while only few cities are still growing and becoming heavily built and densely populated (Aro and Aro, 2019). This phenomenon has a great impact on the landscape resilience in both rural and urban areas.

Helsinki is the country's biggest city with 653,835 residents in 2019 (City of Helsinki, 2020a). Compared to Central European countries, urbanization in Finland, one of most northern countries in Europe, is a relatively modern phenomenon (Eurostat, 2016). In the future, the population of the city of Helsinki is estimated to grow fast, and 260,000 new residents are forecasted to populate the city of Helsinki in 2050. The city scale zoning and land-use policy aims to produce enough apartments for new residents, create new working places, and develop transportation infrastructure in the way it supports sustainable modes of transportation. This has led to densification of the current city structure especially in less compact "forest neighborhoods" and occupying virgin land from green spaces.

Meanwhile, the city of Helsinki aims to be a green city in the future, maintaining its city-wide urban green infrastructure, offering close-to-home green areas for all residents, protecting threatened wildlife and climate adaptation capacity with nature-based solutions. There will be great challenges to achieve these opposite (dense and green city) goals in Helsinki. The total amount of urban green spaces continues to decrease causing biodiversity loss in Helsinki. Due to urbanization especially taxonomic groups sensitive to fragmentation

DOI: 10.4324/9781003266440-12

and anthropogenic pressures such as amphibians and reptiles have decreased at the city scale (McKinney, 2008; City of Helsinki, 2019a). The city of Helsinki has self-assessed its biodiversity values by using standardized City Biodiversity Index (CBI) method. According to the results from previous assessments, the scores were high for governance but poor for protecting native species due to low budget for nature conservation, heavily fragmented green infrastructure and small number of strictly protected nature conservation areas in the city (City of Helsinki, 2020b).

For the provision of the multiple services, urban ecosystems need to be biologically diverse, healthy, and functional in their ecological structures and processes (Alberti and Marzluff, 2004). We want to emphasize that biodiversity loss is simultaneously happening at three interlinked spatial scales having negative consequences to human health and ecosystem functions. First scale is the decrease in total amount of green areas at the city scale. It has been estimated that one-third of new construction sites are developed in green areas during 2016–2018 in fast-growing cities in Finland such as Helsinki (Tiitu, 2021, p. 55). The second scale is the loss of local green areas and biologically diverse vegetation in residential districts. New buildings are either developed in natural landscape, e.g., existing forests or to large private yards replacing existing vegetation with buildings. The third scale is the loss of rich and healthy topsoil at micro-scale, which has a significant impact on ecosystem services and human health. City residents face greater exposure to pathogens and are more susceptible to immune-mediated diseases (Parajuli et al., 2018). There is strong scientific evidence that daily contact to diverse topsoil such as forest floor ("kuntta") enhances immune regulation and resistance toward infections (Roslund et al., 2020) On the other hand, densification increases the proportion of sealed soil and gives way for a human-induced novel biodiversity and high-technological nature-based solutions such as biofilters, rain gardens, or green roofs.

Climate change will bring significant change for seasonal landscape in Helsinki. In the worst scenario of 2050, the mean temperature will rise 2–3°C. We have forfeit snowy winters in Southern Finland. Precipitation has significantly increased, especially in autumn and winter. Short and cloudy days without reflecting light from snow enhance depression among citizens. Runoff water and seawater floodings cause damages for buildings and gray infrastructure. Climate change has an impact on urban biodiversity, environmental conditions of water ecosystems, and natural landscape in Helsinki. Spruce (*Picea abies*)-dominated forests are occupied with newcomers from middle Europe. Invasive alien species and pathogens threaten native urban species. On the other hand, the summer season has become hotter, and especially elderly people are suffering in densely built residential areas (Pilli-Sihvola et al., 2018).

The recent emergency in climate crises and normative response have started to emphasize enhancement of regulating ecosystem services (ES) in cities such as urban temperature regulation, runoff water mitigation or flooding control (Larondelle et al., 2014, Cortinovis and Geneletti, 2019). Sustainable management of runoff water or flooding by nature-based solutions (NbS) can already be seen as a mainstream approach in Europe (EEA, 2021). Urban green infrastructure (UGI) including natural and man-made green areas have a key role in providing multiple ES such as climate change adaptation, local biodiversity conservation, social cohesion, and well-being of residents (Hartig et al., 2014; van den Bosch et al., 2016; Vierikko et al., 2020). However, ES or NbS are not simply a one-directional benefit of ecosystem functioning but rather are coproduction of people and ecosystems (Buizer et al., 2016). Urban planning affects the delivery of ES by designing the spatial arrangement of different functions and land uses (Cortinovis and Geneletti, 2019).

Finally, ideologies, norms, rules, and management practices adopted by planners and decision-makers continuously re-define what sustainable landscape is, and biological and cultural diversity in cities (Buizer et al., 2016; Vierikko et al., 2016). Although more emphasis is on UGI planning, the actual implementation can be challenging because of existing routines, infrastructure, and institutions, which are persistent and highly interwoven (Brown et al., 2013; Pauleit et al., 2019; Lähde, 2020). The lack of collaborative approach and shared understanding what is required for maintaining biologically diverse UGI providing multiple ecosystem services together with professional silos are key barriers that hinder the development of multifunctional UGI as a part of climate smart urban planning (Ahern et al., 2014; Faehnle et al., 2014; Lennon et al., 2016; Lähde and Di Marino, 2018). We come back to this challenge in the following sections, but before that we introduce the role of natural landscape in the urban planning paradigm in Helsinki the past 150 years.

Land-use planning history in Helsinki

Finland is located in north-eastern part of Europe and belongs to Nordic countries (Figure 12.1; Nordic Co-Operation, 2021). Helsinki is situated in the southern part of Finland and belongs to the hemi-boreal vegetation zone. City's administrative boundaries extend to 719 km^2, 2/3 of which is water (land surface 217 km^2). Overall urban biodiversity, based on biotope and species richness (incl. natives and non-natives), is high in Helsinki (Vierikko et al., 2014, Jalkanen et al., 2020). The original landscape of Helsinki before the establishment of urban settlements was a mixture of granite rock hills, small and large forest patches, and shallow open wetlands (reeds) along the coast. There is one river, several urban brooks running through the city. The seashore is 123 km long, and there are 327 islands in Helsinki (City of Helsinki, 2021). Many islands are open for recreational use and can be reached by ferry. The amount of original landscape left in Helsinki is still one of the highest in European capitals. There are about 40 km^2 natural-like boreal forests in the city, and they are the most common type of green areas in Helsinki (City of Helsinki, 2021). The forest network is strongly fragmented, featuring over 800 separate patches, the majority of them smaller than one hectare (Wang et al., 2019). Because forest landscape covers the

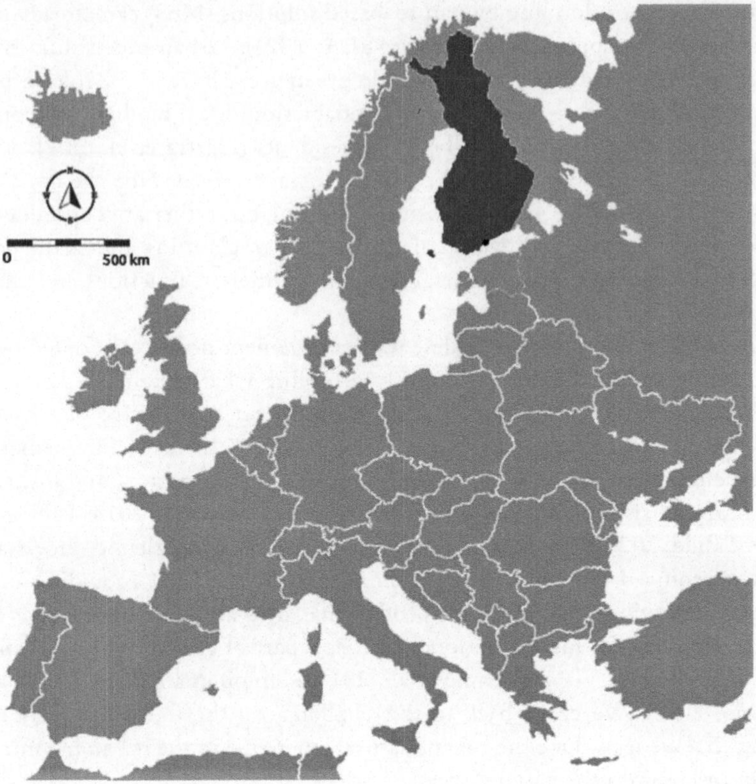

FIGURE 12.1 Finland is located in north-eastern part of Europe. Helsinki, the capital of Finland, is situated along the Baltic Sea and belongs to hemi-boreal vegetation zone.

entire city of Helsinki, almost half of people have walkable access (<500 m) to forests, and small-scale forest patches are commonly valued and visited for nature experience (Harlio and Tuhkanen, 2018). Forests hold high conservation values at local and even at national levels (Vierikko et al., 2014). Citizens also strongly value forests because of their biodiversity values (Wang et al., 2019). Man-made, well-managed public parks (ca. 19 km^2) are the most common type of green area in densely populated city center, and new residential districts developed in the former harbor areas.

The city of Helsinki was founded by the Swedish King on the mouth of river Vantaanjoki in 1550, and Helsinki became a capital in 1812 during the autonomous time under the Russian power. The country received its independence in 1917, and Helsinki kept its capital status. Helsinki borders the Baltic Sea and has always been an important international port and pass-by point between western and eastern European cultures. Today the Helsinki metropolitan region is the economic center of Finland.

The history of urban planning in Helsinki reflects the common international planning paradigms, and one can find representative examples of each era from the city structure. After the city was established in 1812, the historical city center with administrative buildings, still forming the main identity and image of the central areas, was built. The population started to grow at the end of the 19th century due industrialization. Commonly, overcrowding, diseases, and crime were the urban problems of the late 19th century. The solution was to create a wider street network together with proper sanitation and drainage systems. Illumination of housing units and separation of urban functions were further emphasized during functionalism at the beginning of 20th century. The grid structure of the city center of Helsinki leans on these principles.

Rise of motoring allowed the spread of cities and the emergence of independent neighborhood units offering daily services in walkable distances, and where visual connection and recreation opportunities to forests was part of the living standards. One can find several examples (Kannelmäki, Pihlajamäki and Maunula) of these *metsälähiö*, "forest neighborhoods" in Helsinki from 1950s to 1970s when the population of the city was rapidly increasing. "Forest neighborhood" ideology strongly supported direct daily engagement with nature to enhance close human-nature relationships, human health, and multisensory natural experiences as our famous architect argued that 'every man should have the right to walk through forest on their way to work' (Alvar Aalto, 1898–1976). Development of neighborhoods in the middle of natural landscape, conifer-dominated forests with open granite rocks had a negative side effect: sprawling and sparse city structure (Figure 12.2).

As a counter-reaction to urban sprawl, the idea of compact cities was created, and it was further emphasized by sustainable development. Pikku-Huopalahti and Ruoholahti neighborhoods from the 1980s follow the principles of compact city and the eco-Viikki neighborhoods from the 1990s brought up the principles of ecological urban housing and the importance of green infrastructure (pervious soil, diverse vegetation, and restoration of urban creek) for supporting sustainable life in cities. There were huge land-use changes happening in the central area of the city of Helsinki at the beginning of this century when two port areas (Kalasatama and Jätkäsaari) were released from their former use. Both areas were planned and developed as new densely populated residential areas. In the following section, we present the development of the district Kalasatama as an example of the current trend and ideologies about climate smart and sustainable urban planning in Helsinki.

Visions for sustainable urban landscape in Helsinki

Despite the huge stressors described earlier, we see the city of Helsinki has made a great effort for urban sustainability as it aims to achieve carbon neutrality by 2035 and reduce carbon emissions (City of Helsinki, 2018). The city is embedded in the Finnish society, where the baseline situation is promising. Open and

FIGURE 12.2 The ideology of forest neighborhood was introduced in 1950s. Residential areas were developed in the middle of forest connecting built environment with natural landscape. Photo by Kati Vierikko.

participative democratic society with reliable decision making and legislative processes enable common goal setting and sustainability transformation. Furthermore, the city of Helsinki has had a steady growth of welfare and population for almost over 100 years but has managed to control the expanding urban structure. The city still has a relatively high proportion of biologically rich natural (e.g. forests, wetlands, and open rocks) and semi-natural (e.g. meadows and pastures) landscapes and greenery is regarded as one of the city's main identity features (Vierikko et al., 2014). According to the City Strategy (2021–2025), Helsinki wants to be ambitious in climate responsibility, citizens' well-being, economic growth, and nature conservation. In the current mayor's words: "The goal is for this common strategy to take us towards being a better city – towards evolving into a city whose objective is to enable a pleasant, enjoyable and sustainable life for every Helsinki resident." A better and more sustainable city must also be prepared for the impacts of weather events and climate change (City of Helsinki, 2019b).

Helsinki's long-term adaptation vision is to be a climate-proof and safe city in an ecologically and socially sustainable way. The city aims to be known as one of the global forerunners in climate change adaptation (City of Helsinki, 2019b). Adaptation measures are integrated to land use and construction, education and teaching, nature management, recreation, social and health care services, industrial policies, and preparedness and contingency planning (City of Helsinki, 2019b). In the vision, developing green infrastructure and avoiding soil sealing are mentioned as key solutions for stormwater management and avoiding the urban heat island effect. Improving climate resilience would also enhance resilience of natural landscape. The climate adaptation strategy emphasizes that preserving the biologically diverse urban wildlife and nature plays an important role in achieving these goals. The measures of visions include, for example, development and mapping of UGI at the city scale or even at the larger spatial scale. In addition, the vision lists many measures such as increasing green structure, maintaining large trees and planting new street trees and using nature-based solutions such as green roofs at the local scale planning (City of Helsinki, 2019b).

City of Helsinki also aims to protect its wildlife and acts to increase the diversity of the urban natural environment. There are over 1,600 species found in Helsinki including 70% of vascular plants recorded in Finland (Vierikko et al., 2014). The conservation value of forests is estimated to be high even at the national level, which makes its forests potential habitats for native species that could not survive in intensively managed commercial forests in rural landscapes (Vierikko and Jaatsi, 2020). Owned by the city, the urban forests are managed primarily to meet recreational needs while taking biodiversity into account (Saukkonen, 2011). The updated Nature Conservation Programme (2015–2024) doubles the amount of nature conservation areas (ca. 1300 ha) in Helsinki. Furthermore, the city recently launched the 2nd Action Plan for Biodiversity (2021–2028). The aim of the plan is to take biodiversity into account in cross-sectorial decision-making and planning and preserve biodiversity also in the built environment and common green areas, not just in strictly protected areas with a clear vision for nature: "Biodiversity is cherished in the growing and changing city. Important green areas will be preserved and connections between them will be developed to support the well-being of their inhabitant species, as well as that of people."

The role of nature in climate smart city planning

Here we give a more detailed presentation about the vision of Kalasatama district to present the current trends and challenges of urban planning. The area can be seen as a summary of recent aspirations toward a more resilient and climate wise city. History of Kalasatama goes back roughly 90 years. The previously undeveloped site changed through industrial development of the city, as the area provided favorable conditions for port operations and production. In the 1930s, the islands off the coast were annexed to the mainland by landfills. The filling of the beach continued until the 1980s. The landfill contains surplus construction material, such as slag and various construction wastes. The backfilling has been partly unplanned and some of the soil has been originally contaminated. Despite the contamination, nature has spread to the manmade shore over the years. Over time, the site has developed important natural values for urban ecology. Development of the residential and commercial buildings started at the beginning of 2010s. Closest large residential area with natural forests is situated on Mustikkamaa Island that can be accessed along a pedestrian bridge.

Kalasatama is still partly under planning and construction, but both strategy-level goals and practical work aim for a resilient urban system, where well-being of both residents and environmental issues are taken into consideration. The already realized part of the district reflects the ambitions to cut down carbon emissions by dense urban structure that relies on public transport. In Kalasatama, the promotion of UGI has been led by the city's own innovation company, Forum Virium Helsinki. Company's main task is to co-create urban futures with local companies, universities, other public sector organizations and Helsinki residents. In 2019, Forum Virium Helsinki initiated the Virtual Verdure project, where the goal was to support the planning and development of new areas into carbon-neutral city districts that make diverse use of green infrastructure. The northern part of Kalasatama district was the pilot site of the project. We, as one of the partners of Virtual Verdure, got a good insight of the project. Other partners included planners from the city's urban environment division and UGI experts from the private sector. The principal working method were three design sprints during which a certain theme was familiarized, played with and implemented into the pilot site. Addressed themes were design of UGI, building process and maintenance. Particularly, space requirements for ecological processes and provision of ecosystem services in dense urban structures were studied and the main goal was to break professional silos and create joint understanding inside the project group.

Kalasatama district has been part of SMARTer Greener Cities (2020–2023) project where we have analyzed UGI and landscape values of the neighborhood (see more www.smartergreenercities.com). The project adopted social, ecological, and technological system (SETS) approach and analyzed how different components interact. Planning and designing sustainable urban landscape require deeper understanding how social-economical, ecological-biophysical and technological-infrastructural components interact, and together create patterns and processes that influence the character and quality of the city (McPhearson et al., 2016; Grimm et al., 2018). Our landscape analyses revealed that the total vegetation cover has increased in Kalasatama area since removal of harbor and developing the residential (Nyberg et al., 2020). However, vegetation structure is still rather low and large trees are totally missing in public parks and inner private yards (Figure 12.3). Therefore, it is notable that feedback from first residents of Kalasatama has underlined the lack of greenery as a negative aspect despite green areas and vegetation has been introduced in the area. Based on an internet-based survey conducted in 2021, residents preferred the island Mustikkamaa for recreational and biodiversity values (Korpilo et al., 2023). On the contrary, the new public park in the Kalasatama received low biodiversity values, but conflicting negative and positive values. Although the recently established public park is centrally located and accessible for all residents, the overall scenery of public areas is filled with hardscapes: paved areas, walk and driveways with arbitrary street trees, and walls of high-rise buildings (Figure 12.3). This feature will be presumably emphasized in the future through extreme weather conditions creating risk for urban floods and heat waves.

As a result of the Virtual Verdue project, an UGI design manual and guidelines for the design and maintenance of five different types of UGI elements were published. In addition, a method "regional green area factor" developed in the C/O City project (www.cocity.se) to analyze the quality of urban green structures was tested on the pilot site. It was noticed that the method helped to create a common knowledge base regarding UGI and an integrated management

FIGURE 12.3 The public park Kalasatama is centrally located in the district Kalasatama and easily accessible for residents. The landscape is dominating by man-made features. Photo by Elina Nyberg.

of ecosystem services at the district level. After the Virtual Verdure project, the development of the "green area factor for districts" method has been taken further through new pilots in Helsinki. Furthermore, there is another ongoing UGI related project, B. Green, which was initiated based on the results of Virtual Verdure (Forum Virium, 2020).

Based on our insights of the development process in Kalasatama, we claim that although strong emphasize to UGI planning in the Kalasatama district, the progress is still slowed down by path dependency, traditional way of thinking and technical orientated expertise. Even though some advanced efforts for climate adaptation and urban biodiversity enhancement can be identified, the overall process is still led by technical expertise. Social and ecological aspects are used to adjust the overall plan rather than to set fundamentals of it. Analyzing the local solutions in Kalasatama shows that many social-political drivers (e.g. housing policy) affected strongly to the planning while local factors determined that the technical optimization is still dominating in local urban planning (Figure 12.4). Despite co-creation methods and green factor tool were used for the design and implementation of UGI in Kalasatama, we see that lack of holistic thinking and not adopting comprehensive SETS approach at the early stage of planning still hindered the comprehensive enhancement of urban biodiversity and integration of ecological systems into social-technical systems of urban infrastructure.

Crucial question would be how technical solutions could be tailored to enhance local ecological circumstances. This can be reasoned by analyzing the

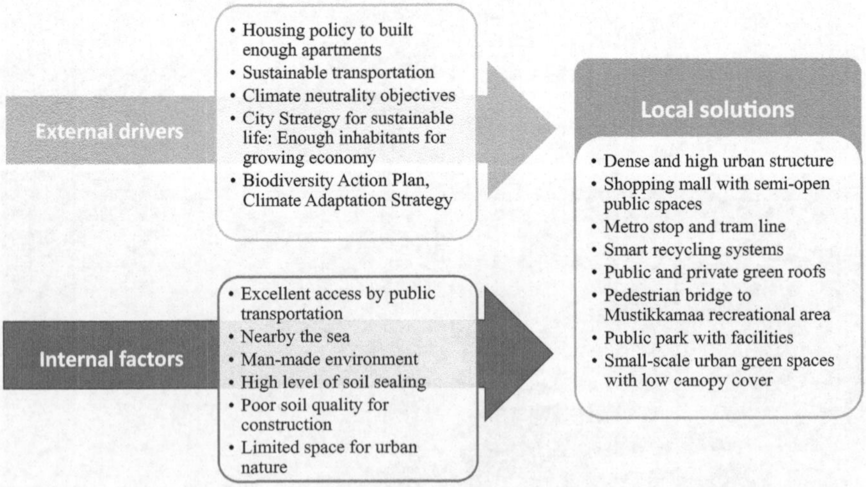

FIGURE 12.4 Pathways of urban planning in our case study area Kalasatama. Drivers are dominated by social (political and economic) aspects, while local environmental factors are seen more as a barrier than enablers implemented local solutions are high technology driven due internal and external factors.

three interlinked spatial scales of biodiversity mentioned earlier. It is possible to increase the total amount of green spaces as the former brownfield areas are scarce of vegetation in Kalasatama. This approach has been emphasized by utilization of green factor tools. However, the existing methods don't advise connecting new green structures into existing ones and thus the local biodiversity of species is not supported, but merely replaced. Secondly, and even more gravely, no attention is put on the loss or replacement of rich and healthy topsoil at micro-scale. In dense urban structure, technical requirements such as parking, flood protection, and water supply and drainage are prioritized leading to elevated and covered courtyards and streets. Furthermore, the design of UGI, for example, in stormwater management solutions is mostly technically and not ecologically driven.

Path to the sustainable future in Helsinki and other cities

In this chapter we have introduced the changing role of natural landscape in urban planning, great visions for climate-proof, biodiversity, and climate smart urban planning in Helsinki and analyzed the Kalasatama district as a forerunner example of climate wise and sustainable urban development. Lessons learned from Kalasatama are that co-creative methods that can lead us toward systems thinking and enlargement of expertise. We strongly believe that Helsinki, and any city in Europe or globally, would benefit to upscale the UGI approach through co-creation and diversified expertise in urban planning. Especially ecological expertise and engagement with local residents needs to be amplified in climate wise solutions and sustainable urban planning (Yli-Pelkonen and Kohl, 2005; Faehnle et al., 2014; Vierikko and Niemelä, 2016). It is also worth remarking that UGI always requires local tailoring.

Based on our case study analyses in Helsinki and more specifically in Kalasatama, we argue that the integrating SETS thinking in urban planning helps to steer future co-creation efforts and planning orientation toward more resilient ones especially regarding human-nature interaction. We see that even the very recent visions of climate adaptation, climate smart district, or sustainable city are rather achievable with the help of mere technical solutions in Helsinki, whereas a successful promoting of UGI and local nature-based solutions requires the integration of social, ecological, and technological dimensions throughout the process, starting from city-scale strategies. Advantages of adopting the SETS approach in urban planning is that it is a multidisciplinary and integrated approach inviting engineers, architects, and landscape designers to interact with scholars of social sciences and urban ecology (Ahern et al., 2014; Lähde, 2020). In addition, monitoring of the results is needed not only to follow up but to communicate if the achieved level is satisfactory when compared to either the starting point or the living prerequisites of different human and non-human dwellers.

The SETS approach brings us towards system thinking, where social actors are just part of a complex and dynamic network where well-being and safety is generated with the ability for adaptation rather than by ruling. Profound enhancement of urban biodiversity would require a post-humanistic approach with deeper understanding regarding interactions between human actions and nature in urban landscape and further the re-evaluation of human needs in the favor of requirements of other species. The history of urban planning paradigms, and most of the western society for that matter, are based on the idea that humankind can increase its well-being by mastery of other natural forces (Grimm et al., 2018). Systemic SETS approach is needed especially if urban biodiversity, climate adaptation with nature-based solutions, and residential well-being is to be enhanced in the district and the whole city.

Acknowledgments

Some results presented here are provided by SMARTer Greener Cities (2020–2023) Nordforsk project number 95377. We would like to thank Leena Kopperoinen and Hanna Nieminen for conducting for the project. We would like to thank Forum Virium and Urban Environment Division for collaboration during Virtual Verdure project (founded by Helsinki-Uusimaa Regional Council).

References

Ahern, J., Cilliers, S., Niemelä, J. (2014). The concept of ecosystem services in adaptive urban planning and design: A framework for supportive innovation. *Landscape and Urban Planning* 125, 254–259.

Alberti, M. and Marzluff, J. (2004). Ecological resilience in urban ecosystems: Linking urban patterns to human and ecological functions. *Urban Ecosystems* 7, 241–265.

Aro, R. and Aro, T. (2019). *Population projections are political dynamite.* MDI blog writing. Retrieved from: https://www.mdi.fi/en/blog-population-projections-are-political-dynamite/

Brown, R., Farrelly, M. and Loorbach, D.A. (2013). Actors working the institutions in sustainability transitions: The case of Melbourne's stormwater management. *Global Environmental Change* 23, 701–718.

Buizer, M., Elands, B. and Vierikko, K. (2016). Governing cities reflexively – The biocultural diversity as an alternative for ecosystem services. *Environmental Science & Policy* 62, 7–13. doi: 10.1016/j.envsci.2016.03.003

City of Helsinki (2018). *Carbon neutral action plan.* Publications of the Central Administration of the City of Helsinki 2018:4. Retrieved from: https://www.hel.fi/static/liitteet/kaupunkiymparisto/julkaisut/julkaisut/HNH-2035/Carbon_neutral_Helsinki_Action_Plan_1503019_EN.pdf

City of Helsinki (2019a). *Tietoa luonnosta – matelijat ja sammakot* (Information about nature – reptiles and amphibians). Retrieved from: https://www.hel.fi/helsinki/fi/-asuminen-ja-ymparisto/luonto-ja-viheralueet/kasvit-ja-elaimet/matelijat/

City of Helsinki (2019b). *Helsinki's climate change adaptation policies 2019–2025.* Publications of the Central Administration 2019:43. Retrieved from: https://www.hel.fi/static/kanslia/Julkaisut/2019/Helsinki-climate-change-adaptation-policies-2019-2025.pdf.

City of Helsinki (2020a). Helsinki facts and figures 2020. Retrieved from: https://www.hel.fi/hel2/tietokeskus/julkaisut/pdf/20_06_18_HKI_taskutilasto2020_eng.pdf

City of Helsinki (2020b). *City Biodiversity Index (CBI) Helsingissä* (City Biodiversity Index in Helsinki). Retrieved from: https://www.hel.fi/helsinki/fi/asuminen-ja-ymparisto/luonto-ja-viheralueet/monimuotoisuus/cbi

City of Helsinki (2021). *Helsingin luontoalueet* (Green areas in Helsinki). Retrieved from: http://www.greenmap.fi/helsingin-luontoalueet/

Cortinovis, C. and Geneletti, D. (2019). A framework to explore the effects of urban planning decisions on regulating ecosystem services in cities. *Ecosystem Services 38*, 100946. doi: 10.1016/j.ecoser.2019.100946

EEA (2021). *Nature-based solutions in Europe: Policy, knowledge and practice for climate change adaptation and disaster risk reduction*. EEA Report No 01/2021. Retrieved from: https://www.eea.europa.eu/publications/nature-based-solutions-in-europe.

Eurostat (2016). *Urban Europe – Statistics on cities, towns and suburbs*. European Commission, 282. doi: 10.2785/91120

Faehnle, M., Bäcklund, P., Tyrväinen, L., Niemelä, J. and Yli-Pelkonen, V. (2014). How can residents' experiences inform planning of urban green infrastructure? Case Finland. *Landscape and Urban Planning* 130, 171–183.

Forum Virium (2020). *B.Green project researchers new solutions for green infrastructure and associated digital tools for supporting city planning*. Retrieved from: https://forumvirium.fi/en/b-green-project-researches-new-solutions-for-green-infrastructure-and-associated-digital-tools-for-supporting-cityplanning-2/

Grimm, N. B., Blasquez, M., Chester, M., Cook, E., Groffman, P., Iwaniec, D., McPhearson, T., Miller, T., Muñoz-Erickson, T. and Redman, C. (2018). *A social-ecological-technical systems approach to understanding urban complexity and building climate resilience*. Proceedings of the IFoU 2018: Reframing Urban Resilience Implementation: Aligning Sustainability and Resilience, Barcelona, 10–12 December 2018, MDPI: Basel, Switzerland, doi: 10.3390/IFOU2018-06044

Harlio, A. and Tuhkanen, H. (2018). *NATOURS 2016–2018 kaupunkiluontohanke Helsinki & Tallinna* (Urban nature project in Helsinki and Tallinn). Oral presentation 3.5.2018.

Hartig, T., Mitchell, R., de Vries, S. and Frumkin, H. (2014). Nature and health. *Annual Review of Public Health 35*, 207–228. doi: 10.1146/annurev-publhealth-032013-182443

Jalkanen, J., Vierikko, K. and Moilanen, A. (2020). Spatial prioritization for urban Biodiversity Quality using biotope maps and expert opinion. *Urban Forestry & Urban Greening* 49, 126586 (open access). doi: 10.1016/j.ufug.2020.126586

Korpilo, S., Nyberg, E., Vierikko, K., Nieminen, H., Arciniegas, G. and Raymond, C. (2023). Developing a Multi-sensory Public Participation GIS (MSPPGIS) method for integrating landscape values and soundscapes of urban green infrastructure. *Landscape and Urban Planning* 230, 104617. doi.org/10.1016/j.landurbplan.2022.104617

Larondelle, N., Haase, D. and Kabisch, N. (2014). Mapping the diversity of regulating ecosystem services in European cities. *Global Environmental Change* 26, 119–129.

Lennon, M., Scott, M., Collier, M. and Foley, K. (2016). Developing green infrastructure 'thinking': Devising and applying an interactive group-based methodology for practitioners. *Journal of Environmental Planning and Management* 59(5), 843–865.

Lähde, E. (2020). *Mission Blue-Green: The Significance of Co-Creation to Promote Multifunctional Green Infrastructure within Sustainable Landscape and Urban Planning and Design in Finland*. Academic dissertation, School of Arts, Design and Architecture, Aalto

University. Retrieved from: https://aaltodoc.aalto.fi/bitstream/handle/123456789/46393/isbn9789526400051.pdf?sequence=3&isAllowed=y

Lähde, E. and Di Marino, M. (2018). Multidisciplinary collaboration and understanding of green infrastructure: Results from the cities of Tampere, Vantaa and Jyväskylä (Finland). *Urban Forestry and Urban Greening* 40, 63–72.

McKinney, M. L. (2008). Effects of urbanization on species richness: A review of plants and animals. *Urban Ecosystems* 1, 161–176. doi: 10.1007/s11252-007-0045-4

McPhearson, T., Pickett, S., Grimm, N., Niemelä, J., Alberti, M., Elmqvist, T., Weber, C., Haase, D., Breuste, J. and Qureshi, S. (2016). Advancing Urban Ecology toward a Science of Cities. *BioScience* 66(3), 198–212. doi: 10.1093/biosci/biw002

McPhearson, T., M. Raymond, C., Gulsrud, N., Albert, C., Coles, N., Fagerholm, N., Nagatsu, M., Olafsson Stahl, A., Soininen, N. and Vierikko, K. (2021). Radical changes are needed for transformations to a good Anthropocene. *npj Urban Sustainability* 1, 5. doi: 10.1038/s42949-021-00017-x

Nordic Co-Operation (2021). *Facts about the Nordic countries*. Retrieved from: https://www.norden.org/en/information/facts-about-nordic-countries

Nyberg, E., Tiitu M., Nieminen H. and Vierikko, K. (2020). Socio-ecological structure in Kuninkaantammi and Kalasatama. Smarter Greener Cities WP3 power report. Finnish Environment Institute, Helsinki. Retrieved from: https://smartergreenercities.com/our-research/publications-and-reports/

Pauleit, S. et al. (2019). Advancing urban green infrastructure in Europe: Outcomes and reflections from the GREEN SURGE project. *Urban Forestry & Urban Greening* 40, 4–16.

Parajuli, A., Grönroos, M., Siter, N., Puhakka, R., Vari, H., Roslund, M., Jumpponen, A., Nurminen, N., Hyöty, H., Rajaniemi, J. and Sinkkonen, A. 2018. Urbanization Reduces Transfer of Diverse Environmental Microbiota Indoors. *Frontiers in Microbiology* 9, 84. doi: 10.3389/fmicb.2018.00084

Pilli-Sihvola, K., Haavisto, R., Leijala, U., Luhtala, S., Mäkelä, A., Ruuhela, R. and Votsis, A. 2018: *Sään ja ilmastonmuutoksen aiheuttamat riskit Helsingissä* (Risks caused by weather and climate change in Helsinki). Kaupunkiympäristön julkaisuja 2018:6.

Roslund, M., Puhakka, R., Grönroos, M., Nurminen, N., Oikarinen, S., Gazali, A., Cinek, O., Kramna, L., et al. (2020). Biodiversity intervention enhances immune regulation and health-associated commensal microbiota among daycare children. *Science Advances* 6(42). doi: 10.1126/sciadv.aba2578

Saukkonen, T. (2011). *Helsingin kaupungin luonnonhoidon linjaus* (Helsinki Nature Management Strategy). Helsinki, Finland: Public Works Department (in Finnish).

Tiitu, M. (2021). *Address urban sprawl and loss of biodiversity*. In Finland's Report on the Implementation of the New Urban Agenda, Ministry of Environment. pp. 54–56. Retrieved from: https://www.urbanagendaplatform.org/sites/default/files/2021-07/NUA%20Report_FINLAND_Min.Env_.%20FINAL%20%283%29_0.pdf

van den Bosch, M., Mudu, P., Uscila, V., Barrdahl, M., Kulinkina, A., Staatsen, B., Swart, W., Kruize, H., Zurlyte, I., and Egorov, A. I. (2016). Development of an urban green space indicator and the public health rationale. *Scandinavian Journal of Public Health* 44(2), 159–167. doi: 10.1177/1403494815615444

Vierikko, K. and Niemelä, J. (2016). Bottom-up thinking – Identifying socio-cultural values and ecosystem services of local urban blue-green infrastructure in Helsinki, Finland. *Land Use Policy* 50, 537–547.

Vierikko, K., Niemelä, J., Salminen, J., Jalkanen, J. and Tamminen, N. (2014). *Sustainable green infrastructure of Helsinki – urban ecological research report and recommendations for the Helsinki master plan 2014*. Helsingin kaupunkisuunnitteluviraston yleissuunnitteluosaston selvityksiä 2014. 113 pp. (In Finnish with English summary) Retrieved from: https://www.hel.fi/hel2/ksv/julkaisut/yos_2014-27.pdf

Vierikko, K., Birgit, E., Niemelä, J. K., Andersson, E., Buijs, A., Fischer, L., Haase, D., Kabisch, N., Kowarik, I., Luz, A. C., Olafsson Stahl, A., Száraz, L., Van der Jagt, A. and Konijnendijk Van-den Bosch, C. (2016). Considering the ways biocultural diversity helps enforce the urban green infrastructure in times of urban transformation. *Current Opinion in Environmental Sustainability* 22, 7–12. doi: 10.1016/j.cosust.2017.02.006

Vierikko, K., Gonçalves, P., Haase, D., Elands, B., Ioja, C., Jaatsi, M., Pieniniemi, M., Lindgren, J., Grilo, F., Santos, M., Niemelä, J. and Yli-Pelkonen, V. (2020). Biocultural diversity (BCD) in European cities – interactions between motivations, experiences and environment in public parks. *Urban Forestry & Urban Greening* 48, 126501. doi: 10.1016/j.ufug.2019.126501

Vierikko, K. and Jaatsi, M. (2020). Designed and engaged urban nature - Biocultural values in Helsinki, Finland. In Cocks, M. & Shackleton, C. (Eds.), *Urban Nature - Enriching Belonging, Wellbeing and Bioculture*. Taylor & Francis Group. p. 51–66.

Wang, Y., Kotze, J., Vierikko, K. and Niemelä, J. (2019). What makes urban greenspace unique – relationships between citizens' perceptions on unique urban nature, biodiversity and environmental factors. *Urban Forestry & Urban Greening* 42, 1–9. doi: 10.1016/j.ufug.2019.04.005

Yli-Pelkonen, V. and Kohl, J. (2005). The role of local ecological knowledge in sustainable urban planning: Perspectives from Finland. *Sustainability: Science, Practice, and Policy* 1(1), 3–14.

13

TRADITIONAL NATURE-BASED ARCHITECTURE AND LANDSCAPE DESIGN

Lessons from Samoa and Wider Oceania

Anita Latai-Niusulu, Susana Taua'a, Gabriel Luke Kiddle, Maibritt Pedersen Zari, Paul Blaschke and Victoria Chanse

Introduction

Positionality: Sina faamatalaga i e na tusia lenei mataupu

An important part of decolonizing research practices is to make the positionality of researchers apparent (Latai-Niusulu et al., 2020; Tuhiwai Smith, 2012). In light of this, we would like to make it clear that the authors of this chapter include Indigenous Samoan academics, as well as researchers who are pākehā (New Zealanders of European descent), and tauiwi (peoples of non-Māori heritage in Aotearoa New Zealand), all of whom are involved in research related to Pacific peoples' well-being and climate change adaptation as part of the Nature-based Urban design for Wellbeing and Adaptation in Oceania (NUWAO) project. NUWAO is funded by the New Zealand government through the Royal Society of New Zealand. Further details about the positionality of each author and NUWAO as a group are available at www.nuwao.org.nz.

Oceania

The islands of Oceania (Figure 13.1), scattered across the vast Pacific Ocean, are dynamic and complex systems. Pacific Island communities, who first inhabited these places more than 3,000 years ago, have adapted ways of surviving amid these tropical environments and the waves of foreign influences that continuously bring change to their shores. In this chapter, Oceania refers to the islands of the Pacific, categorized as Micronesia, Melanesia and Polynesia, the latter including Aotearoa New Zealand. Pacific Island architecture and landscape designs (ALDs) embody Pacific worldviews and a close connection with the environment. They have evolved over centuries and vary from island to

DOI: 10.4324/9781003266440-13

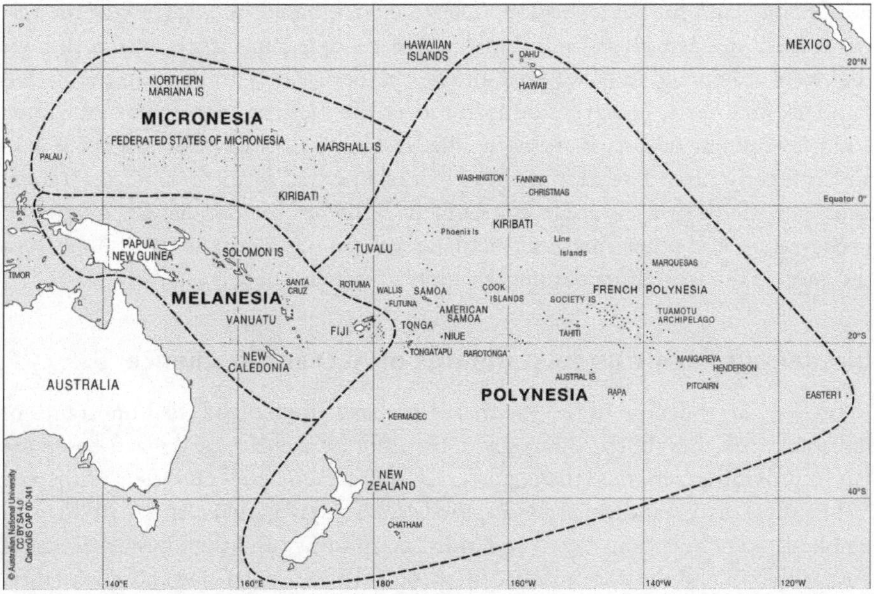

FIGURE 13.1 The islands and regions of Oceania.

Source: CartoGIS Services, College of Asia and the Pacific, The Australian National University under CC BY-SA 4.0 (https://asiapacific.anu.edu.au/mapsonline/base-maps/micronesia-melanesia-polynesia).

island within Oceania. Pacific Island cultures recognize that islands are deeply interconnected socio-ecological systems, in and among themselves, and also in relation to other parts of the world. This knowledge is crucial to the development of nature-based solutions which call for integrated approaches to address climate change, biodiversity loss and other sustainable development challenges (Kiddle et al., 2021a).

This chapter argues that the evolution and continued existence of Pacific Island traditional architecture and the historical designs and structure of their villages, in spite of pressures, demonstrates the resilience of Pacific Island peoples and cultural knowledge. This knowledge is crucial and must be front and center in the development of current and future urban ALD strategies that result in both ecological and human well-being benefits.

This chapter employs a place-based and Indigenous-centric approach to exploring nature-based ALD in Oceania using Samoa as a case study. The discussion begins with a section on Pacific Indigenous understandings of the environment, ALD, and resilience. Secondly, the chapter focuses on Samoa as a case study, specifically the evolution of ALD. The third and fourth sections of the chapter explore the future of Samoan and wider Oceania ALD. Here we draw on lessons from Samoa and make recommendations for future resilient ALD solutions for both Samoa and the wider region. The chapter is based on a review of the relevant literature as well as findings from a series of group interviews/workshops that were conducted in November and December 2021 with Samoan traditional builders, architects, construction lecturers at the National University of Samoa and government officers working in the Samoan Ministry of Transport Works and Infrastructure. The aim of these workshops was to understand traditional and contemporary knowledge on ALD in light of climate change, continued urbanization and other emerging challenges. Findings from the workshops form the basis of the recommendations discussed at the end of this chapter.

Indigenous Pacific understandings of ALD and resilience

Many scholars working in the Pacific region have emphasized the importance of holistic approaches, but also recognize the importance of a local place-based focus in providing a more realistic portrayal of island societies and their resilience. While there is a great diversity of Pacific Island societies and cultures, all share an emphasis on connections between family, community and the physical-spiritual environment, and the core values of respect, reciprocity, service and spirituality. Pacific island concepts and approaches embody the dynamic and complex nature of these relationships, connections and related activities. Many are difficult to define and articulate in English, but exploring these concepts would provide an adequate understanding of the resilience of Pacific island societies. For example, the Fijian concept *vanua* translates as land, home or tribe, but it means much more than that. *Vanua* has physical, social and cultural dimensions inextricably interrelated. *Vanua* can be understood as an extended kinship system. Each member of the *vanua* is considered to be the soul of the geographical land; without the people, the physical *vanua* would be soul-less and incomplete. Flipping the concept the other way around, the people would also be incomplete without the connection to their land (Ravuvu, 1983). Further, Hau'ofa's (1993) conceptualization of Pacific islands as 'Oceania', 'a sea of islands', reiterates how the islands of the region are connected, rather than separated, by ocean. Pacific islanders have always traversed the Pacific Ocean sailing from one island to another to trade, marry and expand social networks.

Pacific Island peoples and their environments are inextricably linked and this is reflected in the rich variety of ALDs in the region. Traditional island buildings, including dwellings, are well-suited to the tropical climate of the Pacific and have enabled these societies to survive in these harsh environments for thousands of years. Roofs, providing shelter from the hot sun and rain, are typically the most important architectural component. The roof is supported by round wooden posts or pillars. Most, if not all, traditional island buildings do not have walls, allowing natural ventilation in an environment where humidity is extremely high. Traditional building materials are within the economic reach of most of the population and building techniques have been highly refined over many centuries. The tropical climate in the Pacific generates lush vegetation, so the most common building materials are organic: for example, timber, bamboo, cane, grasses and reeds. These materials are used for framing, thatch and mats to form and adorn buildings that are organic and blend into the environment. For example, traditional Fijian *bures* (huts) comprise strong wooden corner posts which are set in the ground, and a wooden roof frame erected on top of the posts to create the building frame. These are often supported by small vertical posts which help to reinforce the walls in the center. Traditionally, the *bure* is bound together with ropes made from coconut fiber or other natural materials, but in recent years, construction wire has been increasingly used. *Bures* normally have a thatched roof made from pandanus (or in some cases palm leaves). Almost all roofs are high and use a hipped (four sided) configuration. Traditional buildings in other Pacific islands such as Vanuatu, Tonga, Palau, Kiribati, and the Cook Islands have similar characteristics (South Pacific Disaster Reduction Programme, 1998). For example, the *maneaba* (meeting house) in Kiribati also has a large pandanus thatch roof, resting on short coral (or sometimes wooden) pillars, around which are clustered smaller village houses. Wall and floor coverings such as mats adorn traditional buildings in the region and are beautifully patterned and showcase a varied and artistic range of handicrafts developed through weaving and fine lashing techniques (Whincup, 2010).

In cyclone-prone areas of the Pacific, traditional houses are well adapted to cope with this hazard, demonstrating their resilience. Most traditional Micronesian houses, for example, are timber-framed with thatch roofs. The roofs are designed to be lifted off the posts and placed on the ground and secured during strong winds and cyclones, as well as for repairs. Traditional housing in the Pacific is also well adapted to cope with earthquakes because of their ductility, or ability to bend and sway without collapse. For example, Fijian *bures* are quite strong during shaking as the building is bound together with ropes or wire. The corner posts are normally fairly strong and buried sufficiently to resist uplift. However, when damaged, *bures* are difficult to repair and most residents usually rebuild a new structure. Traditional *bure* construction means that even total collapse of the building is rarely life-threatening. For example, the roof may blow off but the structure can remain intact, and it is common practice that the occupants then rush outside and crawl under the roof for the remainder of calmer winds of the cyclone (South Pacific Disaster Reduction Programme, 1998).

Traditional Pacific structures, however, do have shortcomings as they deteriorate rapidly due to weathering and insect attack, and they also represent a considerable fire hazard. This has resulted in the loss of some very important historic buildings across the region. Across centuries, the tradition of rebuilding with the same standardized techniques ensured continuity and aided the evolution of a traditional Oceanic architecture. This evolutionary process, however, was interrupted from the eighteenth century onwards as missionaries and European planters, traders and administrators introduced new ideas and began to use masonry and other more substantial and durable materials such as galvanized iron for building (South Pacific Disaster Reduction Programme, 1998).

Traditional architecture and landscape design also embody cultural and spiritual values of Pacific islands. For example, the interior design of the roofs of many traditional buildings reflects the intricacies of social relationships and connections with spiritual ancestors that are valued by Pacific islanders. A *maneaba* in Kiribati, for instance, and the social space defined by it provide the key to the identity of each individual as well as the whole community. This view is clearly supported by the enormous effort and care expended on its construction and maintenance and the high regard with which the *maneaba* is viewed by the people. The *maneaba* has been considered as the house of the principal deities of Kiribati cosmology, a non-denominational *bangota* (sacred shrine) for the community at large where, in the presence of gods, all the main civil ceremonies, rituals and debates are conducted. In the *maneaba*, religion, law and politics are inextricably mixed and associated ceremonial feasts are highly ritualized (Saini, 1978; Whincup, 2010).

Case study: ALD in Samoa

The independent state of Samoa is located in the south-central Pacific Ocean. It has a west-northwest to east-southeast orientation (Government of Samoa, 2002), and a total land area of 2,934 square kilometres. The island group comprises four inhabited islands, namely Savaii, Upolu, Manono and Apolima, and several nearby islets adjacent to their coasts (Ministry of Natural Resources and Environment (MNRE): Government of Samoa, 2013; Figure 13.2). Samoa, like many islands, has been exposed to the combined effect of tectonic processes, climate change, colonization and globalization. These processes continue to pose environmental and other risks to the development of the country and the lives of its people (Table 13.1).

The latest census in 2011 indicated that the population of Samoa increased from 176,710 in 2001 to 195,979 (Samoa Bureau of Statistics, 2016), with more than 70% living in 330 villages along the coasts of the two main islands, Upolu and Savaii (Government of Samoa, 2013). Key aspects of Samoan society which are crucial to understanding this society include *faamatai* and the communal nature of resource ownership which is still prevalent today. The majority of Samoa's population are Indigenous Samoans. More than 80% of Samoa's land is still collectively owned (MNRE: Government of Samoa, 2013), and every Samoan has access to customary land, either through their mother's or father's side. 64% of households live on customary land and 28% live on private or freehold land (Samoa Bureau of Statistics, 2016).

Conceptualizing ALD in Samoa: The guiding concept of *va tapuia*

The central Samoan concept of *va tapuia* frames this discussion of ALDs in Samoa, and we argue must be the guiding principle of future ALD in Samoa. *Va tapuia*

FIGURE 13.2 National University of Samoa *fale afolau*. Photo by Anita Latai-Niusulu.

encapsulates the Samoan holistic worldview of the environment and landscape resilience and, despite foreign influences, it continues to feature in Samoan architecture, landscape design, and urban design today as well as in other aspects of society. Tamasese (2007) refers to the literal meaning of *va tapuia*. The words *tapu-ia* (sacred) and *va* (relationship) imply that in relations with all living and non-living things, there exists a sacred life force beyond human reckoning. It can be argued that the sacredness that Tamasese refers to reflects how awareness of *va tapuia* leads to an attitude that accepts and respects the limits of human knowledge in relation to the physical environment. *Va tapuia* is a set of behavioral codes which give rise to the principle of *va fealoai* that permeates human interaction with the environment as a whole and not to just one element of it. *Va tapuia* governs all Samoan interactions, particularly the way decisions are made in the *aiga* as well as in the village *fono*. *Va* is relationships, connections, affiliations, boundaries, difference, separation, space, distance, responsibility, obligation, state of being, position, standing and so much more. The Samoan phrase *'E atoalio le masina'* which equates life to the roundness of the moon illustrates this holistic Samoan worldview (Aiono Le Tagaloa, 1996a and 1996b).

Samoan architecture and landscape environments

Key features of traditional Samoan architecture, such as the circular and open structure of its buildings and houses, must feature more significantly in future ALD. As this section demonstrates, these aspects of Samoan spatial design are

TABLE 13.1 Glossary of Samoan terms

aiga	Family	*matai*	Chief
alii	High chief	*paepae*	Foundation of a house comprising of rock and/or concrete
eleele fou	New land	*poumuli*	*Flueggea flexuosa,* a deciduous, small tree with a compact crown
faamalama laupapa-	Wooden louvres/shutters	*sa'o*	High chief
faamatai	Chiefly system	*tai*	Coastal side or the coast
fale	Traditional Samoan open house	*tala poo le matuatala*	Rounded ends of a *fale tele*
fale afolau	Long rectangular open house with rounded ends for guests	*togovao*	Mangrove wood
fale eli	Outdoor detached toilet	*tufuga*	House builder guilds
fale ese	Toilet	*tulagafale*	Land on which a house has been constructed
faleo'o	Small open house	*tunoa*	Formal word for outdoor detached kitchen
fale tele	Oval or round open house for meetings	*umukuka*	Outdoor kitchen
fono	Village council	*uta*	Inland
itu	Side	*va*	Space
malae	Village green or gathering space	*va tapuia*	Sacred space
Matagi-ala-lua	Splitting of the wind	*va fealoai*	Mutual respect

crucial to the resilience of Samoan people and communities and have been adapted in various ways by local people. The structures themselves are resilient and sustainable and assist in the development and maintenance of peoples' health and well-being. They continue to be seen in the landscape despite a shift toward using foreign construction materials and the building of larger closed European-style houses. Because traditional buildings tend to be without walls, and because of the tropical climate, definition between interior space and external landscape is not as defined as it would be in other parts of the world, meaning many traditional structures are inherent parts of the landscape physically but also culturally.

The Samoan *fale*

The *fale* is an important element of Samoan material culture. The *fale afolau* and *fale tele* are the two main types of traditional buildings that dominate the social-cultural landscape of Samoa (Allen, 1994; Kramer, 1995; Stucky, 2011). Embodied in these two types of buildings are the architecture and aesthetics of traditional Samoan design reflecting customs, social structure and ethos (Engels-Schwarzpaul & Refiti, 2018; Taylor, 1999; Van der Ryn, 2016). According to legends of Samoa, the art of house-building originated from Tagaloa-matua, who created the model for the first *fale tele* (Buck, 1930), inherited by *tufuga*, the house-builder guilds of the Sa-Tagaloa (family of Tagaloa) and Aiga-sa-Sao (Taylor, 1999). Constructing the *fale tele* and *fale afolau* require great skill and knowledge beyond that of the ordinary individual, hence the need for specialized skills of *tufuga*. The construction of these buildings often involves all members of the family and broader assistance from the village community (Latai, 2014). The *fale tele*, and its more modern version observed throughout Samoa in the present, are built on raised platforms or *paepae* of rock or concrete, primarily for protection against flooding and heavy rainfall, but more importantly to reflect chiefly hierarchy in the village (Taylor, 1999). The site on which the *fale tele* is built is called the *tulagafale* (Tuvale, n.d). Every extended family in the village, particularly the founding families, will have a specific *tulagafale* in the village that is connected to the family chiefly title managed by the head *matai/sa'o* of the extended family.

The *fale tele* and *fale afolau* are mainly reserved for extended family and village activities. Activities related to cooking and everyday living take place in two separate houses: the *umukuka* and *faleo'o*. The *fale ese* is located further away from the rest of these buildings. Overall, the architecture of the Samoan *fale* mirrors and constitutes the culture and life of Samoan people, being deeply connected to *va* and the values of kinship, ancestral past, land and community. The spaces outside and inside of the *fale* are pivotal to cultural form, ceremony and ritual. Architectural concepts are incorporated in Samoan proverbs, oratory and metaphors, and linked to other art forms such as boat building and tattooing (Latai, 2014; Van der Ryn, 2016).

Openness is the most outstanding characteristic of Samoan *fale*. All traditional structures from the *fale tele* to the *faleo'o* have no walls, no doors, and a single indoor space. Blinds, woven from coconut leaves, surround the building for protection against the rain or hot sun (Stucky, 2011; Van der Ryn, 2016). In a similar way to the buildings of other islands in Oceania, Samoa's tropical climate dictates the open design of its *fale* to accommodate the hot and humid conditions. The all-year-round warm climate and cool breeze eliminated the need for permanent walls and people can enter and exit from any side of the building. Openness also embodies the relational character of the Samoan person as well as the Samoan family as large and extended. According to Latai (2014), just as the inner space of the house extends outwards, so does the Samoan family and its members in their social relations within the family and the wider community. In short, the family is not an independent entity, but an integral part of the community. Openness has many practical benefits such as ease of cleaning, childcare, and supervision of family chores. Openness also allows for community observation of behavior and shared responsibility in keeping family members' actions in check (Rohinger, cited in Stucky, 2011: 38). The openness of the *fale* does not mean that there are no restrictions associated with certain parts of the *fale* and family life. During family gatherings, the *fale* becomes highly structured and certain spaces within the *fale* such as the front and sides are restricted to chiefs, elders and guests. Children and untitled members are restricted to other areas (Schoeffel, 1995).

Indigenous traditional architecture in Samoa (and Tonga) is also characterized by its circular shape (Van der Ryn, 2016). In Samoa, for example, the rounded ends (*tala poo le matuatala*) of the *fale tele* are where the highest ranking *alii* sits during village *fono*. This greatly enhances the visibility of other people sitting in the *fale*. In addition, from an engineering perspective, rounded ends are also more aerodynamic and can better withstand cyclone force winds (Van der Ryn, 2016).

Modifications and additions to the *fale* and new types of building

The combined effect of many foreign influences (e.g. the arrival of missionaries and other colonizers and also the modern availability of foreign building materials) and natural disturbances (e.g. cyclones and the devastating 2009 tsunami) have seen Samoan *fale* evolving over the years. The *fale afolau* has changed primarily because the distance between the two ends of the building were too far apart for guests and hosts seated on opposite ends of the house to hear each other and conduct a proper conversation, hence the need to reduce the length of the middle straight portion (*itu*), while increasing the width and the height of the roof to compensate for other requirements (Buck, 1930; Figure 13.2). It is now rare to find a *fale afolau* around the country, as cyclones which hit the country in the 1960s and 1990s destroyed most of these traditional buildings. Out of these two traditional structures, the *fale tele* is the more common. In recent decades, a notable use of the *faleo'o* in modern Samoa is related to the development of beach *fale* accommodations in the tourism industry.

Over time, dwellings in Samoa have varied in their size and in the materials used in their construction, often depending on the economic means of the family. These changes have been beneficial but also challenging at times for Samoans. The arrival of Europeans saw the emergence of closed houses (Latai, 2014). The most influential imported building form in Samoa has been the German 'verandah' type houses which were built from the later part of the nineteenth century to the end of World War One. Verandahs, however, which were the most characteristic component of this type of German house, were not introduced because of cost. By 1945 concrete floors and corrugated iron roofs had become standard in Samoa. By this time, the curvilinear end of the traditional *fales* had been replaced by right-angled wooden posts or concrete block ends also (Saini, 1978; Figure 13.3).

The 2006 Samoan Census recorded ten categories of residential buildings, ranging from the traditional open *fale* to multi-story closed European houses (Table 13.2). About 90% of traditional open *fale* in the country are located on customary owned lands. Overall, there is significant variation of residential

FIGURE 13.3 A square open *fale*, which has now become common in Samoa. Photo by Anita Latai-Niusulu.

dwellings located on customary owned lands, while closed European houses are common on privately owned lands. Additionally, the post-2009 tsunami period saw the emergence of various types of pre-fabricated square open *fale* on the landscape funded by the Samoan government and built by various community groups and companies (Government of Samoa, 2011).

Large-scale changes in architecture are evident in Samoa's capital of Apia, located on the island of Upolu. New architecture that now dominates the skyline and landscape mark important developments in the making of modern Apia. New buildings have gradually replaced the colonial buildings that once characterized the township. For example, the Apia Courthouse built in 1902 by the German administration was one of the last surviving colonial structures in the South Pacific, but was demolished in 2020. In addition, the original traditional oval-shaped open parliament house with thatched roof and concrete floor has been replaced by more modern form buildings over the years. The latest building, opened in 2020, is a circular, closed and air conditioned, and is built in a somewhat futuristic-looking European style. Many other multi-story buildings (four to eight floors) have been built in recent decades. However, many of these buildings model European design perhaps better suited to temperate regions. In Apia, all of the buildings constructed in the past 30 to 40 years are closed European structures with significant air conditioning systems. Overall, building for durability and comfort was central to building design, and working in an artificially cooled offices provided a comfortable working environment as far as the Samoan mindset of the time was concerned. However, high-energy costs and impacts were not fully considered in the design of these modern buildings. In addition, the sustainability of the reconstruction of major government buildings has been questioned. For example, a number of these buildings, including

TABLE 13.2 Type of building by land tenure, 2006

Type of building	*Land tenure*							
	Total	*Customary*	*Freehold*	*Leased*	*Government*	*Church*	*Employer*	*NS*
Total	46,048	32,521	9,929	1,259	231	1,843	237	28
Open Samoan *Fale*	12,770	10,396	1,812	199	47	275	37	4
Open Samoan *Fale* with ext.	3,022	2,101	710	90	11	88	20	2
Closed Samoan *Fale*	753	481	205	35	4	16	11	1
Closed Samoan *Fale* with ext.	482	281	149	26	5	16	5	–
Open European *Fale*	10,658	8,705	1,363	181	22	350	33	4
Open European *Fale* with ext.	4,727	3,355	970	138	12	222	25	5
Closed European House	6,883	3,771	2,198	330	81	432	64	7
Closed European House with ext.	5,754	3,131	2,000	197	35	359	30	2
European House 2+ stories	879	219	497	61	11	80	10	1
Samoan *Fale* 2+ stories	75	48	19	2	3	2	1	–
Others	33	25	5	–	–	2	1	–
NS	*12*	*8*	*1*	–	–	*1*	–	2

Adapted from Samoa Bureau of Statistics (2006).

Parliament, are located on Mulinu'u Peninsula on the northern coastline of Upolu within central Apia. But at only one-metre above sea level, the Mulinu'u Peninsula is particularly vulnerable to flooding and storm surge from tropical cyclones, and the risk will increase significantly with rising sea levels (Pacific Australia Climate Change Science and Adaptation Planning, 2013).

Samoan household compounds, village landscapes and the environment

A typical Samoan household compound is made up of a number of discrete buildings deliberately arranged within the landscape to ensure the survival, health and well-being of Samoan families. This diversity of living space becomes crucial during challenging times such as during a funeral or when a cyclone hits. The multiplicity of buildings means there is always space to accommodate visitors or when one structure is destroyed, there is an alternative structure to live in, thus adding to resilience of the family and community. Outdoor kitchens and toilets are common, which helps create healthy living conditions and clean landscapes. Samoan households typically have a *fale tele* or *fale afolau* at the front of the compound. Immediately at the back of the traditional guest house is often a partially closed or closed European house. Further back (or sometimes on the side) one would find a *faleo'o*, a *tunoa* and a *fale eli*. Overall, flexibility and versatility characterize the use of *fale* and domestic space, which is both interior and exterior in Samoa. Eating, weaving, praying, sleeping, meeting and many other activities (other than cooking and toilet related activities) may occur within the same space at different times (Figure 13.4).

The majority of Samoans shifted to the coast in the late 1700s to early 1800s and lived in village settlements where housing was oriented around the village *malae* in a circular/oval shape. The arrangement of villages considered environmental and social sustainability, health and well-being. The majority of villages have *tai* and *uta* sections. Village/extended family *fale tele* typically front on to the *malae,* with smaller *fale* (*faleo'o, umu,* and *fale ese*) situated behind each extended family's *fale tele*, and sprawled out towards the village plantations, cemeteries and bush lands. Figure 13.5 shows a typical example of this layout for one village in northwestern Upolo. A lack of fences and walls in the village mean that the exterior and interior of the *fale* are experienced as one continuum, which walls and fences would disrupt (Shore, 1982; Stucky, 2011). Plantations and cemeteries were mostly located inland and separated from where people lived.

Changes to household compounds and village and town landscapes

The interplay of natural disturbances and human influences have also resulted in the evolution of village landscapes in Samoa, some of which have disrupted human *va* connections with their physical environments. Historic volcanic

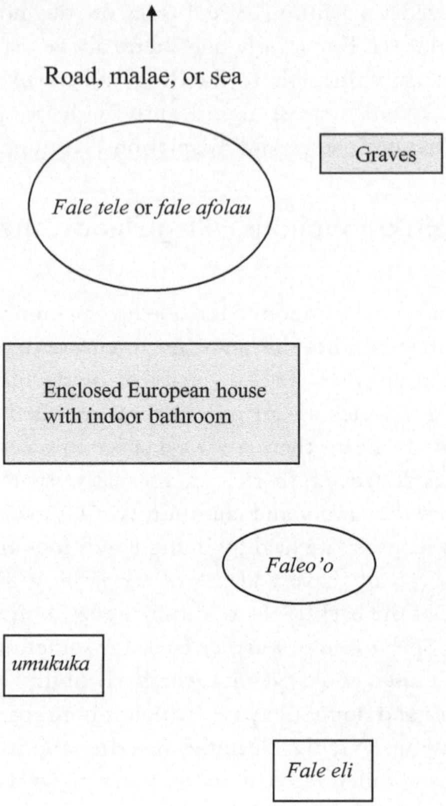

FIGURE 13.4 Layout of Samoan household compounds. Drawn by Anita Latai-Niusulu.

eruptions have influenced human settlement patterns such as shifting entire villages from Savaii to settle on parts of Upolu island. Cyclones have also led to major transformations in village landscapes in Samoa. Cyclones Ofa and Val, for example, in early 1990 and late 1991, are regarded by many in Samoa, including researchers and officials, as the first signs of anthropogenic climate change. After these cyclones, many Samoans started to seriously think about building alternative homes away from the coast. For many living on the northern coast of Savaii, families immediately shifted inland after the cyclones and built houses on lands they had previously used for cultivating food (shown in the area around Saleimoa in Figure 13.7). For many other households and villages in other parts of Upolu, the move to inland parts of the island has been slower and incremental, with some movement not being permanent. For example, younger family members would move to build and live in alternative houses inland but the rest of the family would still occupy the coastal residence and land. Those living inland would often visit the coastal residence and land as demonstrated by the two headed arrows in Figure 13.5.

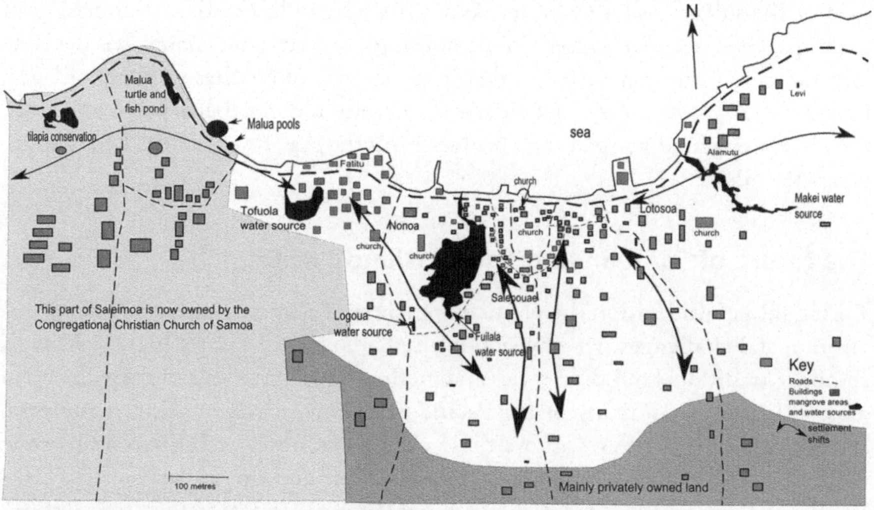

FIGURE 13.5 Village settlement and physical features of Saleimoa village.
Source: Latai-Niusulu (2017).

Many tsunamis have also affected Samoa over the years. The most recent and most devastating in modern times, hit the south coast of Upolu in September 2009, affecting the coastal parts of 20 villages. The tsunami waves reached more than 14 m in height and inundated a zone stretching up to 400 m inland along on the south coast of Upolu. 143 people were killed, and many sections of the coastline were eroded and coastal vegetation destroyed. The majority of people affected have now resettled in inland parts of the affected villages (Government of Samoa, 2011).

There is evidence of growing haphazard development in some parts of the country due to the lack of proper planning and land management thus threatening the resilience of landscapes. The emergence of high-rise buildings in Apia has also been closely related to the loss of green and open spaces. Substantial land reclamation of the Apia waterfront in the early 1970s was necessary to resolve land shortages for urban development. The government building and central bank complex are built on reclaimed land known as *eleele fou*, which was later renamed *Matagi-ala-lua*. In addition, recent developments to upgrade the Apia waterfront in 2018 have resulted in the expansion of impervious concrete pavements for parking and pedestrian footpaths that have contributed to an increased urban heat island effect and thus, the need for increased air-conditioning.

New road developments in Apia in recent years have also dramatically altered the urban landscape, accelerating the path to further inevitable change (Taua'a, 2018). Many roads cut into villages and have impacted the orientation of some houses within. This has sometimes created discord and uncertainty. For example, some families have been left uncertain as to whether their houses should face the road or the village *malae* (Van der Ryn, 2016). In addition, the combination of road upgrades, seawall construction, housing construction, mangrove harvesting, and sand mining have led to the disappearance of beaches in many villages. Further, mangrove areas, particularly in Upolu and northern parts of Savaii, have been separated by roads and bridges from the sea (Boon, 2001; Latai, 2008; Latai-Niusulu, 2017).

The future of Samoan and Pacific Islands' ALD

Taking into consideration the predicted impacts of climate change and other environmental challenges, we envision future Samoan and Pacific Islands' ALD to comprise traditional building styles and methods and variations of these. Samoan society, like those in many of her Pacific neighbors, has evolved in many ways and we cannot separate or ignore these changes in our search for ways to cope with emerging challenges. Due to climate change and the impacts of COVID-19, there has been a renewed desire in many Pacific islanders to look for solutions from within their own cultures and ways of life. This section discusses sustainable and resilient ALDs found in Samoa, which could be adapted to other parts of Oceania with similar physical and socio-cultural conditions.

Traditional Samoan ALDs are crucial aspects of such solutions as they embody Samoan *va* connections to each other as well as to their physical, social and spiritual environment. Finding ways to revitalize traditional Samoan ALDs is crucial to strengthening peoples' *va* connections to the land, sea and all other aspects of the environment. This is key to the sustainability of Samoan society and the well-being of its people. Additionally, future ALDs must also draw on (but not necessarily replicate in all aspects) traditional building styles and methods because these designs and methods are robust, sustainable and reflect a culturally relevant response to the specific ecology and climate of the place. They have been trialed for centuries, and they continue to persist and survive. Traditional building and landscape designs and methods also tend to be more flexible in that they can be made affordable to the local people as the next section demonstrates.

The *faleo'o*

The Samoan *faleo'o* is an innovative, flexible and feasible solution for Samoan people to continue to draw upon, adapt and modify. The *faleo'o* is an affordable option because its size can be determined by the owner's access to resources. It is built from local materials which are readily available such as the *poumuli* and coconut; trees that are commonly found on the plantation lands of most Samoans. Minor modifications which some people are using include the use of tires, rocks, bricks and fishing nets to hold down the roof and tarpaulins to replace coconut blinds (Figure 13.6).

The design of the *faleo'o* is not only climate-appropriate but also energy-effective. These important aspects of building design for Samoa are emphasized

FIGURE 13.6 A family at Lano moved to build this *fale* after their coastal residence was destroyed by the 1990s cyclones. The family maintains both their inland and coastal residences.

Source: Latai-Niusulu (2017).

in the Apia Waterfront Plan 2012–2026 (Ministry of Natural Resources and Environment & Samoa Tourism Authority, 2016) and Samoa's Building Code (Ministry of Works, Transport and Infrastructure, 2017) which are key national documents guiding the current and future development of buildings and landscape design. The Samoan *faleo'o* not only provides natural ventilation due to its open sides, but can be easily raised using posts or moved inland as the sea rises (Figure 13.7). Mechanical systems for cooling air temperature (e.g. air conditioners and refrigeration), solar panels and other weatherproof building materials are foreign, meaning they must be imported at great expense. The *faleo'o* can also be re-built easily when it is destroyed by natural events such as cyclones.

Circular and open designs for large Samoan buildings: Heat reduction designs

Circular and open buildings, such as at the Samoa Tourism Authority village and the National University of Samoa *Fale*, should be encouraged in the architectural design of government buildings in the future as this would result in cooler interior conditions and will be more energy effective. It would also enhance

FIGURE 13.7 *Faleo'o* adapted into what is now commonly known as a beach *fale* seen here at Lano, Savaii.
Source: Latai-Niusulu (2017).

Apia's identity and sense of place as Samoa's urban area continues to grow. The recently built Faleolo International Airport is an example of a large building in open style; fully taking advantage of the sea and land breeze throughout the year (Figure 13.8).

Techniques such as large overhangs, verandahs, high circular roofs, and multiple large side doors, that are kept open when the building is being used, are great heat reduction features of other big buildings such as the Catholic Cathedrals at Leulumoega Tuai and Mulivai (Apia). Samoan art, patterns and stories reflecting values such as family, community and spirituality could be incorporated into the design of all government buildings and public spaces. Other simple heat reduction techniques include the use of *faamalama laupapa* in closed buildings, thermal curtains and the growing of plants around the house that are pruned before the wet/cyclone season begins.

Household compounds and landscape diversification

The multiplicity of Samoan household compounds and family access to land parcels are important adaptive strategies to endure the multidimensional nature of climate change and its impacts. As indicated earlier in the chapter, having many

FIGURE 13.8 The newly built Faleolo International Airport with its high roofs and wide-open ground floor. Photo by Виктор Пинчук (2019) under CC BY-SA 4.0.
(https://commons.wikimedia.org/wiki/File:Faleolo_International_Airport.jpg)

buildings of different sizes, styles and functions means that if one is destroyed, the *aiga* has the other buildings in which they can live. Building and living on two or more parcels of land (often one inland and one more coastal) is another common strategy employed by many Samoan households. This is another way of diversifying places of residence and reducing the risk of being left homeless when a disaster hits.

Green spaces and landscape spatial arrangement

Green spaces should be incorporated in current and future landscape designs as these reflect the Samoan holistic worldview and *va* connections with physical aspects of the environment. These features are still evident in many villages today, but many are also reducing in size due to road and building developments as discussed. For this, and also the development of forested areas, *malae*, mangroves and wetlands, plantation and bush lands, it is important to draw on the traditional village landscape layout and experiences of village councils, who manage village lands and waters. In addition, relevant government ministries and development agencies which are advocating and financing activities to conserve and revitalize these landscapes or think about adaptation to climate change are key stakeholders in nature-based landscape design in Samoa.

Protecting and expanding green spaces is an urgent need for Apia in particular, to make the city cooler during the daytime and provide spaces for peoples' health and well-being. Green spaces could also help rejuvenate urban and coastal biodiversity in an area that is increasingly covered by concrete surfaces and human-made structures. Turning Apia into a 'green' or 'living' city is possible as currently evident in the development of green spaces at *Matagi-ala-lua*, carried out by the Ministry of Natural Resources and Environment and Samoa Tourism Authority as part of the Apia Waterfront Development Project. Further options could be preventing cars from entering the city center, declaring the city center as a park, replanting the area with appropriate native tree species, and ensuring that working with nature, as understood by traditional knowledge, drives new development. Increased urban green space, and indeed culturally relevant nature-based solutions to building and infrastructure challenges, would not only help residents and visitors cope with the heat, but could be part of Samoa's efforts to increase carbon sequestration, support biodiversity and adapt to climate change (Kiddle et al., 2021b). Apia could be developed to encourage a sense of place and well-being for Samoans. Here, Samoa could draw on the development of urban green and blue spaces elsewhere, notably how parks become places that contribute to ecological health in urban settings, but are also for people to meet, socialize, exercise, and generally enjoy the benefits of close proximity to nature. All of these activities contribute to the development of improved physical and mental health and well-being (Pedersen Zari, 2019; Soderland & Newman, 2015).

Moving forward

Ensuring sustainable and resilient building and landscape designs in the Pacific Islands' region is a whole-of-society undertaking that requires collaboration across various government agencies, development and humanitarian actors, and between national and village authorities. The Samoan government has already developed a holistic framework of national governance where all of the 14 core sectors of the government share responsibilities in decision making with non-government groups, communities and development partners (Ministry of Natural Resources and Environment & Samoa Tourism Authority, 2016). This holistic approach is inclusive and enables partnerships and collaborative activities between government, communities and international and local agencies and civil society groups. This framework could enable the mainstreaming of traditional knowledge driven nature-based ALD goals and activities into the plans of key government sectors, local communities and international partners and could be a model for other parts of Oceania.

The role of relevant government agencies in the development of national strategic goals and objectives is crucial to ensuring sustainable building and landscape designs in the Pacific Islands' region. We recommend that any efforts to develop nature-based ALD goals and activities must consider not only vernacular and traditional knowledge-driven approaches to the creation of spaces alone but also how these relate to the existing legislation and national strategic documents in place.

It is also important for governments to consider the significant role of master/traditional builders in the development of national strategies relating to nature-based resilient ALD. These people possess knowledge and skills crucial to these aspects of development. Traditional master builders should also be key consultants in government construction projects and building programs. The education sector also plays a significant role in the sustainability of traditional skills and methods of ALD in Samoa and the Pacific region as a whole. Tertiary institutions should consider traditional ALDs in the development of construction and building programs and courses, particularly in relation to climate change adaptation and resilience initiatives. For example, the Faculty of Technical Education and the Centre of Samoan Studies at the National University of Samoa could consider developing courses that cover Samoan traditional ALDs. To date, the process of building a Samoan *fale* and its significance is taught, at a general level, as part of the Samoan Language and Culture curriculum in schools. However, traditional skills and methods of building *fale* are not part of the construction and joinery curriculum at tertiary level. In 2021, building a model of a *fale Samoa* became a part of the National University of Samoa Culture Week; however, the process and skills are not a formal part of the training of construction workers and builders at the university. Other key government ministries to assist in this area would be those responsible for the environment and forestry. The current activities of the Ministry of Natural Resources and Environment and the Ministry of Agriculture, Forestry and Fisheries in developing awareness of native tree species and their replanting programs have been, and will continue to be, crucial to the development of nature-based traditional ALDs in Samoa, and are key to climate change adaptation and resilience efforts. In addition, future replanting initiatives to increase the resilience of landscapes could consider planting programs for trees, such as *poumuli, togovao*, and certain varieties of coconut and breadfruit, which are necessary to building Samoan *fale*, but also play critical roles in working with ecologies in order to further adapt to the impacts of climate change.

Conclusion

Traditional, nature-based ALDs are tangible aspects of Pacific culture and worldviews. In Samoa, this is encapsulated in the concept of *va tapuia*, the sacred and complex relationship between all living and non-living elements of the socio-ecological realm. In Samoa and the wider Pacific region, buildings and landscapes are interconnected and have evolved in various ways and forms over centuries and showcase the inextricable connection between Pacific peoples and their island and ocean environments. This connection is the foundation of ongoing ability to adapt to an interlinked changing climatic, ecological, political and social context, and must therefore be a critical aspect of future resilient spatial designing which incorporates interconnected architectural and landscape conceptualization, designing and occupation. Using the lens of *va tapuia* from our Samoan case study to analyze these changes helps to identify building types and landscape patterns that are sustainable and therefore likely to be more resilient for the context of Oceania, and those which are not.

It is crucial that models of urban and landscape design for climate change adaptation developed outside of the region are not simply transplanted in the Oceania context. Rather, our Samoan case-study has shown that design driven by traditional ecological knowledge is the most important, localized, starting point for nature-based, resilient, ALD. Such traditional ecological knowledge is the foundation of ongoing ability to adapt to changing contexts in the region. Worldviews and lessons from Oceania, such as the Samoan example of *va tapuia*, can also be used to inform and adapt nature-based ALDs in other parts of the globe.

Acknowledgment

This research was funded by a Royal Society of New Zealand Marsden grant and is part of the NUWAO (Nature-based urban design for Wellbeing and Adaptation in Oceania) project www.nuwao.org.nz.

References

Aiono Le Tagaloa, F. (1996a). *O la ta Gagana*. Apia: Le LamepaPress.

Aiono Le Tagaloa, F. (1996b). *Motugaafa*. Apia: Le LamepaPress.

Allen, A. E. G. (1994). The Ritual architecture: The creation of the Samoan guest fale. *Pacific Arts* 9/10 July: 76–84. Retrieved from: https://www.jstor.org/stable/i23409770

Boon, J. (2001). A socio-economic analysis of mangrove degradation in Samoa. *Geographical Review of Japan*, 74: 159–186.

Buck, P.H. (1930). *Samoan material culture*. Honolulu: The Bernice Pauahi Bishop Museum.

Engels-Schwarzpaul, T. & Refiti, A. (2018). Fale Samoa's extended boundaries: Performing place and identity. In E. Grant, K. Greenop, & A. L. Refiti (Eds.), *Handbook of contemporary indigenous architecture* (pp. 677–696). Singapore: Springer.

Government of Samoa. (2002). *World summit on sustainable development assessment report*. Apia: Government of Samoa.

Government of Samoa. (2011). *Post tsunami Samoa 2009–2011*. Apia: Government of Samoa.

Government of Samoa. (2013) *Samoa post-disaster needs assessment Cyclone Evan 2012*. Apia: Government of Samoa.

Hau'ofa, E. (1993). Our sea of islands. In E. Waddell, V. Naidu, and E. Hau'ofa (Eds.), *A new oceania: Rediscovering our sea of islands* (pp. 2–18). Suva: School of Social and Economic Development, University of the South Pacific.

Kiddle G. L., Bakineti, T., Latai-Niusulu, A., Missack, W., Pedersen Zari, M., Kiddle, R., Chanse, V., Blaschke, P. and Loubser, D. (2021a). Nature-Based Solutions for Urban Climate Change Adaptation and Wellbeing: Evidence and Opportunities from

Kiribati, Samoa, and Vanuatu. *Frontiers in Environmental Science,* 9: 723166. https://doi.org/10.3389/fenvs.2021.723166

Kiddle, G.L., Pedersen Zari, M., Blaschke, P., Chanse, V., Kiddle, R. (2021b). An Oceania Urban Design Agenda Linking Ecosystem Services, Nature-Based Solutions, Traditional Ecological Knowledge and Wellbeing. *Sustainability,* 13: 12660. https://doi.org/10.3390/su132212660

Kramer, A. (1995). *The Samoa Islands: An Outline of a Monograph with particular consideration of German Samoa,* volume I and II. Honolulu, University of Hawaii Press.

Latai, A. (2008). *Toe timata le upega: a critique of coastal governance in Samoa,* Master of Arts thesis, Dunedin: University of Otago. Retrieved from https://ourarchive.otago.ac.nz/handle/10523/9866

Latai, L. (2014). From open *fales* to mission houses: negotiating the boundaries of domesticity in Samoa, *Divine domesticities: Christian paradoxes in Asia and the Pacific.* Canberra: Australia National University Press.

Latai-Niusulu, A. (2017). *Exploring resilience to climate change and other environmental challenges in Samoan communities,* Doctor of Philosophy thesis. Dunedin: University of Otago. Retrieved from http://hdl.handle.net/10523/7461

Latai-Niusulu, A., Nel, E., & Binns, T. (2020). Positionality and protocol in field research: Undertaking community-based investigations in Samoa. *Asia Pacific Viewpoint,* 61(1):71–84. https://doi.org/10.1111/apv.12252

Ministry of Natural Resources and Environment (MNRE): Government of Samoa. (2013). *Samoa's state of the environment report 2013.* Apia: MNRE.

Ministry of Natural Resources and Environment and Samoa Tourism Authority. (2016). *Apia waterfront development project: Waterfront plan 2017–2026.* Apia: Government of Samoa.

Ministry of Works, Transport and Infrastructure. (2017, October 1). *National Building Code of Sāmoa.* Retrieved from: https://www.cip-icu.ca/Files/APE-2017-Projects/NBC-Draft-2017-February15-FINAL-for-STAPLES-8x11.aspx

Pacific Australia Climate Change Science and Adaptation Planning (PACSSAP) (2013). *Samoa LiDAR Factsheet.* International Climate Change Adaptation Initiative, Australian AID.

Pedersen Zari, M. (2019). Understanding and designing nature experiences in cities: a framework for biophilic urbanism, *Cities & Health.* doi: 10.1080/23748834.2019.1695511

Ravuvu, A. (1983). *Vaka iTaukei.* The Fijian Way of Life. Suva, Fiji: Institute of Pacific Studies, the University of the South Pacific.

Saini, B. (1978). *Traditional architecture in the Pacific: An overview.* Brisbane: University of Queensland.

Samoa Bureau of Statistics. (2006). *Population and housing census 2006. Analytical report.* Apia: Government of Samoa.

Samoa Bureau of Statistics. (2016). *2016 Census Brief No. 1.* Apia: Government of Samoa.

Schoeffel, P. (1995). The Samoan concept of feagaiga and its transformation. *Tonga and Samoa: Images of Gender and Polity,* 85–105.

Shore, B. (1982). *Sala'ilua: A Samoan mystery.* New York: Columbia University Press.

Soderlund, J., & Newman, P. (2015). Biophilic architecture: a review of the rationale and outcomes. *AIMS Environmental Science,* 2(4): 950–969.

South Pacific Disaster Reduction Programme. (1998). *Disaster resistant housing in pacific island countries: A compendium of safe low cost housing practices in Pacific Island Countries.* UNDP: New York.

Stucky, L. (2011). *Culture in architecture: Amalau-designing a Samoan village for the future.* Doctorate Project for the Degree of Doctor of Architecture in the School of Architecture, University of Hawaii, Manoa.

Tamasese, T. (2007) 'Bio-ethics and the Samoan indigenous reference', *Sunday Samoan Observer* 13 November: 18–23.

Taua'a, S. (2018). An impact case study of improved road infrastructure on urban communities in Samoa. *Journal of Pacific Studies*, 38(1): 72–98.

Taylor, J. C. (1999). *The Tufuga's Fale Tele: Constructing Social Relations in the Samoan Built Environment.* Master of Arts thesis, Department of Anthropology and Sociology, University of British Columbia, Vancouver Canada.

Tuhiwai Smith, P. L. (2012). *Decolonizing methodologies: Research and indigenous peoples.* Dunedin: Otago University Press

Tuvale, T. (n.d). *An account of Samoan history up to 1918: The Samoan House- O le Pale Samoa.* Retrieved from http://nzetc.victoria.ac.nz/tm/scholarly/tei-TuvAcco-t1-body1-d28.html

Van der Ryn, F. M. (2016). The Measina of Architecture in Samoa-An examination of the Va in Samoan Architecture and Socio-cultural implications of Architectural changes. *The 7th Measina a Samoa Conference* 15th–17th November 2016, Apia, National University of Samoa.

Whincup, T. (2010). TeMwaneabani Kiribati. The traditional meeting house of Kiribati: A tale of two islands, Shima: *The International Journal of Research into Island Cultures,* 4(1): 113–130.

14

ESTIMATION OF SPATIOTEMPORAL VARIATION IN POTENTIAL ECOSYSTEM SERVICES

A case study of Aydın, Turkey

Ebru Ersoy Tonyaloğlu and Birsen Kesgin Atak

History and current context

Aydın province is located on the coastal part of the Aegean Region (37°50′17″N and 27°50′44″E) and surrounded by Denizli in the east, the Aegean Sea in the west, İzmir and Manisa in the north, and Muğla in the south (Figure 14.1). The increase in residential areas and land-use change in Turkey puts an extra pressure on agricultural lands and pastures, and Aydın province follows a similar pattern in terms of land-use change dynamics (Yılmaz Genç et al, 2021).

The region of Aydın province was reshaped many times because of the existence of different civilizations that dominated this region as well as the earthquakes (Uğur, 2003). The province has hosted various civilizations throughout history e.g. Thracians, Ionians and Ancient Romans (Esbah, 2007). In particular, the traces of ancient Tralleis city and settlements around it have a long history that dates back 7,000 years (Akdeniz, 2002).

Aydın province accommodates two important protected areas in Turkey. These are Dilek Peninsula-Büyük Menderes Delta National Park, declared on 19 May 1966, and Bafa Lake Nature Park, registered on 8 July 1994. The Dilek Peninsula is important in that it contains elements of the Mediterranean, Irano-Turanian and European Siberian Flora Region. The Büyük Menderes Delta has the status of "Class A Wetland" of international importance. The Büyük Menderes Delta is rich in different wetland types including lagoons, swamps and lakes. The most naturally important wetlands include the Dilek Peninsula-Büyük Menderes National Park, Bafa Lake Nature Park and Azap Lake. Bafa Lake Nature Park has the status of Special Protection Area due to the breeding populations of the swamp tern and spur lapwing.

The province has a typical Mediterranean climate. These climatic conditions and topographic structure in Aydın province and its surroundings form two

DOI: 10.4324/9781003266440-14

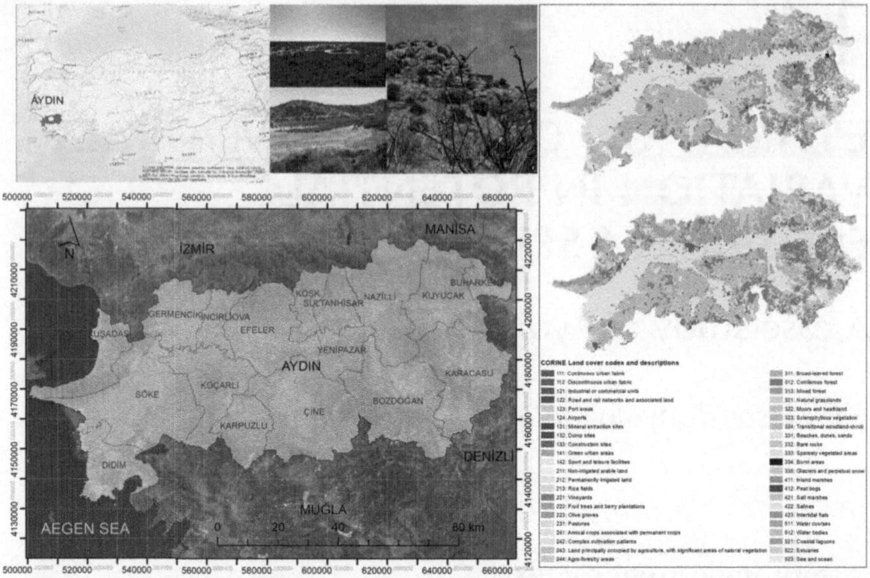

FIGURE 14.1 Location of the Aydın province. Upper left: Aydın province in Turkey @ Basemap Layer Services/Powered by Esri and coastal vegetation types @ Ebru Ersoy Tonyaloğlu. Lower left: Districts of Aydın province with background image from @ ArcGIS 10.5.1 Basemap Layer Services/Powered by Esri. Right: Corine Land Cover Types from @ European Union, Copernicus Land Monitoring Service 2021, European Environment Agency (EEA).

major plant communities, namely forest and maquis. The maquis communities are distributed up to 500–600 m from the coastal areas. Forests and agriculture comprise 80% of the land use in the providence with cultivated agriculture occurring in a valley down the center of the providence, pine-oak forests spanning high mountains, and olive and fruit orchards in the northern quarter of the providence.

Most of the provincial area consists of mountains (63%) and plains (21%) where the most fertile land is mostly located around the Büyük Menderes Meander Basin. The remaining area of Aydın province consists of plateaus (Koday and Akbaş, 2016). Settlements are concentrated on the northern edge of the valley and along the coastal regions.

Aydın province is an agriculture and tourism city where agriculture is the most important economic activity in the province (covering almost 18% of its total area both for arable land and horticulture). Also, with a coastline of 150 km and two important tourism centers, Kuşadası and Didim, Aydın province hosts various year-around active and passive recreational activities as part of domestic and international tourism services.

Current and future stressors

Most of the landscapes that we know today have been shaped by people and their interactions with landscapes through time. Although urban areas currently encompass a small proportion of the world's land surface; most people live in cities due to the facilities and opportunities they provide (Seto et al., 2013). In general terms, cities and their surroundings are regarded as centers of landscape change, anthropogenic disturbances and also among the main drivers of climate change depending on the increase in human population and their needs as well as demand for limited resources (Fedele et al., 2018; Mamat et al., 2018).

While the pattern of land use/land cover (LULC) in cities is influenced by social and ecological processes, it also affects the functioning and pattern of ecological conditions, perceptions and attitudes of citizens against changing landscapes as well as the benefits that we obtain from them. In this context, quantitative assessments of the effects of LULC changes on urban ecosystem services represent a focal scientific research area in the realm of resilient landscapes since there is a direct and clear relationship between ecosystem services and LULC.

Our case study area, Aydın province, stands out with its historical and cultural richness, as well as its natural features and areas. However, the population of Aydın province increased rapidly since the 1990s and this has been associated with many environmental problems due to unplanned urbanization, destruction of agricultural lands and mismanagement of landscapes. Today, having over a million inhabitants, the population density of the province ranks 20th among 81 provinces in Turkey with a total area of 8,007 km^2 (TSMS, 2021). Approximately 54% of the area within the province have been shaped by direct human intervention; consists of agricultural areas, settlement, industrial areas.

In addition to industrial production, mainly developed for the processing of agricultural products, energy production is another important economic resource of the province. However, increasing urban expansion, transportation network, agricultural activities and geothermal energy production are regarded as the most important causes of a range of environmental problems including environmental degradation, loss of ecological integrity as well as local climate change in Aydın province (Çakır, 2005; Umaç, 2019; THHP, 2020).

For example, one of the most important problems encountered within the Aydın province, Büyük Menderes Delta, is the pollution carried by the Büyük Menderes River which is mainly caused by intensive agricultural and industrial activities. Unfortunately, today, many vulnerable plant and animal species are under the threat of this pollution. Being that the Aydın province, which is at the center of important historical and touristic centers such as Izmir, Ephesus, Meryemana, Milet, Didim, Pamukkale, Marmaris and Bodrum, is an important tourism center with its port city feature (KUTO, 2017). Along with all these, there is an increase not only in the settlement pattern but also in road transportation (Deniz et al., 2008). Therefore, mass tourism activities put an extra pressure on natural resources by changing its local land use, increasing pollution and natural habitat loss, and putting more pressure on endangered species. Now, with the increasing effect of tourism, it is seen that the areas suitable for agriculture not only in Kuşadası city center but also in village settlements have left their place to touristic facilities and secondary residences. These adverse effects can particularly be observed on maquis and garig areas (Kahraman, 2011).

Aydın province, with more than 60 power plants today, has an important potential in terms of energy (including geothermal, wind, hydroelectric, natural gas, biogas and solar). Among these, the most important underground resource is geothermal energy resources as it is located in the richest geothermal area of Turkey in terms of its actual potential. There are currently 30 active geothermal power plants within the borders of the province, where a large number of exploration and operation activities are carried out (EA, 2022). Respiratory, cardiovascular, nervous related diseases and cancer are some of the known health problems caused by air, water and noise pollution related to energy and mining sectors (Karagülle, 2018; THHP, 2020).

Moreover, Aydın province is located in one of the riskiest regions against wildfires. In 2021, we had more than 175 wildfires on the western and southern coastal parts of Turkey including Aydın province. At the same time, there are managerial issues that increase the sensitivity to fire in the area. To illustrate, intentional fires were started with the restriction of activities that can be done in some privately owned areas within the borders of the National Park (Deniz et al, 2018). So, the increasing occurrence of wildfires/fires has been a serious threat to ecosystem services in Aydın province, too.

Furthermore, Tonyaloğlu (2019) determined that there was 3.19°C increase in the land surface temperature of the central neighborhoods of the province between 2005 and 2015 just because of the decrease in vegetation with increasing building density. Another study revealed that the urban heat island effect between 1970 and 2004 was gradually increasing in Kuşadası district, which was exposed to intense tourism and construction (Tanrıkulu, 2006). Furthermore, Atak and Tonyaloğlu (2020) examined the relationship between vegetation and urban heat island formation and revealed that urbanization in Aydın has increased the urban heat island effect by creating pressure on the vegetation. Like most of the cities in the world, Aydın province is expected to be affected by climate change including being more vulnerable to natural hazards. In this context, one of the most important aspects is the resilience of the landscape, in other words, its capacity to adapt, reorganize or maintain the processes, structures and functions of the whole system against various disturbances (Folke et al., 2010; Cushman and McGarigal, 2019).

Ecosystem services ensure the continuity of life on Earth including providing many services to citizens, such as access to clean air, water, food and protection of urban environments against natural hazards. Ecosystem services not only provide a healthy environment and increase the quality of life in cities but also increase the resilience of cities by improving their capacity to cope with natural and socio-economic challenges (Gomez et al., 2013). However, this is only possible by protecting and restoring fundamental functions and structures in landscapes through a complete understanding, evaluation and the positive management of landscape change.

Altogether, the population growth, activities for energy production, mining, transportation, anthropogenic pollution and LULC changes not only constitute the most important environmental and health issues in Aydın province but also put pressure on ecosystem services, alter their pattern and may cause irreversible extinction of some ecosystem services. Environmental issues related to changes in the landscape of Aydın province reveal the need for rapid and cost-effective methods and tools that can be useful for understanding spatiotemporal dynamics and patterns of social-ecological interactions and addressing landscape resilience-related questions under different conditions. From this point of view, analysis of multiple ecosystem services provides an effective framework for different conservation, planning and policy-making activities to improve resilience and quality of life in cities (Wratten et al., 2013; Dobbs et al., 2018).

Vision

Environmental issues raised by challenges highlight the need for effective management and response to these challenges, as universally emphasized by many researchers/planners/practitioners, as well as the urgent need for a place-based analysis. Hence, in this chapter, we provide a spatially explicit index for multiple ecosystem services (developed and used by Dobbs et al., 2018) using freely available data in the case of Aydın province, Turkey, which can be applied in similar cases around the world, where rapid changes in LULC threaten the provision of ecosystem services in urbanized landscapes.

We used Landsat satellite images and CORINE land cover maps to determine vegetated/non-vegetated areas (based on Normalized Difference Vegetation Index, NDVI), and to map and estimate the selected ecosystem services. We obtained radiometrically and atmospherically corrected cloud-free Landsat images of the hottest months with the highest photosynthetic activity for 1990 and 2017 from the Landsat archive through the Earth Explorer (USGS, 2021). For the year of 1990, we used Landsat 4–5 TM satellite images dated 13 June, 31 July, 1 September, 17 September and 3 October. For the year of 2017, we used Landsat 8 OLI satellite images dated 7 June, 25 July, 26 August, 11 September and 13 October. Amongst the selected ecosystem services, we used thermal, red and near-infrared bands of Landsat satellite images for the calculation of global climate regulation (carbon storage-CS) and microclimate mitigation (Land Surface Temperature-LST). On the other hand, we assessed ecological integrity (EI) using three indices; Hemeroby index (naturalness of the land cover types using a seven-point scale, see Walz and Stein, 2014), Shannon's diversity index (landscape diversity and heterogeneity) and effective mesh size (landscape fragmentation) using the CORINE land cover maps of 1990 and 2018 (CLMS, 2018a, 2018b). Figure 14.2 represents the spatial distribution of NDVI, carbon

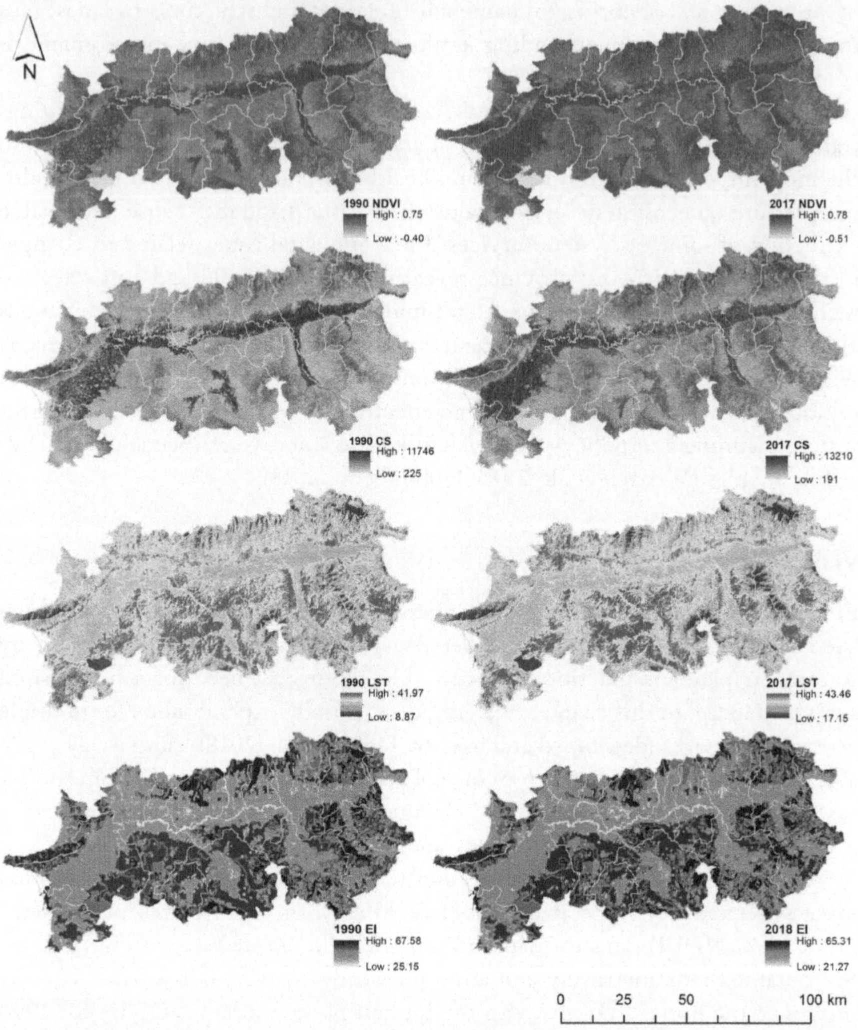

FIGURE 14.2 NDVI, carbon storage (CS), land surface temperature (LST) and ecological integrity (EI) indices for 1990 and 2017–2018. NDVI, Carbon storage and LST were derived from Landsat 4-5 TM and Landsat 8 OLi images @ By Earth Resources Observation and Science (EROS) Center. The EI index was derived from CORINE LC maps from Copernicus/Pan-European/CORINE Land Cover. Maps for 1990 and 2018 @ European Union, Copernicus Land Monitoring Service 2021, European Environment Agency (EEA). Generated by Ebru Ersoy Tonyaloğlu and Birsen Kesgin Atak.

storage (CS), Land Surface Temperature (LST) and EI indices for 1990 and 2017–2018 in Aydın province.

There was no difference in the maximum average NDVI value for Aydın in 1990 and 2017, but the minimum average NDVI was greater in 1990 than 2017, primarily due to thinning of the forest vegetation. Forest cover on high mountains increased in the western and southeast districts, whereas vegetation cover decreased between 1990 and 2017 in the northern plains as drought conditions have reduced yields of arable lands. Even though there was an increase in the vegetation cover (from 48.80% to 60.90%) in the whole province around the peri-urban and rural areas as an extension of the existing vegetated lands (forest and agriculture), the amount of vegetation in urban areas decreased from 1990 to 2017 primarily due to infilling within urban areas. Both in 1990 and 2017, peri-urban and rural areas had the highest total carbon storage in comparison to the urbanized areas. Here, the increase in the Center (Efeler) district is remarkable since it has the highest population and intense urbanization in recent years. This increase has been associated with the increase in urban open green areas and agricultural areas in and around the city, as well as the natural and semi-natural vegetation located and concentrated in the rural areas within the district borders. The district is increasing in population as it is also increasing in the amount of urban green space and the amount of agricultural areas within and around the city, and it is having more natural/semi-natural vegetation in its rural areas. Additionally, there was a decrease in the mean temperature by 1.64℃ between 1990 and 2017. Despite increasing vegetation cover, the microclimate mitigation got worse in our case study area.

Figure 14.3 shows the three indices of EI, namely effective mesh size (MESH), Shannon's diversity index (SHDI) and Hemeroby indices for 1990 and 2018 in Aydın province.

The province came up against a slight fragmentation and isolation of natural land cover types. This is particularly the case in the northern and southern parts of Aydın province where the broad-leaved and coniferous forests became patchier and some of them were transformed into agricultural areas. On the other hand, Aydın province had a more diverse and heterogeneous landscape pattern between 1990 and 2018 with more human-dominated land cover types. This result was confirmed with a decreasing mean Hemeroby index. Particularly, the highway constructed in the mid-1990s has also increased the patchiness of olive groves and complex cultivation patterns. Accordingly, the ecological integrity index was slightly decreased in the district. This is especially related to the increase in continuous urban fabric, industrial/commercial units and road networks. These figures are attributed to the fragmentation of forest and agricultural areas together with the increased diversity of human impact on natural and semi-natural areas.

Figure 14.4 represents the multiple ecosystem services index. In general, the multiple ES index was low in impervious and built-up areas for both dates. The

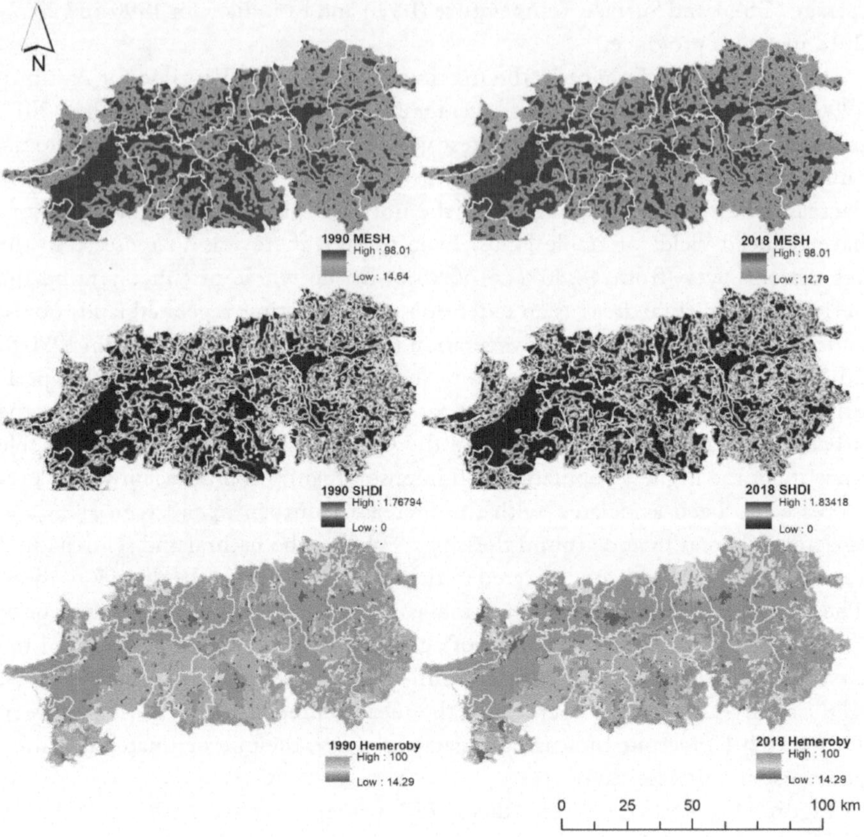

FIGURE 14.3 Effective mesh size (MESH), Shannon's diversity index (SHDI) and Hemeroby indices of Ecological Integrity (EI) for 1990 and 2018. Generated by Ebru Ersoy Tonyaloğlu and Birsen Kesgin Atak.

largest gain in the ESI index is found in forest areas of Kuşadası, Kuyucak, Karacasu, Çine and Bozdoğan districts.

In 2017–2018, ESI index in permanently irrigated agricultural areas is lower than the one in 1990. In both dates, Germencik and Didim districts had the lowest ESI values. This was an expected result because while Didim had experienced a very heavy urbanization process due to second houses, Germencik has geothermal power generation areas. Still, here we can see the direct positive effects of healthy vegetation on the selected ecosystem services.

From all these analyses, it is obvious that the province represents a nice example of how urbanization, agriculture and conservation can all work together to increase the production of ecosystem services. However, there is still a need for a holistic landscape management roof, which is carried out in coordination among different stakeholders. Today, a growing body of literature and experience reveal the effectiveness of the nature-based solutions (NBS) to adapt and mitigate

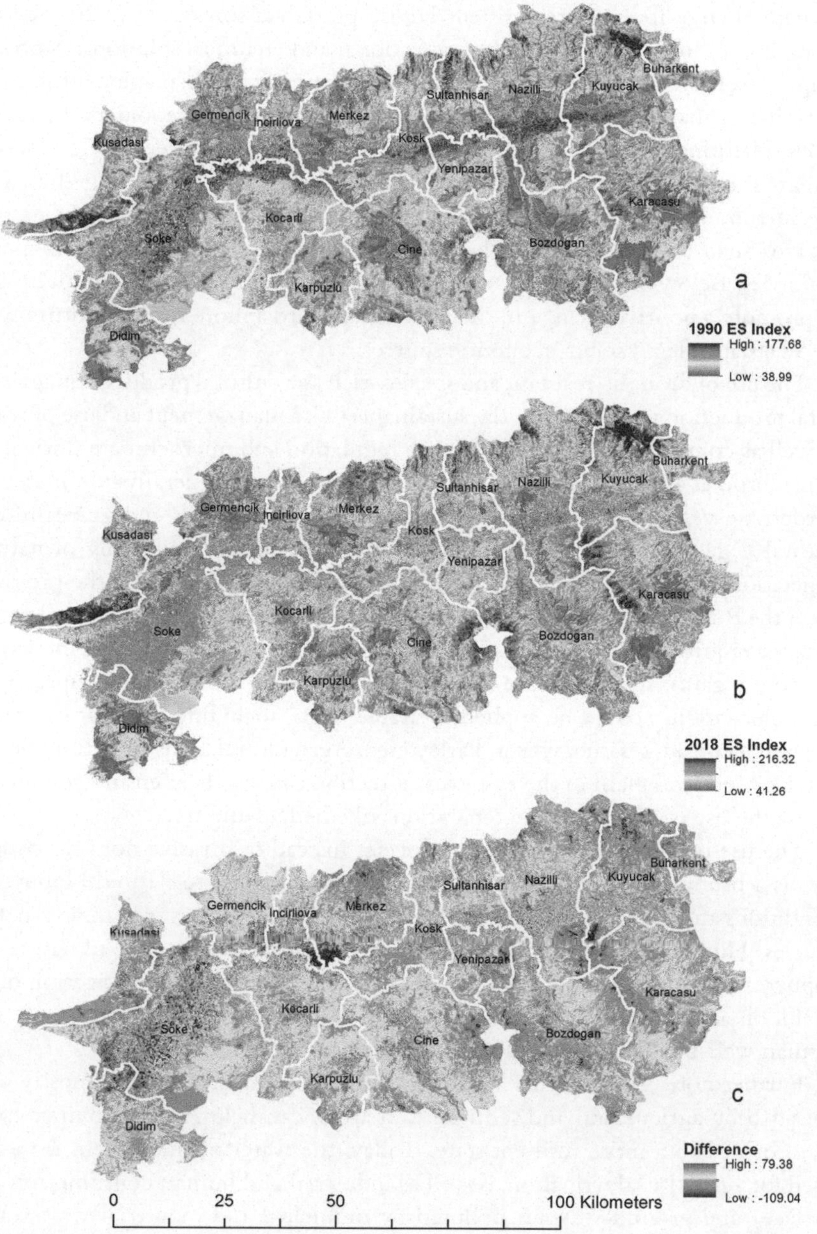

FIGURE 14.4 Multiple ecosystem services (ESI) indices for 1990 (a), 2018 (b) and their difference (c). In the difference map, while negative values represent the gains in ESI index, positive values represent the reverse conditions. The mean multiple ecosystem services index value for Aydın province was increased by 21.01% between 1990 and 2018. Generated by Ebru Ersoy Tonyaloğlu and Birsen Kesgin Atak.

climate change for a more resilient landscape (Chausson et al., 2020; Seddon et al., 2020). In comparison with conventional and technical solutions, NBS can help to promote a top-down and bottom-up planning and management strategy as well as enhancing communication between different stakeholders. In recent years, planning and management approaches that support green infrastructure integrate ecosystem services and NBS into city plans, plan cities together with the surrounding rural areas and increase the resilience of cities against natural hazards such as floods and earthquakes have come into prominence in many countries. Below are some cost-effective and long-term NBS, which might help to promote a positive change in the province toward a more resilient future with its own natural and socio-economic sources.

The use of drought-resistant and species-rich agricultural products for agricultural production would ensure the sustainable/wise management and use of water as well as contributing to global climate regulation and microclimate mitigation (Eigenbrod et al., 2009; Power, 2010). Aydın province is generally dominated by xerophytic vegetation where we have fig trees, citrus, olives and grapes in our natural vegetation. Therefore, by promoting horticulture consisting of natural vegetation elements, where the foothills of mountains surrounding the province meet the Büyük Menderes plain, it will be possible to reduce irrigation and maintenance requirements, prevent soil erosion and salinity issues, as well as increasing economic gains, social welfare and resilience of these types of landscapes. Such an understanding could be applied to arable lands (including drought resistance species of cotton, sesame, wheat, barley etc.). Agricultural lands stretching on the Büyük Menderes plain in the east-west direction can also help ensure connectivity by the use of natural parcel separations like hedges and tree rows.

The use of indigenous tree/shrub species instead of invasive non-native species is a promising approach since they are well-adapted to natural biophysical conditions and their reliance on water and soil resources is lower than non-native species (Hufnagel and Garamvölgyi, 2014). So, promoting the use of native xerophytes within urbanized areas would increase carbon storage, mitigation of air pollution and urban heat islands, quality of water sources, and biodiversity and human well-being (Forman, 2014; Mirsanjari et al., 2020).

Furthermore, the riverine system of Büyük Menderes River, mostly surrounded by agricultural and semi-natural areas, can help Aydın Province to be transformed to a more resilient state. Today, the water of this system is mostly used for agricultural irrigation as well as industrial and human consumption and surface- and ground-water is polluted by municipal and industrial wastewater. However, the river channel riparian zone and wetlands, if it can behave naturally by creating buffers around it, would create a more connected and biodiverse environment, mitigate flood risks and store carbon.

Path to the future

The general results of this research suggest that the potential multiple ecosystem services supply and their spatiotemporal dynamics have changed for the better because of the increase in forest vegetation cover and agricultural land extension. However, the negative change in microclimate mitigation and ecological integrity emphasizes that the constructive contribution of vegetation cover cannot be enough to compensate for the subtle ecological consequences that the whole province may face in the near future because of its demographic and economic structure. As urbanization, the fragmentation of natural vegetation and the energy production sector may cause greater challenges in Aydın province, which is also compounded by the effects of economic priorities; NBS can secure the production of multiple ecosystem services now and into the future. It is important to understand the type and magnitude of change and the ability of landscapes to accommodate/adapt to change across space or time if we want to respond to various challenges created by extensive human activities.

Landscape-scale approaches to measure and evaluate different attributes of an ecosystem can recognize the interactions and interdependencies of urban and rural areas at the landscape scale and provide planners/decision-makers/managers with opportunities to direct limited financial resources to opportunity areas of greatest benefits as well as areas of apparent deficiency in functioning of ecosystems (Dobbs et al., 2018; Cushman and McGarigal, 2019; Tonyaloglu, 2020). Mapping and assessment of urban ecosystem services allow the use of this information in spatial planning.

The method used in this chapter is flexible, rapid, and cost-effective and can be applied in similar contexts to assess and map the potential for multiple ecosystem services. In addition, other potential ecosystem services can be integrated to this method to expand the scope of intended purposes. Despite the fact that the method used in this chapter simplifies the complex processes and synergies in landscapes, the approach is still promising for providing spatially explicit, economic and timely estimations of the relationships between landscape change and potential ecosystem services over time. In particular, the use of time series over a long time period would yield better estimations of regional/national trends and inform planning policies and management decisions on supporting ecosystem services and landscape resilience for the benefit of people, nature and biodiversity.

In our case, these indices provide valuable information on which areas we should intervene in for landscape planning and climate change adaptation /mitigation, such as urban planting and the protection of forest shores if we intend to improve local microclimate, ensure ecological integrity and get multiple ecosystem service benefits. Each indicator used to form the ESI index in this chapter can be adapted and used for monitoring in different ecosystems. In our case, the NDVI index is used to understand the change in state and health of vegetation together with some other associated ecosystem services. As a general indicator of vegetation productivity, NDVI might be used to monitor the health and vitality of forests as it is able to represent the location and extent of a forest spatially. In the case of Aydın province, it was obvious that even though the density and extent of forest had expanded between 1990 and 2018 around the existing forests, we also had losses from their core. The CS index is also useful to assess changes in forest carbon as an indicator of the ecological condition of forest ecosystems (stable, growing or declining) and therefore in determining the spatial strategies for forest management.

Altogether the indices that we used to assess the multiple ESI specify the fraction of vegetated/green areas compared to impervious surfaces at different time periods and pave the way for decision-makers and planners that can be exploited in different cases for transforming landscapes into a more resilient state. These indices can be applied in any context around the world not only to historical landscape changes but also to simulate future landscape changes. These might include determining the state and change in urban green areas and forests that need conservation and improvement (e.g. parks, urban trees and forests), opportunity areas where we can increase the vegetated areas and their connectivity (e.g. vegetated roofs, urban agriculture, hedgerows in agricultural lands, rivers) and evaluating potential/suitable areas, if required, for urban expansion and other human-related land uses.

References

Akdeniz, E. (2002). Neolitik ve Kalkolitik Çağlarda Büyük Menderes Havzası'ndaki Kültürel Yapılanma ve Orta Kalkolitik Çağ Problemi. *OLBA* 5, 59–76.

Atak, K.B. & Tonyaloğlu, E.E. (2020). Alan kullanım/arazi örtüsü ve bitki örtüsündeki değişimin arazi yüzey sıcaklığına etkisinin değerlendirilmesi: Aydın ili örneği. *Turkish Journal of Forestry, 21*(4), 489–497.

Çakır, M. (2005). *Aydın ili ilçelerindeki büyük ölçekli işletmelerin sorunları ve Türkiye ekonomisine katkısı* (Doctoral dissertation, Adnan Menderes Üniversitesi).

Chausson, A., Turner, B., Seddon, D., Chabaneix, N., Girardin, C.A., Kapos, V., Key, I., Roe, D., Smith, A., Woroniecki, S. & Seddon, N. (2020). Mapping the effectiveness of nature-based solutions for climate change adaptation. *Global Change Biology, 26*(11), 6134–6155.

CLMS (2018a). Copernicus Land Monitoring Service, CORINE Land Cover. https://land.copernicus.eu/pan-european/corine-land-cover/clc-1990 (Accessed 10 January 2018).

CLMS (2018b). Copernicus Land Monitoring Service, CORINE Land Cover. Available online: https://land.copernicus.eu/pan-european/corine-land-cover/clc2018 (Accessed 10 January 2018).

Cushman, S.A. & McGarigal, K. (2019). Metrics and models for quantifying ecological resilience at landscape scales. *Frontiers in Ecology and Evolution*, 7(440). https://doi.org/10.3389/fevo.2019.00440

Deniz, B., Eşbah, H., Küçükerbaş, E.V. & Şirin, U. (2008). Kentsel Alan Kullanımlarındaki Vejetasyon Yapısının Analizi: Aydın Kenti Örneği. *EkolojiDergisi*, 17(66), 55–64.

Dobbs, C., Kendal, D. & Nitschke, C.R. (2014). Multiple ecosystem services and disservices of the urban forest establishing their connections with landscape structure and sociodemographics. *Ecological Indicators*, 43, 44–55.

EA (2022). Enerji Atlası. Available online: https://www.enerjiatlasi.com/jeotermal/ (Accessed 7 Apriy 2022).

Eigenbrod, F., Anderson, B.J., Armsworth, P.R., Heinemeyer, A., Jackson, S.F., Parnell, M., Thomas, C.D. & Gaston, K.J. (2009). Ecosystem service benefits of contrasting

conservation strategies in a human-dominated region. *Proceedings of the Royal Society B: Biological Sciences, 276*(1669), 2903–2911.
Esbah, H. (2007). Land use trends during rapid urbanization of the city of Aydın, Turkey. *Environmental Management, 39*(4), 443–459.
Fedele, G., Locatelli, B., Djoudi, H. & Colloff, M.J. (2018). Reducing risks by transforming landscapes: Cross-scale effects of land-use changes on ecosystem services. *PLoS One, 13*(4), e0195895.
Folke, C., Carpenter, S.R., Walker, B., Scheffer, M., Chapin, T. & Rockström, J. (2010). Resilience thinking: Integrating resilience, adaptability and transformability. *Ecology and Society, 15*(4). https://www.jstor.org/stable/26268226
Forman, R.T. (2014). *Land Mosaics: The ecology of landscapes and regions (1995). The Ecological Design and Planning Reader.* Island Press: Washington, DC, 217–234.
Gómez-Baggethun, E., Gren, Å., Barton, D.N., Langemeyer, J., McPhearson, T., O'farrell, P., Andersson, E., Hamstead, Z. & Kremer, P. (2013). Urban ecosystem services. In: Elmqvist et al. (Eds.), *Urbanization, biodiversity and ecosystem services: Challenges and opportunities* (pp. 175–251). Dordrecht: Springer.
Hufnagel, L. & Garamvölgyi, Á. (2014). Impacts of climate change on vegetation distribution No. 2-climate change induced vegetation shifts in the new world. *Applied Ecology and Environmental Research, 12*(2), 355–422.
Kahraman, C. (2011). Kuşadası'nda Arazi Kullanımı. *Coğrafya Dergisi, 1*(22), 27–44.
Karagülle, D. (2018). Aydın ilinde çevre sağlığı eylem planı çalışmaları-çevresel risk faktörleri araştırması.
Koday, Z. & Akbaş, F. (2016). Aydın İlinin İdari Coğrafya Analizi/Administrative Geography Analysis of Aydin Province. *Atatürk Üniversitesi Sosyal Bilimler Dergisi*, 57, 63–91.
KUTO (2017). Kuşadası Çevre Yolu Raporu, Kuşadası Ticaret Odası. Available online: http://www.kuto.org.tr (Accessed 10 April 2022).
Mamat, A., Halik, Ü. & Rouzi, A. (2018). Variations of ecosystem service value in response to land-use change in the Kashgar Region, Northwest China. *Sustainability, 10*(1), 200.
Mirsanjari, M.M., Zarandian, A., Mohammadyari, F. & Visockiene, J.S. (2020). Investigation of the impacts of urban vegetation loss on the ecosystem service of air pollution mitigation in Karaj metropolis, Iran. *Environmental Monitoring and Assessment, 192*(8), 1–23.
Power, A.G. (2010). Ecosystem services and agriculture: Tradeoffs and synergies. *Philosophical Transactions of the Royal Society B: Biological Sciences, 365*(1554), 2959–2971.
Seddon, N., Smith, A., Smith, P., Key, I., Chausson, A., Girardin, C., House, J., Srivastava, S. & Turner, B. (2021). Getting the message right on nature-based solutions to climate change. *Global Change Biology, 27*(8), 1518–1546.
Seto, K.C., Parnell, S. & Elmqvist, T. (2013). A global outlook on urbanization. In: Elmqvist et al. (Eds.), *Urbanization, biodiversity and ecosystem services: Challenges and opportunities* (pp. 1–12). Dordrecht: Springer.
Tanrıkulu, M. (2006). *İzmir'de Şehirleşmenin Sıcaklık ve Yağış Üzerine Etkisi* (Doctoral dissertation, Yüksek Lisans Tezi. Ankara Üniversitesi, Sosyal Bilimler Enstitüsü, Coğrafya (Fiziki Coğrafya) Anabilim Dalı. 127s).
THHP (2020). Temiz Hava Hakkı Platformu, Kara Rapor 2020: Hava Kirliliği ve Sağlık Etkileri. *Printworld Matbaa San. ve Tic. A. Ş, İstanbul.*
Tonyaloğlu, E.E. (2019). Kentleşmenin kentsel termal çevre üzerindeki etkisinin değerlendirilmesi, efeler ve İncirliova (Aydın) örneği. *Türkiye Peyzaj Araştırmaları Dergisi, 2*(1), 1–13.

Tonyaloğlu, E.E. (2020). Spatiotemporal dynamics of urban ecosystem services in Turkey: The case of Bornova, Izmir. *Urban Forestry & Urban Greening, 49*, 126631.

TSMS (2021). The Turkish State Meteorological Service. Available online: http://kuraklikizle.mgm.gov.tr/ (Accessed 20 April 2021).

Uğur, A. (2003). Aydin Şehrinin Kuruluşu ve Gelişme Evreleri (Establishment and Development Phases of the Aydın City). *Coğrafi Bilimler Dergisi/Turkish Journal of Geographical Sciences, 1*(2), 41–62.

Umaç, B. (2019). *Aydin İl Merkezinde Seçili Bölgelerde Biyofiziksel Ölçümlerin Karşılaştırılması* (Master's thesis). University of Aydın Adnan Menderes.

USGS (2021). U.S. Geological Survey. Available online: https://earthexplorer.usgs.gov/ (Accessed 20 January 2021).

Walz, U. & Stein, C. (2014). Indicators of hemeroby for the monitoring of landscapes in Germany. *Journal for Nature Conservation, 22*(3), 279–289.

Wratten, S.D., Sandhu, H., Cullen, R. & Costanza, R. (Eds.). (2013). *Ecosystem services in agricultural and urban landscapes* (p. 217). Oxford: Wiley-Blackwell.

Yılmaz Genç, S., Behradfar, A., Castanho, R.A., Kirikkaleli, D., Naranjo Gómez, J.M. & Loures, L. (2021). Land use changes in Turkish territories: Patterns, directions and socioeconomic impacts on territorial management. *Current World Environment 16*(1), 105–122.

15

SCENARIO-BASED THINKING TO NEGOTIATE COASTAL SQUEEZE OF ECOSYSTEMS

Green, blue, grey and hybrid infrastructures for climate adaptation and resilience

Scott Hawken, Kaihang Zhou, Luke Mosley and Emily Leyden

Introduction: scenario-based thinking for resilient coastal landscapes

Coastal landscapes are under threat globally due to climate change, in particular due to rapidly rising sea levels (IPCC, 2021). These dynamic coastal environments are valuable for a range of land uses including tidal and supratidal ecosystems, human settlement, agriculture and various built infrastructures (Small & Nicholls, 2003). Together these land uses form what we call herein "anthropogenic coastal landscapes". In describing them in this way, we acknowledge that there is a continuum of more natural and more built or artificial conditions (Hobbs, Higgs, & Hall, 2013; Hobbs et al., 2014; Lundholm, 2015; Sutton-Grier, Wowk, & Bamford, 2015; Mangone, 2016). The anthropogenic coastline has been formed through a complex combination of processes. For instance, artificial or grey infrastructure has often been designed to arrest or defend against geomorphological change to prioritize land uses such as housing and agriculture yet coastlines keep evolving regardless with landscape types such as beaches and wetlands in a continual state of flux. In many coastal suburbs, infrastructure, such as stormwater systems, housing and coastal wall defenses, have become defunct through climate-related forces including sea level rise, tidal flooding, saltwater intrusion through and across built infrastructures and storm surges. As a consequence, we need to consider how to address climate change threats. The most immediate and obvious approach to defending built and settlement assets is a linear, defensive approach epitomized by dikes and seawalls. However, this can have disastrous outcomes for all land uses, even those which are intended to be protected (Zevenbergen, Rijke, Van Herk, & Bloemen, 2015). An ultimately more powerful approach is through adaptation and making space for the integration of various natural and

DOI: 10.4324/9781003266440-15

built systems to manage climate change in a resilient way (De Bruijn, De Bruijne, & Ten Heuvelhof, 2015). In particular, integrating land use such as green and blue infrastructure is critical (Hawken et al., 2021). Green infrastructure is any land uses which delivers ecosystem services (Hawken, Rahmat, Sepasgozar, & Zhang, 2021). Typically, green infrastructure involves terrestrial ecosystems such as forests or other vegetation whilst blue infrastructure includes aquatic land use such as estuaries and lakes and rivers.

Such an integrated resilient approach requires the reconceptualization of simple monofunctional land uses as multifunctional or hybrid land uses to achieve win-win outcomes for both human- and non-human-dominated environments. A key to such an approach is the by now well-known concept of "green infrastructure". Although it is in popular usage the concept of green infrastructure is complex to define and not always transparent. Silva et al. (2017, 2020) argue that green infrastructure encompasses a range of natural, semi-natural and artificial multifunctional strategies to solve ecological and socioeconomic challenges simultaneously.

To study alternative coastal landscape "futures", we use a scenario-building approach applied in the Millennium Ecosystem Assessment (2005) to systematically and creatively think about alternative courses of action. Scenarios are not about identifying correct or incorrect courses of action although they may suggest benefits and drawbacks of alternative approaches. They are most useful in generating creative choices that might not immediately be apparent using conventional, business-as-usual conditions. In scenario-based approaches, stakeholders are not tied to one outcome but can experiment and develop alternatives freeing up negotiations and avoiding adversarial politics which can stifle action (Carpenter, Bennett, & Peterson, 2006; Pettit et al., 2019; Small & Nicholls, 2003). The approach is particularly helpful in generating win-win outcomes or making choices in complex situations.

Scenario building generally involves a number of key steps. In this chapter, a multidisciplinary group of authors discuss key areas of importance in ecosystem and resource management, and environmental and urban planning for a local district that is already being affected by sea level rise. The authors then went through a series of steps to generate and design scenarios, with multidisciplinary input from local specialists, that can inform future approaches to resilience building within coastal areas. Our approach is consistent with that used by others including specialists in conservation (Peterson, Cumming, & Carpenter, 2003) and planning (Fisher, Orland, & Steinitz, 2020). We follow a series of methodical steps to define the analytic framework and the scenarios themselves. This process involves recording the stresses, pressure points, uncertainty, impacts and comparative outcomes that result for each scenario. The steps are outlined in Figure 15.1. We suggest such an approach is important in building a consensus around what resilience is in relation to urban landscapes. As many scholars have contested, *resilience* is a difficult term to define. We point to Meerow and co-authors 2016 definition that:

> urban resilience refers to the ability of an urban system—and all its constituent socioecological and socio-technical networks across temporal and

spatial scales—to maintain or rapidly return to desired functions in the face of a disturbance, to adapt to change, and to quickly transform systems that limit current or future adaptive capacity.

(Meerow, Newell, & Stults, 2016, p. 39)

Notable in this definition is the need to best understand what functions are desired and the need to assess what land uses or infrastructure limit adaptability in

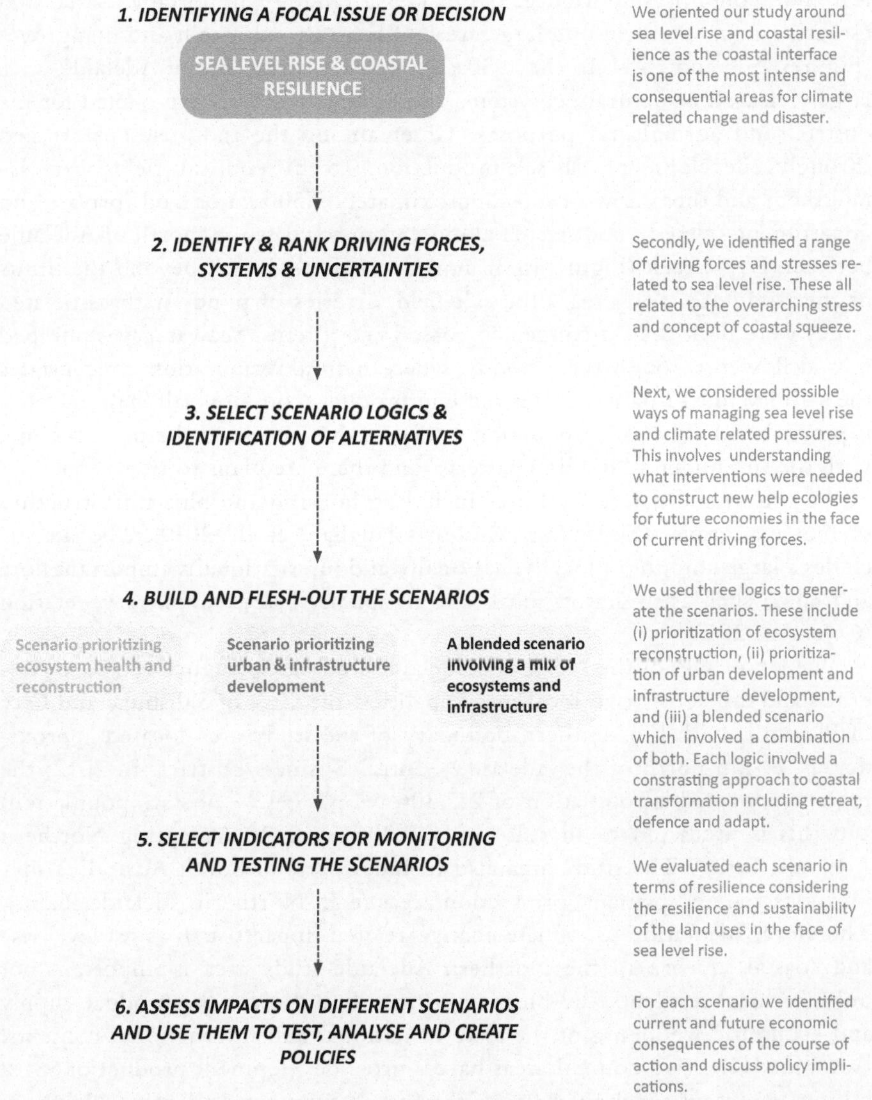

FIGURE 15.1 The six steps used in the scenario building and testing process. Generated by Scott Hawken and Kaihang Zhou. For the color figure, see the e-book.

the face of future pressures. This strong element of uncertainty and openness in relation to resilience makes the scenario building approaches useful.

History and current context: The challenge of coastal squeeze in anthropogenic coastal landscapes

The Northern Adelaide study region (South Australia) was selected as an ideal area to assess resilience to climate-related impacts due to its social, ecological and economic importance. It includes a range of important coastal and estuarine habitats, including large areas of seagrass, saltmarsh and mangroves (Figures 15.3 and 15.4). In the 1930s, the coastal landscapes of Adelaide were largely cleared of natural ecosystems and comprehensively remodeled for industrial and agricultural purposes. Chief among the industries prioritized through redevelopment was salt production. To this end, salt fields were established, and these now occupy approximately 5,500 ha of land (private and government-owned), and stretch along 28 km of coastline, north of Adelaide between Dry Creek (Figure 15.3), north to Middle Beach beyond the limits of the study area. To create the salt field, a series of ponds with associated bunds were built that disconnected coastal ecosystems. Seawater was pumped in and flowed through these ponds, where natural evaporation concentrates the salinity until finally the desired end product (common salt/halite/NaCl) is precipitated. The salt production operation has ceased at the present time, with the site under a "holding pattern" and there are plans to transition parts of the site to alternative land uses including housing and blue infrastructure such as mangroves (Dittmann, Mosley, Clanahan, et al., 2019). The site includes a large amount of locally, nationally and internationally important flora and fauna such as migratory birds and nationally listed samphire vegetation (Coleman, 2013).

The landscapes of the Northern Adelaide study area (Figure 15.3) are governed and fall across two local municipalities: the City of Salisbury and City of Playford, with the southern boundary of the study area located approximately 10 km north of the Adelaide Central Business District. In 2012, the region supported a population of 217,306 people (~13% of state population) and this is predicted to increase by 169,000 by 2040 (Adapting Northern Adelaide, 2016). The study area also includes parts of South Australia's major industrial and primary production areas (e.g., Northern Adelaide Plains). Therefore, adaptation to climate change-related impacts such as sea level rise and coastal squeeze in the Northern Adelaide study area is important not only for local residents and business owners but also for the produce supply and economic wellbeing of the state in national and international contexts. Combined, the two council areas have a gross development product of $8.12 billion, representing about 8.9% of the State's economy (Adapting Northern Adelaide, 2016).

Current and future stressors: Types of climate related stressors in coastal landscapes

As mentioned in the current study, we focus mostly on the "coastal squeeze" aspects associated with climate change in the Northern Adelaide study area. Coastal squeeze occurs when built or natural topographic conditions constrains the landward migration of ecosystems as sea levels rise, with likelihood of ecosystem loss (Borchert et al., 2018; Figure 15.2). Pressure can be placed upon ecosystems by infrastructures such as roads, embankments, natural rises or seawalls amongst other topographic features (Doody, 2004; Pontee, 2013).

The presence of the salt field stretching along a large area of coast in northern Adelaide area is currently constraining the ability of coastal vegetation to migrate, and mangroves are currently being squeezed up against artificial bund walls and roads (Figures 15.2–15.5), whereas natural sites have a much larger salt marsh transition area. There are also large hazards present at the site (i.e., hypersaline water and salt crystals) arising from salt production and maintaining the bunds and their integrity is proving challenging in the context of salt mine closure. There is potential for catastrophic failure and leakage of hypersaline brine to the coastal ecosystems, recently there has been a leakage event at St Kilda associated with localized (approx. 24 ha) mangrove and salt-marsh death.

There is also intense pressure to transform the salt fields and coastal areas into new urban developments. The Dry Creek Salt Pans have been identified in the state government's 30-Year Plan for Greater Adelaide as a Future Urban Growth Area. Whilst the precise development has not been determined, it is projected that some 10,000 homes will be accommodated in this area along with other possible uses such as shopping centers, education facilities and marinas (James, 2019). Such urban developments will consume large quantities of resources for filling low lying land and also place additional pressure on coastal ecosystems, water security and drainage systems. There are also questions of long-term social equity associated with such developments should climate change impacts cause future developments to fail (Hawken et al., 2021a).

In regard to future stressors, in the northern Adelaide area climate change will mean warmer and drier conditions, increased risk of climate hazards, such as extreme heat, fire and flooding, and changing conditions in Gulf St Vincent, includes under a high emissions scenario (Adapting Northern Adelaide, 2016):

- Annual rainfall is projected to decline by about 11% while rainfall intensity could increase by 16%
- Annual maximum temperatures are projected to increase by 2.3°C while annual minimum temperatures could increase by 2°C and extreme heat days per year (35°C or higher) could increase by 76% to 82% (up to 44 days).

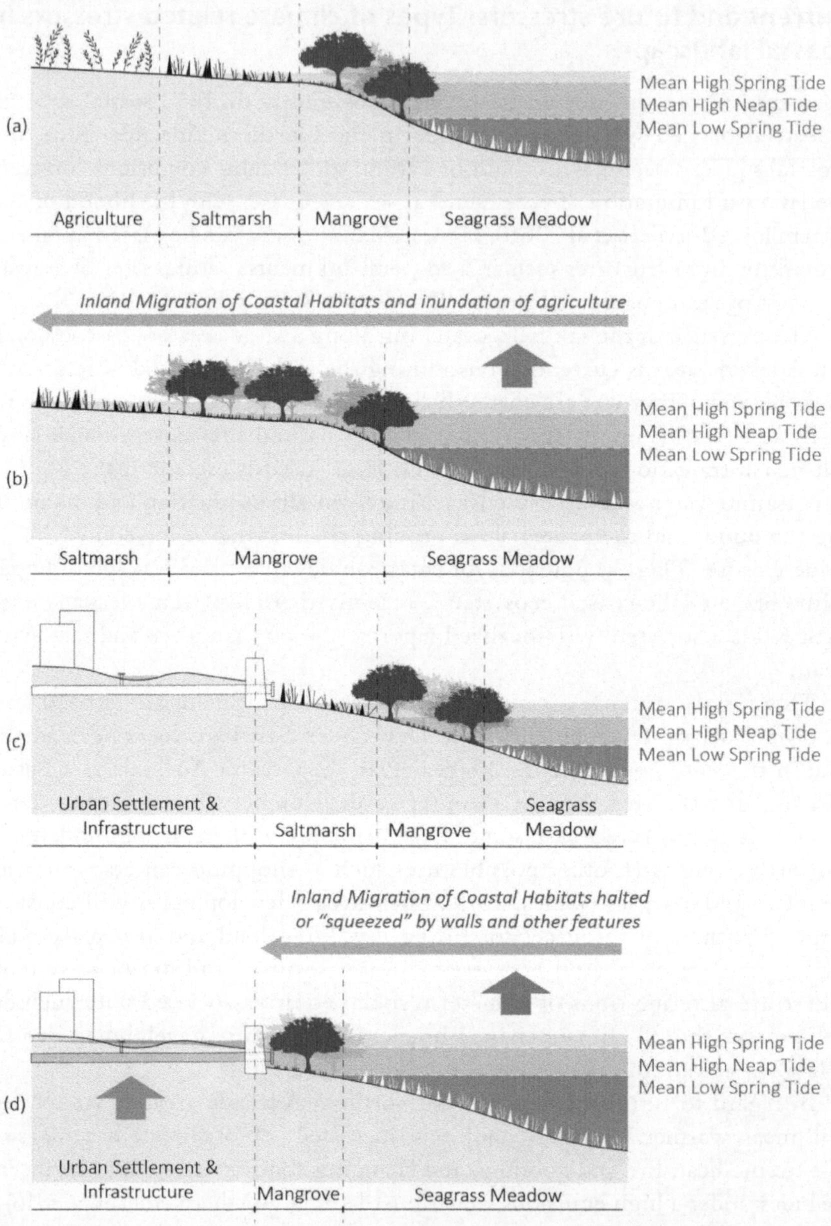

FIGURE 15.2 The coastal squeeze concept illustrated for both agricultural-coastal interfaces and urban-coastal interfaces. The first (a, b) sees tidal ecosystems migrate inland and the destruction of agricultural landscapes as they become saline. The second (c, d) sees the elimination of tidal ecosystems as sea levels rise and "squeeze" tidal ecosystems against built elements such as roads and seawalls. Tidal ecosystems largely are eliminated in this second condition. Generated by Scott Hawken and Kaihang Zhou. For the color figure, see the e-book.

FIGURE 15.3 Study area map. Major infrastructure barriers such as the salt pan bunds, roads, levees associated with wastewater plants and sea-walls will cause coastal squeeze and limit the dynamic migration of ecosystems as sea levels rise. Generated by Scott Hawken and Kaihang Zhou. For the color figure, see the e-book.

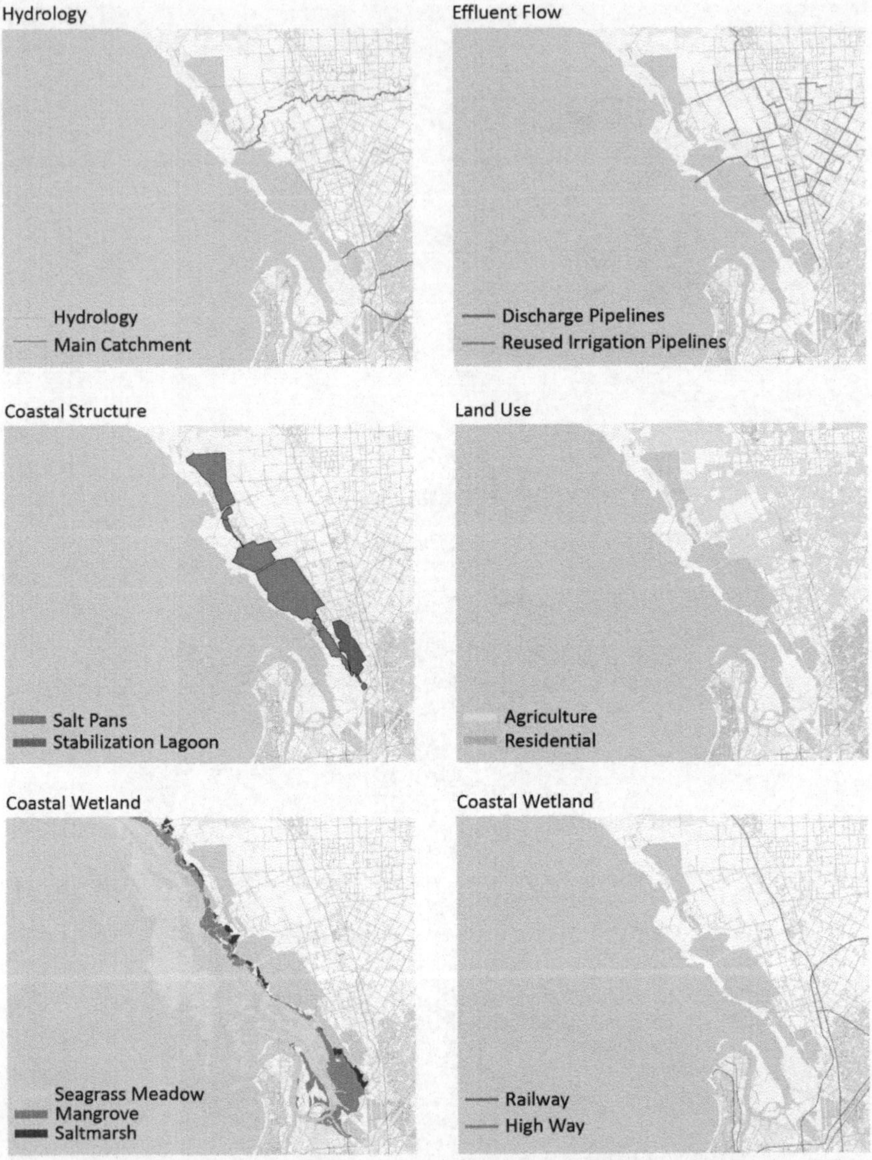

FIGURE 15.4 The tidal ecosystems, including green and blue infrastructure such as sea grass meadows and mangroves, are a fraction of the original systems prior to European settlement. Other land uses such as salt production and wastewater treatment, agriculture and suburban settlement have replaced such tidal systems. The threat and pressure placed by sea level rise may force a renegotiation of the land uses and an opportunity to reallocate land for new uses. Generated by Scott Hawken and Kaihang Zhou. For the color figure, see the e-book.

FIGURE 15.5 Photographs showing (top) coastal squeeze created by the salt field seawalls/bunds, and (bottom) a subsequent tidal restoration trial to open up one salt field pond, thus allowing drainage and flushing of hypersaline waters, and restoration and migration of coastal ecosystems. Photos by Luke Mosley and Emily Leyden. For the color figure, see the e-book.

Sea level rise is the key stressor we consider and driver of coastal squeeze in this area. Latest global sea level rise predictions (IPCC, 2021) are shown in Figure 15.6. The mid-range scenarios show approximately 0.6 m of sea level rise with higher emission scenarios approaching 1 m of sea level rise, or much higher if there is ice sheet instability. Sea level rise will likely lead to progressive inundation of the coastal ecosystems outside of the salt field bunds and begin to impact residential and agricultural land uses at parts of the site but different scenarios are explored below on how this may occur (Figure 15.7).

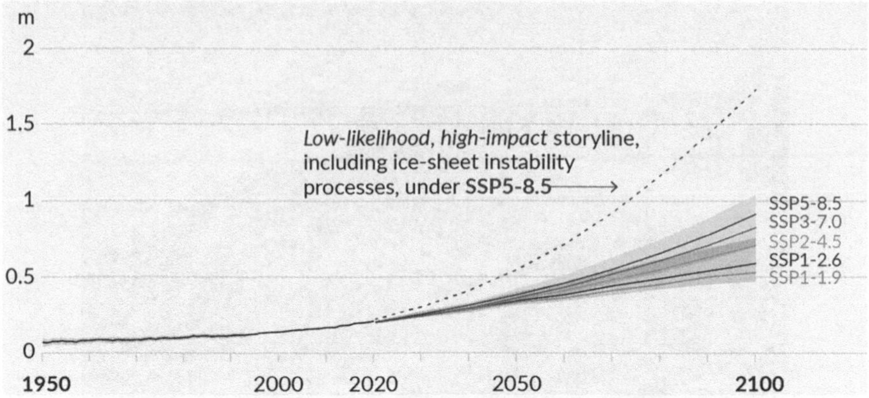

FIGURE 15.6 Global mean sea level change relative to 1900. The different colored lines represent different greenhouse gas emissions scenarios. For the color figure, see the e-book.

Source: Figure SPM.8, Panel (d) from IPCC (2021): Summary for Policymakers. In: *Climate Change 2021: The Physical Science Basis. Contribution of Working Group I to the Sixth Assessment Report of the Intergovernmental Panel on Climate Change* [Masson-Delmotte, V., P. Zhai, A. Pirani, S.L. Connors, C. Péan, S. Berger, N. Caud, Y. Chen, L. Goldfarb, M.I. Gomis, M. Huang, K. Leitzell, E. Lonnoy, J.B.R. Matthews, T.K. Maycock, T. Waterfield, O. Yelekçi, R. Yu, and B. Zhou (eds.)]. Cambridge University Press, Cambridge, United Kingdom and New York, NY, USA, pp. 3–32, doi:10.1017/9781009157896.001.

Vision: blue-green infrastructures for coastal adaptation

Although many strategies are available for climate adaptation in coastal environments, we consider three major approaches to explore impacts on ecosystems and infrastructure. The first is a "retreat" approach that allows and even facilitates the dynamic migration and expansion of natural ecosystems such as seagrass, mangroves, saltmarsh and supratidal environments. The second is a "defensive" approach that allows and facilitates the protection of current and future land uses through a range of approaches but most typically through seawalls and grey infrastructure engineered works. In contrast to the "retreat" approach, the "defensive" approach fixes land uses in place based on an ideal location. Finally, the "adaptive" approach integrates various strategies to create and integrate both green and grey and blue infrastructure. This approach is characterized by a greater degree of spatial and design complexity. We discuss each of the three scenario "visions" below. For each of these scenarios, the changing land uses set up a different set of questions and expectations regarding resilience. Scenarios A and B prioritize the resilience of particular land uses whilst scenario C adopts a more complex "win-win" approach. In this regard, we suggest that in any scenario, it is necessary to prioritize resilience for "whom, what, when, where, and why?" in the words of Meerow et al. (2016).

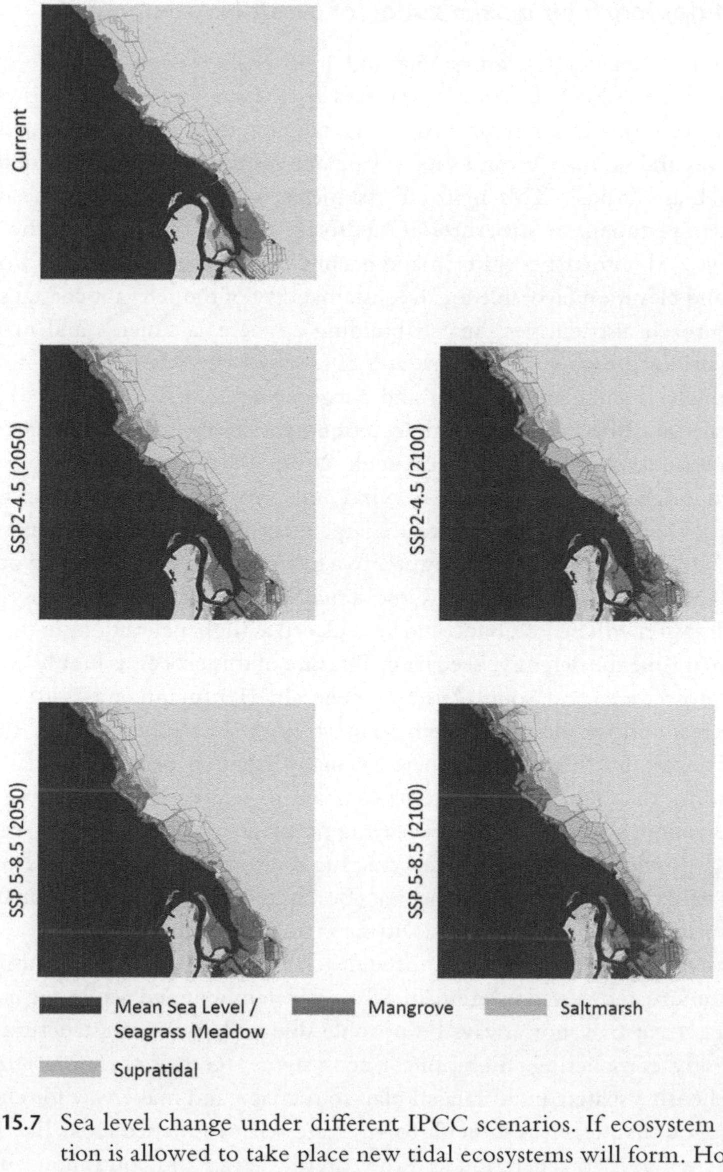

FIGURE 15.7 Sea level change under different IPCC scenarios. If ecosystem migration is allowed to take place new tidal ecosystems will form. However, in many instances, seawalls, roads and other topographic features, often associated with land tenure patters, will constrain this migration and tidal and supratidal ecosystems will instead disappear. Generated by Scott Hawken and Kaihang Zhou. For the color figure, see the e-book.

Retreat approach for maximization of coastal ecosystems

The "retreat" approach (Figures 15.8 and 15.9) enables ecosystems the space to move upwards in land elevation as sea level rises. In our study area, this requires the large-scale conversion of post-industrial salt pans to various ecosystems such as mangroves and saltmarshes and supratidal plant communities local to South Australia such as salt bush. This approach assumes ecosystems can adapt and migrate without impediment by structures or landforms. Under this scenario, the area of mangrove and saltmarshes shifts inland occupying a similar niche or hydrological profile and elevation favorable for the sustainability of the relevant coastal species. This approach also enables "marsh building", where salt marsh and mangrove ecosystems keep pace with current rates of sea level rise where sediment and organic supply is sufficient (Kirwan and Megonigal. 2014; Lovelock et al., 2015; Schuerchet al. 2018). This vertical accretion process also enables "blue carbon" sequestration to continue (Dittmann et al., 2019). This appears to the case for the northern Adelaide study area where wind and wave action on the shallow Gulf waters provides sediment, and there are large quantities of seagrass organic matter "wrack". Connecting urban stormwater to a reconfigured salt field area would also provide sediment supply benefits. Without marsh building processes, ecosystems at lower elevation will be lost, but could be replaced at higher elevations in the retreat scenario (unlike the defensive scenario). The rate of this process is highly uncertain for the project area and South Australia generally (Dittmann et al., 2019). Landward retreat options allow ecosystem connectivity to be maintained (i.e., seagrass-mangrove-salt march gradient). There are many other co-benefits associated with maintaining the ability of ecosystems to adapt to sea level rise including shoreline protection, sequestering, and ensuring permanence of blue carbon, reducing flooding, providing fisheries habitat, creating recreational opportunities and supporting valuable fish and wildlife habitat (Borchert et al., 2018; Temmerman et al., 2013), including in our study area (Dittmann et al., 2019a, b).

However, a retreat approach is difficult to implement due to current patterns of land tenure and vested interests in urban development and economic land use. Further retreat may not always be possible due to legacy infrastructure which is currently constricting migration of ecosystems. Retreat therefore requires a comprehensive spatial and financial plan to manage and make way for changing coastal geodynamics. It is difficult to envisage what resilience looks like in such contexts without consideration and the careful design and continuous management of such dynamic systems to support the evolution of novel and constructed coastal ecologies. This scenario is already visible in parts with selective breaches constructed in existing salt works bunds to allow the reconstruction of mangrove and saltmarsh ecosystems (Figure 15.5 bottom). It is calculated such constructed blue infrastructure systems can store as much as 110 tons of "blue" carbon per hectare (Dittmann, Mosley, Beaumont, et al., 2019; Dittmann, Mosley, Clanahan, et al., 2019; Dittmann, Mosley, Jones, et al., 2019; Jones et al., 2019; Sandhu et al., 2018).

Defensive approach for maximization of coastal urban and agricultural development

The "defensive" approach (Figures 15.8 and 15.9) envisages the salt marshes to be converted into settlement landscapes and high-value agricultural landscapes. It assumes the high level of investment required to convert low-lying land can be recompensated by even higher levels of real estate income through subsequent property sales. As mentioned, such assumptions are implicit in state planning policy for South Australia and other jurisdictions around Australia. Coastal settlements around Australia and globally are currently experiencing recurring and increasing frequency of flooding. To protect property and communities, there is therefore a great push to develop defensive structures such as seawalls and embankments. Retrofitting existing low-lying settlements is, however, difficult as areal flooding and surcharge can occur behind the walls and render houses and drainage infrastructure unfit for those purposes. In some instances, novel ecosystems are developing in such settlements in low-lying swales, depressions and even kerbs and medians of streets. However generally such defensive approaches result in a major loss of coastal habitats (Cooper et al., 2020).

Building walls or other engineering structure to defend from sea level rise essentially over time compresses the tidal ecosystems into a smaller area. Defensive walls that protect against storm surge or sea water incursion prevent salt marsh ecosystems and mangrove ecosystems from migrating inland (Figures 15.8 and 15.9). They are therefore progressively lost (i.e., "drowned out") and replaced by submergent vegetation such as seagrasses. A defensive approach also requires the building and maintenance of costly coastal infrastructure. Such costs will likely increase over time in the face of increased wave energy as sea levels rise. Furthermore, although surface waters may be able to be defended against, groundwater intrusion and rise on the landward side of structures could cause many issues including salinization of surface soil, salt damage to houses and other infrastructure. Conventional urban developments are not recommended for such landscapes and should take into consideration worst case scenarios involving storm surges, environmental degradation of materials and changing geomorphological conditions. Although such defensive approaches are envisaged as "fixed", building systems in such developments can incorporate adaptive features that can allow the development to evolve over time. This might include built features such as pontoons that allow developments to rise and fall with the tide or green buffer zones to dissipate storm surges (Barsley, 2020). It might also include policy mechanisms such as incentives (Fang, Howland, Kim, Peng, & Wu, 2019; Wang, Lu, Lu, & Nie, 2021) to permanently move should environmental conditions becomes extreme. Long-term uncertainty is associated with this scenario's ability to defend itself against changing environmental conditions. Long-term insurance from 50 to 100 years against sea level rise and climate-related impacts maybe prohibitively high and render the development unfeasible unless realistic assumptions about flooding are built into the development.

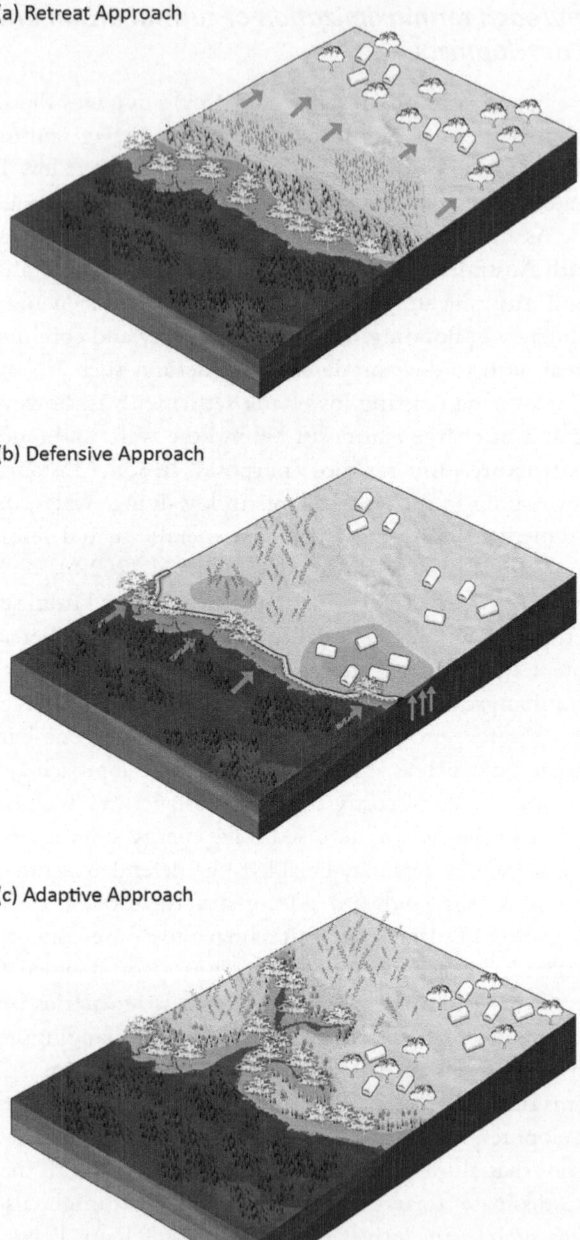

FIGURE 15.8 Strategic approaches used in the three scenarios: (a) retreat, (b) defensive, (c) adaptive. Generated by Kaihang Zhou. For the color figure, see the e-book.

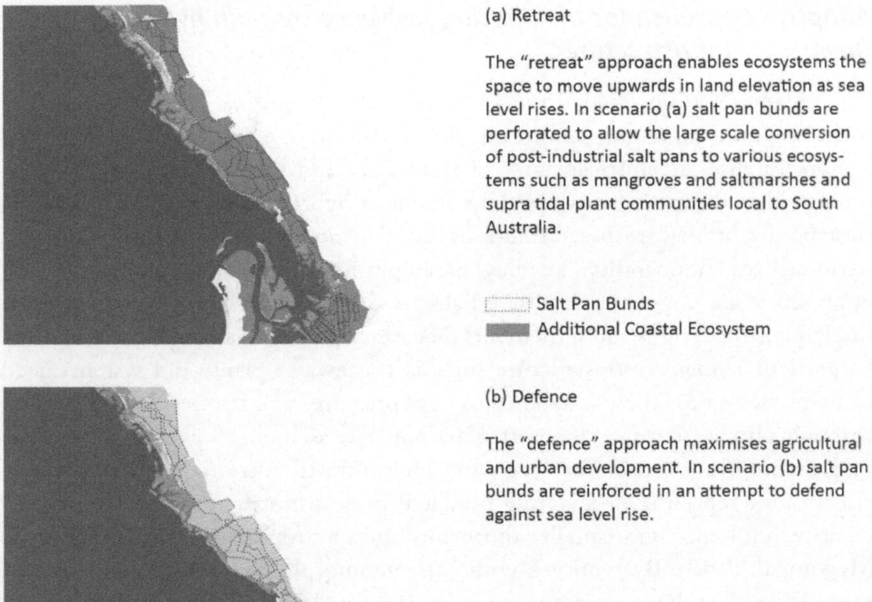

FIGURE 15.9 Three scenarios (a) retreat, (b) defense, (c) adapt, each follow a different decisive development process that either inhibits or facilitates ecosystem migration and evolution. The first shows a massive expansion of coastal ecosystems, the second shows diminishing ecosystems as sea levels rise against seawalls and infrastructure squeezing the ecosystems out, and the last shows an evolution of diverse ecosystem and settlement systems in a hybrid, incorporating a mix of urban settlements and green infrastructure in close proximity that can adapt and strengthen resilience as the climate pressures increase. Generated by Scott Hawken and Kaihang Zhou. For the color figure, see the e-book.

Adaptive approach for maximizing resilience through hybrid green-grey infrastructure

The "adaptive" (Figures 15.8 and 15.9) approach integrates a variety of strategies including creation of new landforms, the modification of existing landforms and a more intense and intricate mix of residential and tidal ecosystem land uses, often within the same landscape. The intent of hybridization is to use green infrastructure or "nature-based solutions" to generate greater resilience. This scenario utilizes "floodability" as a new paradigm for urban development (Ernstson et al., 2010; La Loggia et al., 2020; Palazzo, 2019). Cities around the world are implementing designs and settlements that are resilient and adaptable to flooding. In particular, major infrastructure such as wastewater plants and systems need to be reassessed for their suitability to new pressures and stresses that are arising through climate change. In particular, our case demonstrates the importance of providing more space for green and blue infrastructure to manage sea level rise and for replacing or adapting outdated grey infrastructure with more efficient technologies with smaller footprints and environmental costs (Kirwan & Megonigal, 2013). By working with such geomorphological and nature-based processes rather than against them in a defensive way, the urban landscape is inherently more resilient (Hawken et al., 2021b). Long-term spatial corridors dedicated to tidal ecologies allow ecosystem migration to occur. Built infrastructure such as access roads and causeways are configured to facilitate water flow and ecosystem migration whiles being orientated to deflect and disperse storm surge impacts and other coastal environmental pressures. Residential landscapes are either adapted as floating pontoon landscapes or built up as marsh islands which use the landform generating abilities of mangroves and marshlands to "grow" terrestrial landscapes over time. Such approaches involve a dynamic changing approach more in tune with the realities of climate change than any fixed masterplan. Lease arrangements are used to ensure that long-term problems do not become the problem of residents but are managed at the larger scale and remain the responsibility of developers and government.

Such an approach can also accommodate stormwater functions necessary for the treatment of runoff from adjacent urban areas (Hawken et al., 2021b). A further example of such an approach is illustrated in Figure 15.10 which communicates a proposal for the Bolivar Wastewater Plant in Adelaide, which replaces the aging technology for new advanced dual membrane and ultraviolet technologies to allow for more extensive water reclamation and circular resource use. The smaller footprint of the approach allows for the industrial polishing ponds to be turned over to tidal ecosystems such as mangroves, salt marshes and freshwater constructed wetlands. Whereas the defensive approach largely follows a business-as-usual development paradigm, the adaptive approach uses inherent dynamic aspects of green infrastructure to generate and shape coastal environments.

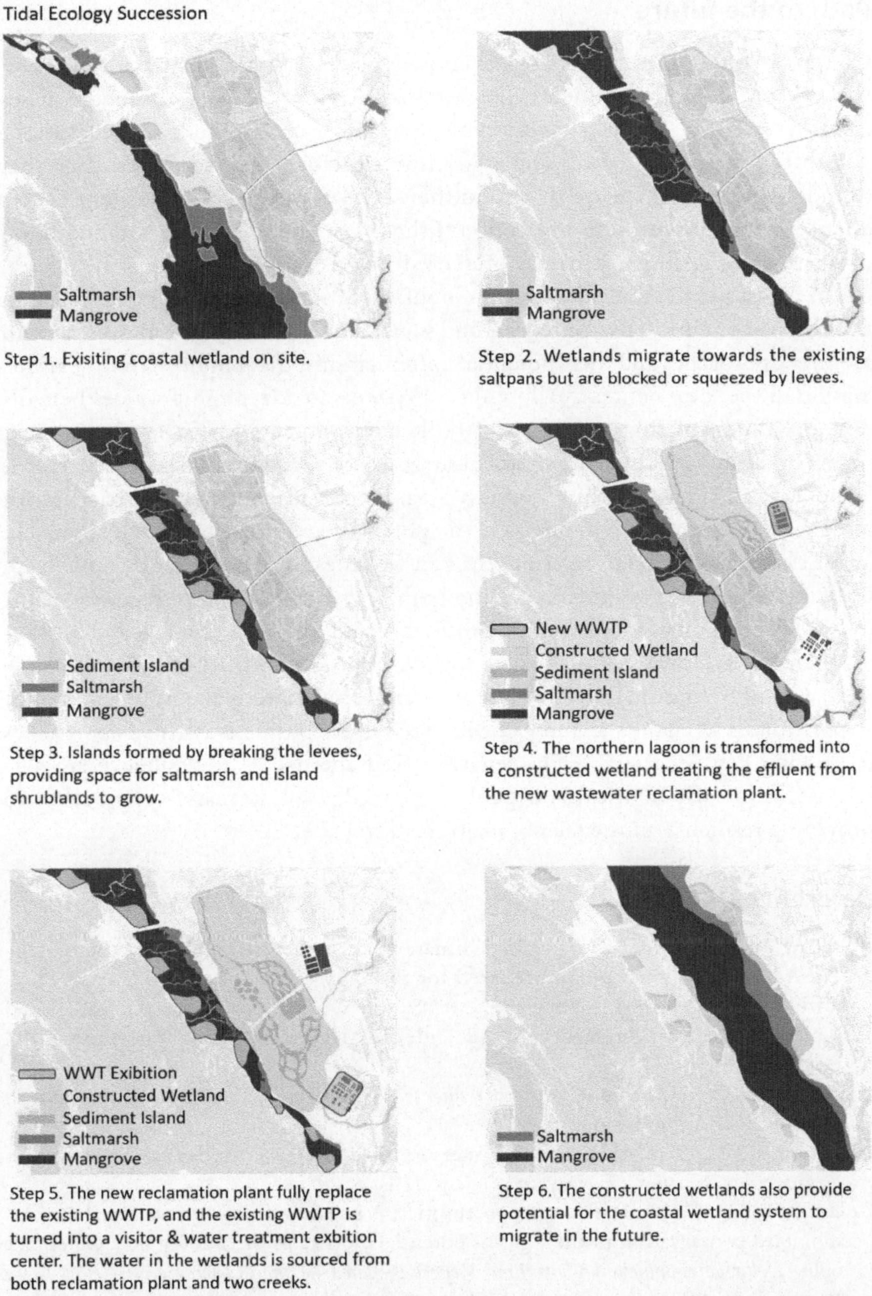

FIGURE 15.10 A hypothetical scheme for the Bolivar Waste Water Plant demonstrates how grey and green infrastructure can be designed together to facilitate win-win outcomes. Generated by Kaihang Zhou. For the color figure, see the e-book.

Path to the future

Climate change and associated sea level rise is placing unprecedented pressures on natural and built environments in coastal zones globally. Coastal squeeze arises from the inability of coastal ecosystems to adapt or migrate to higher elevations due to the presence of walls and other infrastructure, and is a major issue that cannot be ignored if we are to sustain the values of these ecosystems (e.g. recreation and tourism, erosion protection, fisheries, carbon sequestration and flood mitigation). The three scenarios presented in this chapter for the same coastal urban landscape in Adelaide offer different ways of thinking about resilience in uncertain economic, environmental and social context. There are complexities in all three approaches and their potential combinations, that require careful consideration in the local context. Allowing ecosystems to adapt also provides benefits essential for increasing resilience of built environments to sea level rise (e.g., marsh building and buffering of storm surges). Localized examples of scenarios 1, 2 and 3 are all evident within the study area. By considering such approaches on a large scale, we can begin to envisage the potential and impacts of each approach. Whilst each approach has its strengths and weaknesses, we suggest that a hybrid, utilizing adaptive strategies integrating both grey and green infrastructure, is superior when it comes to maximizing resilience and a balance of social and ecological outcomes. However, to advance such an approach in the Northern Adelaide area, a greater consultation across sectors and disciplines is required to identify opportunities to challenge and critically reconsider business-as-usual approaches in making key decisions. Such scenario-based approaches to resilience are also important to raise political and public knowledge of alternative approaches to providing resilience in the face of major sea level rise.

References

Adapting Northern Adelaide. (2016). Climate change adaptation plan for the Northern Adelaide region. A plan prepared for the City of Salisbury and City of Playford by Seed Consulting Services, URPS, the Workplace Innovation and Social Research Centre (University of Adelaide), CSIRO Land and Water and FMG Engineering.

Barsley, E. (2020). *Retrofitting for flood resilience: A guide to building & community design.* London: RIBA Publishing.

Carpenter, S. R., Bennett, E. M., & Peterson, G. D. (2006). Scenarios for ecosystem services: An overview. *Ecology and Society, 11*(1), p. 29.

Coleman, P. S. J. (2013). Risk and opportunities: A briefing paper on coastal habitat and shorebird conservation in the light of potential closure of the Ridley Dry Creek salt field. *Delta Environmental Consulting Report prepared for the Adelaide and Mount Lofty Ranges Natural Resources Management Board.*

De Bruijn, J. A., De Bruijne, M. L. C., & Ten Heuvelhof, E. F. (2015). The Politics of Resilience in the Dutch 'Room for the River'-project, in: *Procedia Computer Science, 44, 2015; Conference on Systems Engineering Research.* Elsevier.

Dittmann, S., Mosley, L., Beaumont, K., Clarke, B., Bestland, E., Guan, H., et al. (2019). From Salt to C; carbon sequestration through ecological restoration at the Dry Creek Salt Field. *Goyder Institute for Water Research Technical Report Series* (19/28).

Dittmann, S., Mosley, L., Clanahan, M., Quinn, J., Crooks, S., Emmer, I., et al. (2019). Proof of concept for tidal re-connection as a blue carbon offset project. *Goyder Institute for Water. Research Technical Report Series* (19/29).

Dittmann, S., Mosley, L., Jones, A. R., Clanahan, M., Beaumont, K., & Gillanders, B. M. (2019). Research agenda for Blue Carbon in South Australia. *Goyder Institute for Water Research Technical Report Series* (19/31).

Doody, J. P. (2004). 'Coastal squeeze'—an historical perspective. *Journal of Coastal Conservation, 10*(1), pp. 129–138.

Ernstson, H., Van der Leeuw, S. E., Redman, C. L., Meffert, D. J., Davis, G., Alfsen, C., et al. (2010). Urban transitions: On urban resilience and human-dominated ecosystems. *Ambio, 39*(8), pp. 531–545.

Fang, L., Howland, M., Kim, J., Peng, Q., & Wu, J. (2019). Can transfer of development rights programs save farmland in metropolitan counties? *Growth and Change, 50*(3), pp. 926–946.

Fisher, T., Orland, B., & Steinitz, C. (2020). *The international geodesign collaboration: Changing geography by design*. Redlands, CA: ESRI Press. Retrieved from https://www.esri.com/en-us/esri-press/browse/the-international-geodesign-collaboration-changing-geography-by-design

Hawken, S., Avazpour, B., Harris, M. S., Marzban, A., & Munro, P. G. (2021). Urban megaprojects and water justice in Southeast Asia: Between global economies and community transitions. *Cities, 113*, p. 103068.

Hawken, S., Rahmat, H., Sepasgozar, S. M., & Zhang, K. (2021). The SDGs, ecosystem services and cities: A network analysis of current research innovation for implementing urban sustainability. *Sustainability, 13*(24), p. 14057.

Hawken, S., Sepasgozar, S., Prodanovic, V., Jing, J., Bakelmun, A., Avazpour, B. S., et al. (2021). What makes a successful sponge city project? Expert perceptions of critical factors in integrated urban water management in the Asia-pacific. *Sustainable Cities and Society*, 75, 103317.

Hobbs, R. J., Higgs, E., Hall, C. M., Bridgewater, P., Chapin III, F. S., Ellis, E. C., et al. (2014). Managing the whole landscape: Historical, hybrid, and novel ecosystems. *Frontiers in Ecology and the Environment, 12*(10), pp. 557–564.

Hobbs, R. J., Higgs, E. S., & Hall, C. M. (2013). Defining novel ecosystems. In: Hobbs, R. J., Higgs, E. S., & Hall, C. M. (Eds.) *Novel ecosystems: Intervening in the new ecological world order*, Wiley-Blackwell, pp. 58–60.

IPCC. (2021). Summary for Policymakers. In: Climate Change 2021: The physical science basis. contribution of working group I to the sixth assessment report of the intergovernmental panel on climate change [Masson-Delmotte, V., P. Zhai, A. Pirani, S.L. Connors, C. Péan, S. Berger, N. Caud, Y. Chen, L. Goldfarb, M.I. Gomis, M. Huang, K. Leitzell, E. Lonnoy, J.B.R. Matthews, T.K. Maycock, T. Waterfield, O. Yelekçi, R. Yu, and B. Zhou (eds.)]. Cambridge University Press, Cambridge, and New York, NY, pp. 3–32, doi:10.1017/9781009157896.001

James, C. (2019). Soil dumped at Dry Creek to test salt fields for housing. *The Advertiser*. Retrieved November 27, 2021, from https://www.adelaidenow.com.au/messenger/north-northeast/housing-concept-for-salt-fields-at-dry-creek-near-globe-derby-park/news-story/dece7a60feaa33e181802f43993bfd24

Jones, A. R., Dittmann, S., Mosley, L., Clanahan, M., Beaumont, K., Waycott, M., et al. (2019). Goyder Institute blue carbon research projects: Synthesis report. *Goyder Institute for Water Research Technical Report Series* (19/30).

Kirwan, M. L., & Megonigal, J. P. (2013). Tidal wetland stability in the face of human impacts and sea-level rise. *Nature, 504*(7478), pp. 53–60.

La Loggia, G., Puleo, V., & Freni, G. (2020). Floodability: A new paradigm for designing urban drainage and achieving sustainable urban growth. *Water Resources Management, 34*, pp. 3411–3424.

Lundholm, J. (2015). The ecology and evolution of constructed ecosystems as green infrastructure. *Frontiers in Ecology and Evolution, 3*, p. 106.

Mangone, G. (2016). Constructing hybrid infrastructure: Exploring the potential ecological, social, and economic benefits of integrating municipal infrastructure into constructed environments. *Cities, 55*, pp. 165–179.

Meerow, S., & Newell, J. P. (2016). Urban resilience for whom, what, when, where, and why? *Urban Geography, 40*(3), pp. 309–329.

Meerow, S., Newell, J. P., & Stults, M. (2016). Defining urban resilience: A review. *Landscape and Urban Planning, 147*, pp. 38–49.

Palazzo, E. (2019). From water sensitive to floodable: Defining adaptive urban design for water resilient cities. *Journal of Urban Design, 24*(1), pp. 137–157.

Peterson, G. D., Cumming, G. S., & Carpenter, S. R. (2003). Scenario planning: A tool for conservation in an uncertain world. *Conservation Biology, 17*(2), pp. 358–366.

Pettit, C. J., Hawken, S., Ticzon, C., Leao, S. Z., Afrooz, A. E., Lieske, S. N., et al. (2019). Breaking down the silos through geodesign – Envisioning Sydney's urban future: *Environment and Planning B: Urban Analytics and City Science, 46*(8), pp. 1387–1404.

Pontee, N. (2013). Defining coastal squeeze: A discussion. *Ocean & Coastal Management, 84*, pp. 204–207.

Sandhu, H., Clarke, B., Baring, R., Anderson, S., Fisk, C., Dittmann, S., et al. (2018). Scenario planning including ecosystem services for a coastal region in South Australia. *Ecosystem Services, 31*, pp. 194–207.

Silva, J. M. C. da, & Wheeler, E. (2017). Ecosystems as infrastructure. *Perspectives in Ecology and Conservation, 15*(1), pp. 32–35.

Silva, R., Martínez, M. L., van Tussenbroek, B. I., Guzmán-Rodríguez, L. O., Mendoza, E., & López-Portillo, J. (2020). A Framework to manage coastal squeeze. *Sustainability, 12*(24), p. 10610.

Small, C., & Nicholls, R. J. (2003). A global analysis of human settlement in coastal zones. *Journal of Coastal Research, 19*(3), pp. 584–599.

Sutton-Grier, A. E., Wowk, K., & Bamford, H. (2015). Future of our coasts: The potential for natural and hybrid infrastructure to enhance the resilience of our coastal communities, economies and ecosystems. *Environmental Science & Policy, 51*, pp. 137–148.

Wang, H., Lu, S., Lu, B., & Nie, X. (2021). Overt and covert: The relationship between the transfer of land development rights and carbon emissions. *Land Use Policy, 108*, p. 105665.

Zevenbergen, C., Rijke, J., Van Herk, S., & Bloemen, P. (2015). Room for the river: A stepping stone in adaptive delta management. *International Journal of Water Governance, 3*(1), pp. 121–140.

16

UTILIZATION OF FOREST LANDSCAPES FOR BIODIVERSITY CONSERVATION IN A MEDITERRANEAN ECOSYSTEM

A case study of Greece

Alexandra D. Solomou

Current context

The Mediterranean Basin is a biodiversity hotspot and one of the world's most outstanding regions due to its unique biological richness and landforms. It can be considered as a transitional zone, spanning from South Europe's temperate temperature to Africa's warm/dry regions, where multiple ecosystems coexist (Myers et al., 2000; Solomou and Sfougaris, 2021). Greece, which is within the Mediterranean Basin, is situated in southeastern Europe at a crossroads between Europe, Asia and Africa.

The diversity of Greek flora comes as a result of the fruitful interaction of several elements (Strid and Tan, 1997; Georghiou and Delipetrou, 2010; Korakis, 2015), including:

a Greece's geographical location, which is a geological crossroads and a place where basic flora units can be discovered. The Mediterranean, Iran-Caspian, and even tropical/ subtropical flora regions are flora regions that enhance the Greek flora with their elements to varying degrees.
b The relief and geomorphology of Greek landscapes. The presence of mountain ranges, peninsulas, and island clusters, which represent the region's geological history, has had a significant impact on population fragmentation, migration, and plant species isolation.
c The presence of previous geological ages' flora (Tertiary flora), and the site's role as one of the most important refuges for European species throughout glacial periods.

Although civilization, including urbanization and agriculture, has widely changed landscapes of Greece over time, Greek forests have been sustainable for

DOI: 10.4324/9781003266440-16

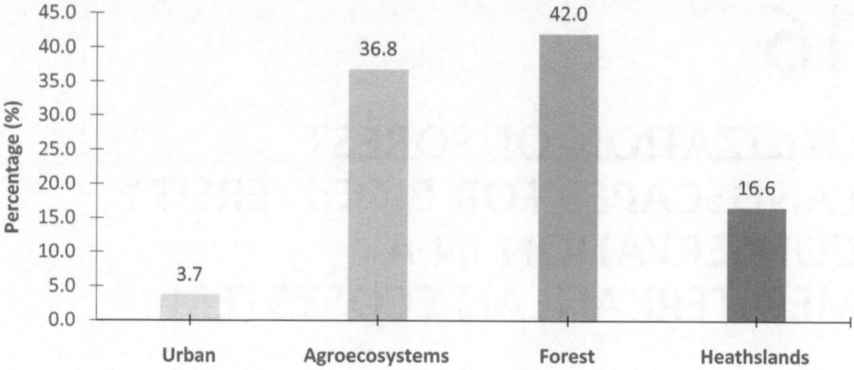

FIGURE 16.1 Cover (%) of ecosystem types of Greece. Generated by Alexandra D. Solomou.

thousands of years. Today, forest is the most common natural terrestrial ecosystem type of Greece, accounting for 42% of the country's land area (Figure 16.1; Solomou et al., 2016; Topalidou et al., 2021). Forest biodiversity in Greece is among the highest in Europe (in terms of species richness and quality). Because of the high biodiversity of the forests, they are legally protected, and significant efforts have been made to identify and conserve their biodiversity values.

There are three main types of forests in Greece: old-growth forests, natural forests, and managed forests. Old-growth and natural forests are exceptionally valuable for biodiversity (EEA, 2016), but intensively managed forests can also have high conservation values. Old-growth and natural forests provide major services through their regulating role in the environment and their ability to provide valuable opportunities for natural life and evolution for many species and populations of the flora and fauna (Sheil, 2018). In managed forests, the biodiversity is dependent on the continuity of habitat, the variability of site conditions, and tree species and forestry that resembles natural dynamics. Managed forests provide wood (e.g., lumber, firewood) and also a rich variety of other products, such as fruits, seeds, mushrooms, snails, and medicinal/herbal plants. The forests are also used for grazing, hunting, mountain tourism, and other activities that are very important for the economy of local communities (Thompson et al. 2009; Solomou et al., 2016).

Current and future stressors

Greek forest ecosystems are threatened both by natural causes (e.g., fires, erosion, and desertification) and by human activities (e.g., overgrazing, land clearing, and deforestation) (Ristić et al., 2019). Fires have always been a potential threat to Greek forest ecosystems, having devastating implications which affect both forest health and human life (Xanthopoulos, 1998, 2007). The destruction caused by fires in Greece over the past two decades has been described as "biblical" in terms of their intensity (Tedim et al., 2015).

It is certain that in the future, due to climate change, fire will continue to be a determining factor in the development of Mediterranean ecosystems because fires are not only almost uncontrollable in this region, but in some cases, they should not be eliminated. Many forest ecosystems in fire-affected areas, including those in Greece, are fire adapted; developing biological mechanisms of defense, adaptation, and survival.

Although most Greek forests are fire adapted, they are at increased risk of long-term damage immediately following a fire. Because of this, actions are taken to protect the ecosystem from the risk of further degradation and destruction following a fire and to help the burnt forest ecosystem to recover (i.e., to help it return to its previous structure and condition). These actions include a series of institutional measures such as the declaration of the burnt area as a reforestation area.

Case study

To provide an example of post-fire management in Greece, Mount Penteli (Ntaou Pentelis – Lyreio, Attica) as a case study, was used (Figure 16.2). Mount Penteli rises like a pyramid in the northeast of Attica and together with the mountains Parnitha, Ymittos, and Aigaleo embraces the basin of Athens, the

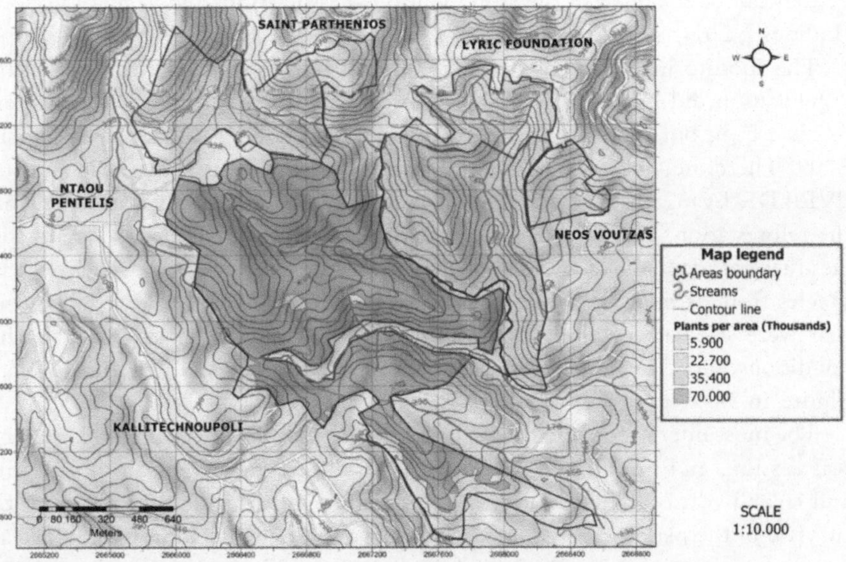

FIGURE 16.2 Map of the study area. Generated by Alexandra D. Solomou.

capital of Greece. In Mount Penteli, there are several historical and cultural monuments (monuments of the classical era, monasteries, churches, etc.), which give evidence of the historical periods from the Neolithic times until today, which testifies to the long-term presence of humans in the area. This presence is mainly connected with the proximity of the mountain to Athens, which is the largest urban center in Greece.

Mount Penteli belongs to the zone of the warm Mediterranean formations of the eastern Mediterranean (Oleo-Ceratonion plant alliance). The forest areas (Ntaou Pentelis – Lyreio, Attica), which were most recently burned in 2018, had been burned many times in recent years. Although a small part of them, on the borders of the coastal zone, had not been burned until recently. Most of the areas that burned in 2018 (approximately 60%) were residential areas with scattered patches of forest. The rest (about 40%) were forests and woodlands.

The main tree species of the burned forest was the Aleppo pine (*Pinus halepensis*) in dense stands or mixed with broad-leaved trees mainly holly (*Quercus coccifera*) and spruce (*Picea abies*). The subcanopy consisted of holly, schinous (*Pistacia lentiscus*), myrtle (*Myrtus communis*), and wildflower (*Olea oleaster* var. Silvestris) as well as various brushwood such as Montpellier cistus (*Cistus monspeliensis*) and Attic thyme (*Thymus atticus*).

Kailidis (1981) classified the various plant communities of Greece according to their flammability, considered the combinations wild olive (*Olea ceratonion*) and holm oak (*Quercus ilex*) as highly flammable, and the plant alliance Ostryo-Carpinion as flammable. In the Mediterranean region, the overwhelming percentage of fires occurs in the Thermo-Mediterranean and the Mediterranean vegetation floor (Georgiadis, 2004). Almost all the plant communities of Mount Penteli can be included in the aforementioned associations and vegetation floors (Baliousis, 2011).

The monitoring of vegetation after the 2018 fire showed that the natural regeneration did not have satisfactory results due to the fact that Mount Penteli had been burned many times in the recent past: in 1995, 1998, 2005, and 2009. Therefore, the area was declared as reforestable. In 2020, the company ilVERDE-Evangelia Andrianopoulou and Co. EE (https://ilverde.gr/) began the reforestation. The criteria for the reforestation included selecting plant species that are adapted to the site conditions and using and sourcing endemic plant species from as close to Mount Penteli as possible. The tree and shrub species that were selected for the reforestation were tolerant of the extreme weather conditions (e.g., frost and heat) that occur on Mount Penteli (Table 16.1 and Figure 16.3).

The most important threats to the reforestation of Mount Penteli are grazing, soil erosion, and fire. The risk of damage from grazing, especially in the first and second year of the restoration, is high and remains a serious problem for the survival of the plants. The issue of grazing was more easily addressed by declaring the area reforestable, which prohibits grazing with the reforested areas. Soil erosion is a threat after fires because the ground is not protected by living plants,

TABLE 16.1 Tree and shrub species used for the reforestation of Mount Penteli

Species	%
Aleppo pine (*Pinus halepensis*)	28.41
Mediterranean cypress (*Cupressus sempervirens*)	30.89
Spanish broom (*Spartium junceum*)	9.51
Oleander (*Nerium oleander*)	12.55
Tree medick (*Medicago arborea*)	2.43
Laurel (*Laurus nobilis*)	3.53
Holm oak (*Quercus ilex*)	1.94
Carob (*Ceratonia siliqua*)	0.12
Amur privet (*Ligustrum amurense*)	1.12
Late cotoneaster (*Cotoneaster lacteus*)	5.22
Mastic tree (*Pistacia lentiscus*)	0.16
Myrtle (*Myrtus communis*)	2.61
Holly (*Quercus coccifera*)	0.15
Mediterranean buckthorn (*Rhamnus alaternus*)	1.34
Total	100.00

and heavy rainfall exacerbates the soil erosion risks. Soil loss within this mountainous landscape reduces the site's potential for plants to establish, which could result in long-term loss of biodiversity. Fire protection is based on an organized system of prevention, suppression, and repression and is dealt both at the level of planning and at the level of implementation by established State services (e.g., civil protection and fire brigade).

Path to the future

Following a fire, it is necessary to spend at least two years to assess the possibility of natural regeneration. If natural regeneration does not have satisfactory results (e.g., a sufficient number of plants do not establish), reforestation becomes the best option. Reforestation improves ecological services and biodiversity and increases the quality of life of residents who are close to such projects. Regeneration and reforestation both contribute to climate change mitigation by sequestering carbon in the wood of the trees and shrubs.

Although natural regeneration reduces the costs of landscape restoration, sometimes it cannot be applied on the ground. Reforestation offers a solution when regeneration is not possible. Reforestation also creates the possibility to alter the plant species that are re-established, which offers opportunities to shift the plant communities in response to the changing climate of the region. Reforestation is more time, labor, and financially intensive, but increasingly organizations and institutions are becoming aware of the tremendous value of the forests and the need to ensure their protection.

FIGURE 16.3 Reforestation of burnt areas in Mount Penteli. Photos by Alexandra D. Solomou.

Acknowledgments

Thank you to ilVERDE-Evangelia Andrianopoulou and Co. EE (https://ilverde.gr/) that made this project possible.

References

Baliousis, E. (2011). *The flora and vegetation of Mount Pendelikon (East Attiki, Greece).* Ph.D. Thesis, School of Science – Faculty of Biology, Department of Ecology and Systematics, National and Capodistrian University of Athens.

EEA. (2016). *European forest ecosystems – state and trends.* EEA Report No. 5/2016.

Georghiou, K., & Delipetrou, P. (2010). Patterns and traits of the endemic plants of Greece. *Botanical Journal of the Linnean Society, 162*(2), 130–153.

Georgiadis Th. (2004). *Vegetation ecology.* Patra: University of Patra Publications. http1://ec.europa.eu/environment/nature/legislation/habitatsdirective/index_en.htm (accessed on 30 December 2021).

Kailidis, D. (1981). *Forest fires.* Thessaloniki: Yahoudi-Yapouli.

Korakis, G. 2015. *The flora and vegetation of the forests of Greece.* Athens: Association of Greek Academic Libraries.

Myers, N., Mittermeier, R. A., Mittermeier, C. G., Da Fonseca, G. A., & Kent, J. (2000). Biodiversity hotspots for conservation priorities. *Nature, 403*(6772), 853–858.

Ristić, R., Malušević, I., Radić, B., Milanović, S., Milčanović, V., & Polovina, S. (2019). The role of forest ecosystems in the process of mitigation and adaptation to the effects of climate change. *Contributions, Section of Natural, Mathematical and Biotechnical Sciences, 40*(1), 25–32.

Sheil, D. (2018). Forests, atmospheric water and an uncertain future: The new biology of the global water cycle. *Forest Ecosystems, 5*(1), 1–22.

Solomou, A. D., & Sfougaris, A. (2021). Contribution of agro-environmental factors to yield and plant diversity of olive grove ecosystems (*Olea europaea* L.) in the mediterranean landscape. *Agronomy, 11*(1), 161.

Solomou, A. D., Karetsos, G., Skoufogianni, E., Martinos, K., Sfougaris, A., & Tsagari, K. (2016). Assessment of Greek forests protection and management. In: Zhelezov, G. (Ed.), *Sustainable development in mountain regions* (pp. 141–166). Cham: Springer.

Strid, A., & Tan, K. (1997). *Flora hellenica.* Königstein: Koeltz Scientific Books.

Tedim, F., Xanthopoulos, G., & Leone, V. (2015). Forest fires in Europe: Facts and challenges. In: Shroder, J.F. and Paton, D. (Eds.), *Wildfire hazards, risks and disasters* (pp. 77–99). Amsterdam: Elsevier.

Thompson, I., Mackey, B., McNulty, S., & Mosseler, A. (2009). Forest resilience, biodiversity, and climate change. *Secretariat of the Convention on Biological Diversity, Montreal. Technical Series no. 43. 1–67.* (Vol. 43, pp. 1–67).

Topalidou, E., Solomou, A. D., Santos, S. S., Krystallidou, E., Kakara, S., & Mantzanas, K. (2021). Dynamic role and importance of multi-kingdom communities in Mediterranean wood-pastures. *Sustainability, 13*(18), 10179.

Xanthopoulos, G. (1998). Forest fires in Greece: Past, present and future. Epikentra (Foundation for Political Research and Advanced Education), *6*, 62–71. (Special issue titled "Greek forests: myths and reality - Proposals for the present and for the future) (In Greek).

Xanthopoulos, G. (2007). Forest fire policy scenarios as a key element affecting the occurrence and characteristics of fire disasters. In *Proceedings of the 4th international wildland fire conference*, Sevilla.

17

CREATING RESILIENT LANDSCAPES

From a hopeful vision to a long-lasting existence

Amin Rastandeh and Meghann Jarchow

A vision from the twenty-second century

The twenty-first century was a turning point in the history of human-nature interactions. Today, in the twenty-second century, the principles of creating resilient landscapes are an indispensable part of environmental planning and management at multiple scales. To design and manage landscapes, top-down decision-making approaches are not acceptable anymore, even in less-developed regions of the world. Today, there is more of a focus on managing for processes and functions rather than managing to maintain prior/static conditions. In comparison with the past, the role of governments is more functional, as one component of a complex system of stakeholders of different interests. The science of landscape resilience is a borderless concept providing opportunities for the co-generation of knowledge and co-operation in multi-national projects. Our landscapes are now resilient to the impacts of climate change. Thanks to our collective effort during the twenty-first century, we are now better able to flourish within the new range of environmental conditions caused by climate change. Our landscapes are now climate-adapted (i.e., drought- and flood-tolerant). The inherent interconnectedness among all life – human and beyond – is widely recognized and valued. Fragmented habitats have been restored and locally extinct wildlife species (i.e., flora and fauna) have been reintroduced into their historical ranges after decades of absence. The number of biosphere reserves has sharply increased since the 2020s. Areas of biodiversity importance are protected by NGOs and local people. Environmental/climate justice is widespread and considered by all as an integral part of human rights. People of different socioeconomic classes have equal access and choose to connect to nature. Food and energy production have generally shifted to serve local people and meet local needs. Traditions linked to agriculture are preserved as part of our cultural heritage.

DOI: 10.4324/9781003266440-17

The generation of clean energy is among the most desired human activities in our landscapes. In areas dominated by human populations, land-sharing/sparing and reconciliation ecology are now two widely accepted norms for supporting the coexistence of humans with wildlife. The voices of Indigenous Peoples are heard and heeded in the process of decision-making on landscapes. Indigenous knowledges have been incorporated into dominant worldviews thereby affecting the resulting actions toward landscapes. Women, more than ever, lead the efforts to connect to and learn from the land at all levels from small farms to countries. Reforestation and re-perennialization projects have created sustainable job opportunities for local people that function at longer times scales.

These are just examples of what can be seen in the world's landscapes of the twenty-second century. Just 100 years ago, these achievements were thought to be impossible, and therefore, unachievable. But, we, the travelers of the twenty-first century, employed hope as a humane motivational driving force to create a better world for those who live in the twenty-second century.

Lessons learned from the case studies

One result of climate change is that the predictability and stability of the weather that we had previously experienced has been replaced by a period where unprecedented and extreme events are becoming common. Increasing ecological resilience, which is "the magnitude of disturbance that can be absorbed before the system changes its structure… and processes" (Holling 1996, p. 33), is needed in these new systems rather than engineering resilience, which is "stability near an equilibrium steady state, where resistance to disturbance and speed of return to the equilibrium are used to measure the property" (Holling 1996, p. 33). Ecological resilience helps us to identify what processes and functions that we want to maintain, but it does not hold tight to the past, which is what we need to do now. We need to have a clear vision for the processes and structures that we want to maintain, but we have to be flexible and innovative in thinking about how we might achieve that. More importantly, resilience helps to refocus on the how rather than the end point.

The 15 case studies presented in this book have a wide range of implications for applied research, strategic policymaking, and practice in the real world that are both context specific and extend beyond the specific contexts of the places. Overall, lessons learned from these case studies can help us to expand our understanding of the diversity of problems, as well as the practicality of solutions in the real world. Applying a future-facing and solution-oriented perspective, the contributors addressed the concept of landscape resilience from different angles at various scales. The issues discussed in this book reflect a diversity challenges and opportunities for creating landscape resilience worldwide. In short, Præstholm et al. (Chapter 2) addressed how multifunctional land consolidation can help us to find new paths to landscape resilience in Denmark. Potthoff and Dramstad (Chapter 3) studied the links between landscape diversity, heterogeneity, and diversity of productions and discussed how these components may affect landscape resilience in Norway. Jarchow (Chapter 4) explained how we can create resilient landscapes that allow us to become sustainably self-sufficient in food, energy, and water in the United States. Streit Krug et al. (Chapter 5) argued that re-perennialization and diversification of landscapes supported by scientific knowledge and social movement are required to create resilient landscapes and accomplish a social perennial vision for the Great Plains in the United States. Martinez-Cruz et al. (Chapter 6) provided multilevel recommendations for developing resilience strategies of women and Indigenous Peoples in Guatemala and Mexico. Orcasberro et al. (Chapter 7) discussed why grazing management is fundamental to landscape resilience in the Campos grasslands of South America. MacKenzie (Chapter 8) explored opportunities and challenges for mitigating the vulnerability of the poor in the face of climate change through gardening in Vanuatu. Mathenjwa et al. (Chapter 9) showed how green infrastructure and open spaces in urban areas can contribute to landscape resilience and human well-being in South Africa. Kilbane (Chapter 10) presented a solution-oriented method for enhancing landscape resilience in the face of biodiversity loss, climate change, and land-use conversion in Australia. Fukamachi (Chapter 11) indicated that how human nature symbiosis can contribute to landscape resilience in Japan. Vierikko et al. (Chapter 12) addressed the relationship of applying high-tech nature-based solutions to landscape resilience in Finland. Latai-Niusulu et al. (Chapter 13) showed how traditional nature-based solutions can help to enhance landscape resilience in Samoa. Ersoy Tonyaloğlu and Kesgin Atak (Chapter 14) developed a methodology to better understand the potential role of ecosystem services in creating resilient landscapes in Turkey. Hawken et al. (Chapter 15) discussed how the application of scenario-based approaches can help to find win-win solutions for resilience and adaptation to the impacts of climate change in Australia. Solomou (Chapter 16) examined how forest restoration and fire management for biodiversity conservation can contribute to landscape resilience in Greece.

Taken together, these case studies suggest that there is no quick-fix response to the challenges caused by climate change and anthropogenic development. Instead, to create resilient landscapes, we need to embrace innovation and change while maintaining continuity in ecosystem functioning. We need a comprehensive reform in our land-use policies, food and energy systems, biological conservation strategies, recreation patterns, travel behaviors, economic activities, and political relations in our landscapes. To achieve this, we need to find common-ground solutions that increase synergy and unity among us. In the real sense of the word, we need to change our landscapes. Before changing our landscapes; however, we first need to change ourselves (i.e., our views and lifestyles). We, ourselves, should become resilient and for this, what we necessarily need in this era is hope.

Irrespective of the socio-economic, cultural, ecological, and political differences that characterize our landscapes, we learned from the case studies that the creation of resilient landscapes is associated with a broad range of challenges, many of which have similar root causes. We learned that landscape resilience cannot be achieved in isolation. It is synergy, combined with cooperation, integration, and multifunctionality, that can help us to depart from this point toward a more resilient future. To tackle the interconnected and interrelated socio-ecological challenges, we learned that one-dimensional solutions should not be among our options anymore. We learned that we need to re-develop and re-organize our landscapes in ways that concurrently support abiotic, biotic, and cultural resources and preserve them into the future. We learned that the concept of landscape resilience is inherently multi-dimensional and cross-disciplinary. The plastic, complex, evolving nature of this concept requires us to incorporate the scale of time as one important dimension into the process of creating socio-ecological resilience in our landscapes. We learned that the creation of resilient landscapes is a process that is sensitive to socio-ecological parameters and spatiotemporal scales. And finally, we learned that there are still many questions for and uncertainties about the future.

Conclusion

The idea of writing this book was crystallized during the summer of 2020, when the global community was grappling with mega-droughts, historical wildfires, sudden floods, and a global pandemic. The reality is that we are currently facing unprecedented, and unfortunately, unpredictable socio-ecological challenges, mainly triggered by climate change and widespread anthropogenic development across the Earth. In this context, our vision of the twenty-second century may seem to some people ambitious, unorthodox, and ultra-optimistic. Nonetheless, we believe that the effectiveness of hope for making a better world, as previously underscored by others (e.g., Rosenzweig 2003; Ojala 2012; Stern 2012; Wilson 2016; Li and Monroe 2019; Strazds 2019; Geiger et al. 2021; Lewandowsky et al. 2021), should not be omitted when moving toward the creation of resilient landscapes. As referenced in Chapter 1, there are already many hopeful signs around the world to support the notion that the creation of resilient landscapes in the face of climate change is today more possible than the past. Hope, if combined with wise science, can lead to the change we envisioned in this chapter.

In this book, we put hope at the center of our arguments to build up a platform for change: from inaction to proaction; from dysfunctionality to multifunctionality; and from brittleness to resilience. This overarching goal, however, cannot be achieved overnight. It takes a great deal of time and requires a substantial amount of effort, not in isolation, but together. It is why in the beginning of this chapter, we described the whole twenty-first century just a turning "point" in the history of human-nature interactions, as if it is just one chapter in an old, large book that its next chapters will be written by us, our children, and their children. It is under the umbrella of hope that we can create resilient landscapes that effectively support ecosystem services in ways that help us to flourish in a future altered by climate change.

This work is one collective step toward a deeper and more comprehensive understanding of the problems and solutions associated with the process of creating resilient landscapes. This step was taken by 44 subject-matter experts from across the world. The solutions provided in this book should not be considered as a panacea to all of the socio-ecological issues we are currently facing on the ground. Perhaps, the practicality and effectiveness of these solutions are directly correlated with a large number of factors, including the willingness of stakeholders of different interests to change their landscapes toward a more resilient future. More importantly, transformability, as discussed in Chapter 1, can pave the way toward this goal.

The informative, provoking, and insightful questions raised in this book open up new opportunities for addressing less-studied, oft-neglected aspects of the science of landscape resilience in an era of climate change. To take the next step, we invite interested people of different backgrounds to address these questions and re-test, re-evaluate, and re-frame the issues discussed in the hope that all of us consolidate and enrich our developing knowledge of creating resilient landscapes.

Acknowledgment

This work was supported by the National Science Foundation through the EPSCoR Track II cooperative agreement OIA-1632810 and NSF DBI-1560048.

References

Geiger, N., Swim, J. K., Gasper, K., Fraser, J., & Flinner, K. (2021). How do I feel when I think about taking action? Hope and boredom, not anxiety and helplessness, predict intentions to take climate action. *Journal of Environmental Psychology*, 76, 101649.

Holling, C. S. (1996). Engineering resilience versus ecological resilience. In: *Engineering within ecological constraints* (Editor, Schulze, P. E.). Washington DC: National Academy Press, pp. 31–43.

Lewandowsky, S., Facer, K., & Ecker, U. K. (2021). Losses, hopes, and expectations for sustainable futures after COVID. *Humanities and Social Sciences Communications*, 8(1), 1–17.

Li, C. J., & Monroe, M. C. (2019). Exploring the essential psychological factors in fostering hope concerning climate change. *Environmental Education Research*, 25(6), 936–954.

Ojala, M. (2012). Hope and climate change: The importance of hope for environmental engagement among young people. *Environmental Education Research*, 18(5), 625–642.

Rosenzweig, M. L. (2003). *Win-win ecology: how the earth's species can survive in the midst of human enterprise*. Oxford: Oxford University Press.

Stern, P. C. (2012). Fear and hope in climate messages. *Nature Climate Change*, 2(8), 572–573.

Strazds, L. M. (2019). Radical hope: Transforming sustainability. *Journal of Sustainability Education*, 21. Retrieved from http://www.susted.com/wordpress/wp-content/uploads/2019/12/Strazds-JSE-December-2019-General-Issue-PDF.pdf

Wilson, E. O. (2016). *Half-earth: Our planet's fight for life*. New York: Liveright.

INDEX

Note: **Bold** page numbers refer to tables; *italic* page numbers refer to figures and page numbers followed by "n" denote endnotes.